COVID COUP:

"The Rise of the Fourth Reich"

LEONARD G. HOROWITZ
DMD, MA, MPH, DNM (Hon.), DMM (Hon.)

Medical Veritas International, Inc.

COVID COUP: "The Rise of the Fourth Reich"
Copyright © Leonard G. Horowitz, 2021
All rights are reserved. No part of this text may be reproduced or transmitted in any form, by any means, without the written consent of the author or publisher —Medical Veritas International, Inc.
Cover design: Leonard G. Horowitz
Library of Congress Cataloging Preassigned
Horowitz, Leonard G.
The Ayahuasca Cult p. cm.
Includes bibliographical references.

1. Popular Works;
2. COVID, Deep State, Politics
—Health science—MKULTRA, MKNAOMI, MindWar, Neuroscience, Vaccinology, Virology
—Genetic Engineering—Creationism, DNA, Bio-electronics — Conspiracy—Social Engineering, Depopulation, Transhumanism — Intelligence industry—Artificial Intelligence (AI), CIA, MI6, Mossad, FBI —Drug crimes—Bioterrorism, Treason, mRNA vaccine Racketeering
3. Public Health
4. Natural Immunity and Spiritual Healing,

Card Number: Pending
Additional cataloging data is pending.
ISBN: 9798764321981
Additional copies of this book are available for bulk purchases.
For more information, please contact:
Medical Veritas International, Inc.
5348 Vegas Dr., Suite #353 • Las Vegas, NV 96815;
1-888-508-4787; E-mail: tetra@tetrahedron.org,
URL web site: http://www.tetrahedron.org

OTHER BOOKS BY DR. LEONARD G. HOROWITZ

The Ayahuasca Death Cult
The Las Vegas Deep State Massacre
The Book of 528: Prosperity Key of Love
Healing Codes for the Biological Apocalypse
Death in the Air: Globalism, Terrorism & Toxic Warfare
Healing Celebrations: Miraculous Recoveries Through Ancient Scripture, Natural Medicine & Modern Science
Emerging Viruses: AIDS & Ebola—Nature, Accident or Intentional?
DNA: Pirates of the Sacred Spiral
Walk on Water
LOVE The Real da Vinci CODE
Choosing Health for Yourself: A Clear & Practical Guide to Motivating Self-Care

DEDICATION

I dedicate this book to my beloved deceased partner, the legendary investigative journalist, activist, and injustice avenger, Sherri Kane, whose manslaughter by corrupt law enforcers and bribed judges in Hawaii demands retribution.

Table of Contents

DEDICATION	7
FOREWORD	11
CHAPTER I	17
INTRODUCTION AND BACKGROUND INTELLIGENCE	17
CHAPTER II	63
THE WUHAN OUTBREAK	63
CHAPTER III	83
COVID-19's LAB ORIGIN & WHO's ACCOUNTABLE	83
CHAPTER IV	117
PANIC AT THE NIH, WELLCOME TRUST,	117
WHO AND HARVARD	117
CHAPTER V	175
WHY THE FBI LIED ABOUT	175
CHARLES LIEBER'S CHINESE ESPIONAGE	175
CHAPTER VI	211
CONSOLIDATING GLOBAL CONTROL by MINDWAR	211
CHAPTER VII	273
THE COVID COUP WITH FINANCING FROM HELL	273
CHAPTER VIII	333
INCITING VIOLENCE AND CIVIL WAR	333
AGAINST THE UN-VACCINATED	333
CHAPTER IX	367
BILL GATES, THE CLINTONS, AND THE WHO'S COMPLICITY IN THE COVID COUP	367
CHAPTER X	379
COVID, The Biden Crime Family,	379
and Deep State Corrupted Science	379
CHAPTER XI	401
COVID GENOCIDE, THE LAW &	401
THE RISE OF THE FOURTH REICH	401
CHAPTER XII	425
TRANSHUMANISM AND THE GREAT GLOBAL RESET	425
CHAPTER XIII	441
THE TOP TEN MOST DAMAGING CORONAVIRUS LIES	441

CHAPTER XIV	469
THE COVID COUP, NEW WORLD RELIGION, AND SPIRITUAL WARFARE	469
CHAPTER XV	489
WHAT WE MUST DO TO SAVE CIVILIZATION	489
SUMMARY & CONCLUSION	505
ABOUT THE AUTHOR	507
Appendix 1	509
THE HOROWITZ COVID PROTOCOL	509
FOR PERSONAL HEALTH & IMMUNITY	509
References and Notes	519

FOREWORD

This book sounds alarms that rang before the holocaust of World War II, the Nazi concentration camps, and human experiments that resulted in the Nuremberg trials. Thereafter, powerful political and financial agents survived and thrived in the Rockefeller, IBM, IG Farben petrochemical-pharmaceutical cartel forged from that holocaust, which is now evidenced in the "COVID Coup."

This book reminds me of my earliest memories of my grandfather speaking about the horror of der Khurbn (Yiddish for Holocaust). His accounts filled me with questions as to how and why Adolph Hitler could have been responsible for so much slaughter? And so, at age 21, I decided to visit the former Axis countries to view the physical evidence for myself.

My trip included a pilgrimage to Auschwitz, in memoriam to my Dutch cousins, whose family line traced back to the venerable rabbis and scholars of Constantinople. They, along with three-quarters of the Jewish population in the Netherlands, were annihilated during the holocaust.

After arriving in Amsterdam, my ancestral home, I turned south toward Oswiecim. But as I approached Auschwitz, I was overwhelmed by a sense of dread that prevented me from completing my journey. Dread is a gut-wrenching breath-stealing feeling. Even today, more than four decades later, I can still feel the sense of foreboding that enveloped and incapacitated me on the outskirts of the Nazi's most notorious extermination complex.

Auschwitz was an industrial complex containing over 40 concentration and extermination camps, which consumed more electricity than the entire city of Berlin. Slaves were warehoused in horrific conditions and forced to work at IG Farben (Standard Oil's partner), but only so as long as their output exceeded input, which averaged 2 months. Those who could not produce more than they ate were categorized as "useless eaters" and incinerated to keep German blood free of infection.

This fear of infection was pushed by Nazi doctors who falsely claimed Jews were primarily responsible for outbreaks of typhus. Hitler's henchmen advised military authorities to quarantine Jews to isolate them from the population. It was these reprobate doctors that provided the public health justification to transport Jews and others of "unclean" blood to slave camps, so they could be worked to death in support of Germany's war effort, then exterminated in "showers" during chemical "disinfection."

James Pool's *Who Financed Hitler* documents the German industry's financial contributions to the early Reich. Pool showed Hitler was neither a lone madman nor a modern pied piper. "Der Fuhrer" did not singularly lead Germany into the inferno.

The historical evidence also reveals that without the support of Germany's aristocracy and international financiers, Hitler's ascent to chancellor in 1933, and dictatorship in 1934, would not have been possible.

A decade later, as the Third Reich reached its nadir, Nazi Party officials and their collaborators, in accordance with the "Nero Decree," destroyed Germany's remaining infrastructure along with many of the incriminating records documenting the regime's crimes against humanity. Yet, even with this near-total destruction, the surviving documents present clear and convincing proof IBM, GM, Standard Oil, and Ford Motor Company, among others, did not just violate "The Trading with the Enemy Act of 1917," but aided-and-abetted the Reich in its ability to conduct global warfare, terrorize civilization, and kill millions.

Dr. Horowitz presents stunning parallels in this book.

Through its German subsidiary Dehomag, Tech giant IBM was the main provider of computer expertise and equipment to the Nazis. IBM's automated census search tool gave the Gestapo a way to rapidly and efficiently identify Jews and other non-Aryans for arrest and transportation to the concentration camp complex. Slave laborers upon arrival at the complex were then tattooed with an IBM number code for SS tracking, and when in the interest of optimum production, exterminated.

These work/death slave complexes formed the backbone of Germany's war machine and created economic possibilities, that according to historian Charles Cheape, "literally bewitched" American industrialists like IBM's CEO, Thomas Watson. In 1938, four months after German stormtroopers invaded Austria, Henry Ford traveled to

Germany to receive the Grand Cross of the German Eagle— the highest honor Germany could bestow on a non-German. Hitler publicly acknowledged he was inspired by Ford's business philosophy and kept a life-size portrait of Ford next to his desk. When American boys landed at Normandy and fought their way inland, they were met by Nazi armored vehicles built by Opel - a 100% GM-owned subsidiary.

America's industrialists claimed their interest in Germany was strictly pecuniary, and that they never knowingly supported the Nazi war effort. Yet, even after the Reich's murderous intent became evident, the owners and directors of these great transnational corporations continued to do business with the Reich, claiming afterwards, they had no idea what their German subsidiaries were up to. Assuming this defense were true, should IBM's willful blindness absolve them of their war crimes? And what of Henry Ford, who supplied trucks and engineered parts to transport Jews to the death camps, or Standard Oil's provision of tetraethyl lead, a fuel additive, without which the Luftwaffe could not have conducted the aerial bombardment of America's allies?

Additionally, America's national media had ample evidence of the holocaust but failed to provide sufficient coverage to break through the noise of more 'pressing' events. To the extent it was covered, the truth was occluded by mixed messaging. As early as 1933, *TIME* magazine wrote that Dr. Joseph Goebbels explained "away all Germany's defeats and trials in terms of 'the Jew.'" Yet, in this same article, *TIME* described Hitler as a vegetarian superman and Nazi ruler who uplifted the German soul. *TIME*'s coverage in the interwar period is a salient example of how criticism intermingled with praise was used to obscure Nazi horror.

Compare *TIME*'s equivocation on the architects of the Holocaust with today's corporate-controlled media attacks on "anti-vaxxers." When the national media is motivated, they are quite capable, through repetition, narrow messaging, and confusion, to obliterate the credibility of any target. Though some argue no one could have foreseen what Germany would become, even after the German army goose-stepped into the Rhineland in 1936, American industry and media continued to facilitate Nazi aggression and its resultant annexation of Austria in 1938.

Today, legions of "fact-checkers" and censors bar any mainstream analysis of the structural factors (profit being only a part) that caused the national media to cover up American collaboration with the Third Reich.

Paradoxically, in order to effect this cover-up, Nazi technique is employed. Free speech is suppressed and opposition to the dominant narrative proscribed. To the extent any analogy to Nazi Germany arises, it is made clear the comparison is strictly *verboten*.

To look at the "final solution" and say we are not there yet, and therefore the analogy improper, conflates the endpoints of National Socialism with its emergent processes. Nazi Germany did not begin with the holocaust and world war, but like the United States had a democratic form of government. This ended when the Nazi Party engineered the 'false flag' of the Reichstag fire in 1933, much like 9/11 justified the "Patriot Act" and its subsequent curtailment of America's Constitutional guarantees securing citizens' rights.

Given our media gatekeepers failed to report the truth about the Reichstag fire and the subsequent rise of the Third Reich, what makes us believe they would warn us today if another pretext was created in furtherance of these same structural interests? The national media's prior culpability in enabling the instigators of World War II is revealing and unsettling, as is their concealment of the alliance between American industrialists and the Nazis leading up to the conflagration.

This alliance and its attendant world war greatly advanced the technology of mechanized mass murder and enabled the massacre of 75,000,000 people. In the post-world-war period, these same entities killed 5 million in Korea; between 1965-75 dropped 7.5 million tons of bombs were dropped on Southeast Asia, twice the number of bombs dropped in WWII.

Despite these grand efforts to reduce the global population, Defense Secretary McNamara (and former President of Ford Motor Company) became acutely aware of mechanized warfare's targeting limitations and inefficiencies. Hamstrung by the collateral damage which would predictably result to the Übermensch's bloodlines from nuclear or chemical munitions, the merchants of death led by SS bio-weapons lab chief Erich Traub (secreted into the US under "Operation Paperclip"), turned their lethal gaze to bio-warfare, and the weaponization of nano-genetics. On April 24, 2020, after decades of "vaccine" trials on the African continent, Bill Gates, in an interview with Stephen Colbert, giddily proclaimed the Covid vaccine was the "Final Solution."

Foreword

Gates and his media savvy handlers knew this was the code name for the systematic and deliberate genocide of all European Jews, formulated by the Nazi leadership, as the answer to the "Jewish question." The "Final Solution" culminated in the holocaust, which saw the killing of 90% of Polish Jews and two-thirds of the Jewish population of Europe. On December 11, 2020, in follow-up to Gates' promised "Final Solution," a blitzkrieg of novel biologics were unleashed against an unsuspecting and fearful world.

Dr. Leonard G. Horowitz, in this meticulously documented work, has compiled the evidence as to why this *new and improved* "Final Solution" has been long-planned. In this treatise, the *COVID Coup: The Rise of the Fourth Reich*, Dr. Horowitz maps the historical roots and ultimate purpose of the Third Reich and its metamorphosis into the "Fourth Industrial Revolution."

Dr. Horowitz's COVID Coup is dangerous to the mainstream's *mise en scene*. It names and connects the dots. The Covid Coup neither adheres to the dominant narrative nor pays homage to the historical revisionism of the Nazi era, which portrays Hitler as the causative agent. This book's focus is the financial, military, industrial network out of which a Fuhrer emerged to become an instrument of modern fascism, which now approaches its denouement with the COVID Coup. Excoriated for his uncannily accurate prognostications during the past quarter-century, Dr. Horowitz is that rare individual who has the heart and guts to stand in the middle of the killing grounds and report exactly what he sees. His intolerance of falsehood, and his fearlessness of those who produce it, make him the ideal reporter for our times.

Few have identified and analyzed the COVID Coup's precursors more adroitly than Dr. Horowitz. His 1996 prophetic work —*Emerging Viruses Aids & Ebola, Nature, Accident or Intentional?* — sent out a clarion call of what would come today. According to Dr. Horowitz, the COVID Coup now unfolding is not a discrete event but part of a historical continuum dating back to the Russian Pogroms of the late 19th century. Since then, its progenitor structure has always been present, and each time the world is readied for war, elements of this structure are propped up to do the requisite work.

The World Economic Forum (founding member, Nazi banker Hermann J. Abs) is the current iteration of that ascendant structure, and its current work is to ready the world for the "4th Industrial Revolution" and transhumanist conversion that Horowitz evidences while reporting solid science. Dr. Horowitz's warning in COVID Coup is unequivocal. Failure to expose the underlying bulwark of this coup and the Fourth Industrial Revolution will metastasize into the Fourth Reich.

This warning is not alarmist. In a 2016 video entitled, "What is the 4th Industrial Revolution,"[1] the World Economic Forum (WEF) enunciated its intent to destroy not just humanity but the very meaning of what it means to be human.

"Now a Fourth Industrial Revolution is building on the Third," the WEF reported. "It is characterized by a fusion of technologies that is blurring the lines between the physical, digital, and biological spheres…The very idea of a human being, some sort of natural concept, is changing. Our bodies will be so high tech we won't be able to really distinguish between what's natural and what's artificial."

Though the means by which the WEF seeks to achieve its intention were unavailable to the Third Reich, the ends are the same. And like the Reich, while its ends are not subject to rational comprehension, the dystopian vision is plainly stated. The 4th Industrial Revolution is a world where monsters seek out and destroy the divinity of our genome and twist us into their own grotesque image. It is a world where the natural has been annihilated. A hellish nightmare, where we will no longer be able to distinguish ourselves from their synthetic devices, is overtaking us.

With the vantage of historical perspective, we can look back at the destruction of German democracy in the 1930s and see its parallels in the present. Not to desecrate the memory of the victims but as prophylactic against its recurrence; so that the sacrifice of millions will not have been in vain.

Who among us, after reading the words of the WEF above, would argue pathological evil has been extinguished, or that vigilance against its re-ascendance is misplaced?

On the sacred memorials of the holocaust victims is emblazoned the words, "Never Forget, Never Again."

We will not forget; not so long as we have intrepid chroniclers like Dr. Leonard G. Horowitz standing at the ramparts, beseeching us to stand and face all those who act to enslave or destroy us.

JTK deCordova, October 2021, Barbados WI

CHAPTER I
INTRODUCTION AND BACKGROUND INTELLIGENCE

Aldous Huxley, the author of *Brave New World*, wrote about the ideal political and religious drug "soma" in 1958 in <u>Brave New World Revisited</u> (p. 77). He covered the topics of "Over-Population," society's "Quantity, Quality and Morality," "Propaganda in a Democratic Society," the "Arts of Selling" and "Brainwashing," "Chemical Persuasion," "Hypnopaedia" predicting *Wikipedia*, and recommending "Education for Freedom."

Huxley's brother, Julian, was the pre-eminent transhumanist—Sir Julian Sorell Huxley. Transhumanism (abbreviated as 'H+' or 'h+') is "an international philosophical movement," states *Wikipedia*, "that advocates for the transformation of the human condition by developing and making widely available sophisticated technologies to greatly enhance human intellect and physiology."

"Transhumanist thinkers study the potential benefits and dangers of emerging technologies that could overcome fundamental human limitations and the ethical limitations of using such technologies. The most common transhumanist thesis is that human beings may eventually be able to transform themselves into different beings with abilities so greatly expanded from the current condition as to merit the label of *posthuman beings*."

Today, Huxley's prophesy is most forcefully advanced by World Economic Forum Founder and Executive Chairman Klaus Schwab. They, with Bill Gates and Johns Hopkins University officials, sponsored "Event 201" —a coronavirus pandemic 'predictive programming' exercise that took place on October 18, 2019, in New York City. That was approximately six weeks before the press began reporting the first cases of COVID-19 at the Wuhan seafood market. According to Event 201 co-sponsor Schwab, what is unfolding is a "Fourth Industrial Revolution[2] ["4IR"; that] will affect the very essence of our human experience."

COVID COUP: "The Rise of the Fourth Reich"

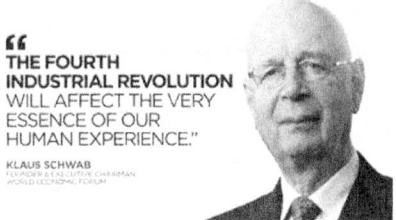

> **THE FOURTH INDUSTRIAL REVOLUTION WILL AFFECT THE VERY ESSENCE OF OUR HUMAN EXPERIENCE."**
>
> KLAUS SCHWAB

This book explains the role of the COVID Coup in this 4IR. The facts detailed herein evidence the "Fourth Industrial Revolution" is the long-planned corporate-elite financed "Rise of the Fourth Reich."

Regarding the COVID pandemic, Schwab is alleged[3] to have said, "The COVID-19 outbreak is the first big step towards unprecedented control over mankind." Is there any basis for this quote? Fact-checkers say "No." Facts, however, say "Yes."

Schwab is a champion for global governance to take control over the 'human condition.' Schwab lectured, "The problem that we have is not globalization. The problem is a lack of global governance." Accordingly, the COVID pandemic has imposed global governance by devastating the economies of virtually all nation-states, *other than leading Communist nations*, China and Russia. What does this evidence?

Continuing his 'predictive programming,' Schwab reportedly explained that, "In the New World [Order] it is not the big fish which eats the small fish, it's the fast fish which eats the slow fish."[4]

With Big Tech's super-fast 5G telecommunications and expanded artificial intelligence computing (such as IBM's Watson computer that simulates "Hal" in *2001 A Space Odyssey*), you will be 'eaten' up by the transhumanists. Your brain will not be able to comprehend, integrate, or keep up with the speed of fiber optics, data-mining your every move, word, and metabolic function. Your brain will need to be connected to the Cloud for you to operate in this 'Brave New World.' Otherwise, you will be a "useless eater," unworthy of life as a burden to the 'National Security State' and global governance. To be sure, the bioelectronic technology to accomplish this brain-Cloud union is currently heralded by many experts, including Dr. James Giordano, in *Battlescape Brain: Military and Intelligence Use of Neurocognitive Science*.[5]

Introduction and Background Intelligence

Julian Huxley, who lived from 1887 to 1975, foresaw this transformation of civilization. As a distinguished British evolutionary biologist, eugenicist, and internationalist, he advocated for natural selection. "Survival of the fittest," applicable in the financial world, is also applicable to being a "slow fish" in Big Tech's fast pond. As a propagandist for evolution by natural selection, in 1959, Huxley received a Special Award of the Lasker Foundation in the category Planned Parenthood – World Population. Huxley was a prominent member of the British Eugenics Society and was its president from 1959 to 1962.

Curiously today, Klaus Schwab's partner in financing the Fourth Industrial Revolution, Big Tech's dominance over politics, the media, and our culture—Bill Gates—is like-minded. Gates's father was an outspoken official in the Planned Parenthood organization promoting eugenics, 'racial hygiene,' 'ethnic cleansing,' and 'population management' (i.e., depopulation). Bill Gates's mother worked for IBM and helped secure IBM's financing of Bill's Gates's start-up called Microsoft. IBM administered the Rockefeller-IG Farben partnership in 'data-tracking'—directing holocaust laborers and death camp victims to meet their destinies during World War II.

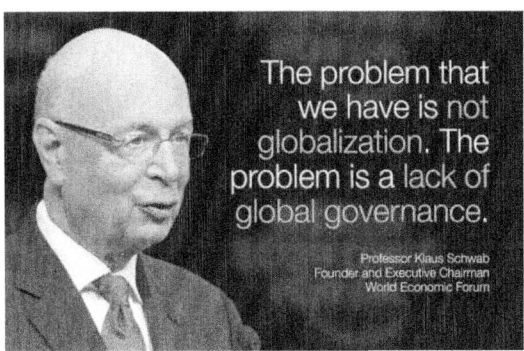

For those who may not know, Planned Parenthood-World Population is the leading legislative lobbying group promoting abortions as a remedy for 'overpopulation.' This organization emerged from Margaret Sanger's efforts. She opened the first birth control clinic in the U.S. in 1916. Sanger founded the American Birth Control League in 1921, which changed its name to Planned Parenthood in 1942.

It is no 'coincidence' that Planned Parenthood began virtually the same time that Pearl Harbor was bombed in Hawaii. America used that event as an excuse to engage the Germans and Japanese in that war. Those proceedings were largely prompted by the Rockefeller family and their lawyers, John Foster and Alan Dulles —the first directors of the CIA. They operated for commercial gain. Their arrangements favored their Standard Oil Company partnership with IG Farben —Germany's leading industrial racketeering enterprise. CEO Hermann Schmitz, head of IG Farben, partnered with Carl Duisberg, director of the Bayar Company. Today, the 'Bayer' aspirin company owns Monsanto and is the co-owner of the "CureVac" company with Pfizer's parent, GlaxsoSmithKleinBeecham and Wellcome that plays a major role in the COVID Coup. This syndicate advertises an alternative mRNA COVID-19 vaccine supposedly competing against Pfizer's and Moderna's vaccines. Despite lower reported efficacy and questionable safety, the CureVac version is predicted to supply lesser-developed countries defending against newly emerging viral variants, such as the "D-strain" of coronavirus that emerged from the UK and Africa.

During a trip to the United States in the spring of 1903, Duisberg visited several large American trusts, including the Rockefellers' Standard Oil. In 1904, after returning to Germany, he proposed a nationwide merger of the producers of dyes and pharmaceuticals in a memorandum to Gustav von Brüning, the senior manager at Hoechst. The Rockefellers and Merck officials formed the world's ongoing virtual monopoly over the drug, chemical, genetic (or eugenic) engineering, and later biotechnology industries.

To help with these matters, Aldous Huxley's and Julian Huxley's close cousin, Sir Andrew Fielding Huxley, was appointed. This world-famous nerve physiologist is considered the most distinguished pioneer of research in "Energy Medicine."

As you will learn later in this book, energy medicine and neuroscience includes "Frequency Therapeutics," a company started by Pfizer's leading biomedical scientist. You will also learn that the 'nano-bioelectronics' coating enabling mRNA "vaccine" delivery into cells is mediated by electromagnetic energy forces super-conducted through body water. The future of medicine, connecting human brains to the Cloud and data-mining virtually everything ongoing in vaccine recipients, is based on such energy medicine and the electro-dynamics of this "hydrogel" device.

Historians report that Aldous Huxley was very interested in brain psychopharmaceuticals. Many people viewed him as "Mr. LSD." His enthusiasm for the transformative properties of psychedelics led him into friendships with Richard Alpert and Timothy Leary while Huxley was a visiting professor at MIT. Alpert (a.k.a. Alan Watts) and Leary were conducting psychedelic research at Harvard at that time. As you will read later in this book, Harvard and MIT play crucial roles in developing mRNA vaccines, vital nano-bioelectronic hydrogel devices therein, and the alleged criminal manipulation of civilization leveraging COVID Labs at Harvard and MIT have enabled the drug syndicate's population control and transhumanist agendas.

The Sydney Morning Herald[6] detailed Andrew Huxley's early contributions in this field of brain science. "Their insight, known as the ionic hypothesis, solved one of the leading questions in brain science at the time." Huxley's hypothesis helped explain the numbing effects of anesthetics, which block the activity of ions, and it made possible devices that harness electricity from brain cells to operate prosthetic limbs.

Today, the Transhumanist Movement celebrates brain implants, nano-bioelectronic nerve sensors, and prosthetic limbs. These are advertised as helping "war heroes" who lost their arms or legs in combat.

Andrew Huxley's research also led to the identification of genetic diseases known as channelopathies. These result from defects in energy (electron) flow in ion gates, or ion channels as they are known today in science and medicine, including how mRNA vaccines operate.

The Huxley boys came from the well-to-do Huxley family. However, biographers have often noted the occurrence of mental illness in their family. Their father was writer Leonard Huxley, and his paternal grandfather was Thomas Henry Huxley. Thomas was a friend and supporter of Charles Darwin and a proponent of the Evolution of Species theory. This is the basis of White Supremacy, Black Hate, anti-Semitism, 'racial hygiene,' eugenics, and the political 'scientific' agenda fueling Hitlerian/Rockefeller/IG Farben (i.e., Big Pharma) 'national socialism' overtaking American politics, economics, and society at the time of this writing.

Where the 'rubber meets the road' on this topic interlaces with religious conflicts, theological uprisings, and social engineering regarding vaccination mandates and the 'global elite.' As Bill Gates openly represents, vaccines offer the 'Final Solution" to diseases *and* the "lower races." Vaccines are forecast in the Transhumanist Era to deliver disease-resistant super-humans—the 'Ubermensch'—genetically bred to "Be All That You Can Be."

The first major commercial vaccines were produced by this cartel — the IG Farben/Rockefeller/Merck & Co/Bayer/Wellcome alliance. Today, these companies are similarly situated and financially empowered through their 'successors-in-interest.' They spun-off enterprises that resemble honest businesses engaging honorable science and commerce. This illusion disappears upon careful investigation, as you will read below. 'Follow the money,' the partnerships, and institutional investors. You will see from today's winners on Wall Street, Hitler's partners actually won WWII.

Today, what is called 'globalism,' advanced by multinational corporations, is precisely what Hitler's administration called the 'Neuordnung' —the "New World Order"— the Rise of the Fourth Reich, or "Fourth Industrial Revolution."

Aldous Huxley shared his concerns about these geopolitical and scientific proceedings and their risks through his writings. The problem for investigators is discerning Huxley's *Brave New World* as simply a work of fiction (as it has been declared) or a brilliant 'psychological operation' —a 'PSYOPS' for socially engineering the 'market' for globalist control.

Substantial evidence proves Huxley's *Brave New World* was a British Secret Service-financed PSYOPS. Such evidence was found in the activities of Huxley's cousin, Gervas Huxley (1894-1971). Quoting the 'CIA's mouthpiece,' *Wikipedia*, "Gervas was recruited in 1939 to help set up the wartime Ministry of Information (MOI)."

The MOI initially organized the "Press Relations" group that was responsible for both the issue of news and the censorship of information for England in alliance with America. The MOI became the chief allied propaganda service that evolved to this day to direct and censor news on behalf of its 'Deep State' controllers. Currently, this is the main reason

Introduction and Background Intelligence

there is such coordinated disparity between "left" and "right" major news networks, and gross censorship of citizens by Big Tech.

The PSYOPS administers a divide-to-conquer strategy. That's why, for example, *FOX News* slams mandatory vaccines that are heavily promoted on competing, "liberal networks." *FOX* takes a dim view of 'infanticide' in opposition to the Population Council's 'Family Planning' and 'Planned Parenthood' organizations. Yet, hypocritically, *FOX* encourages vaccinations and censors the massive number of deaths and diseases vaccines are causing. They back your freedom to abstain from vaccinations yet promote toxic "immunizations" nonetheless. That's a 'double-sided message' schemed to create confusion and submission.

Generally, people don't realize the Population Council and intelligence agencies are behind the propaganda persuading viewers that their favorite embattled television network, such as CNN, Bill Gates's MSNBC, Britain's BBC, or Rupert Murdoch's FOX, are actually playing on the same Deep State directed team. They're all <u>*controlled*</u>.

The CIA's online propaganda mill, *Wikipedia*, which regularly favors censorship, does so with, for example, The Population Council.

This organization is written up as "an international, nonprofit, non-governmental organization. The Council researches biomedicine, social science, and public health and helps build research capacities in developing countries. One-third of its research relates to HIV and AIDS; while its other major program areas are in reproductive health and its relation to poverty, youth, and gender."

Wikipedia censors that the Population Council heavily promotes vaccines, including the hepatitis B vaccine that triggered HIV/AIDS.[7] Nelson Rockefeller's protégé, Henry Kissinger—a top consultant and business director for Merck & Co.—actually ordered the development of the immune system destroying viruses HIV and Ebola for germ warfare[8] and global *depopulation,* as you will learn more about later.

After the war [that instantly saw the Rockefellers' lawyers, the Dulles Brothers,' establish the U.S. Central Intelligence Agency (CIA) in precisely the IG Farben building that was spared allied bombings by Rockefeller-direction], Gervas Huxley sat on the Executive Committee of the British Council and became a successful author of biographies. Today, the British Council promotes a propaganda film[2] favoring the Wellcome Trust. The current director of the Wellcome Trust, Jeremy "James" Farrar, plays a major role in fraudulently concealing the genetic engineering of the AIDS-laced COVID-19 lab virus, as you will read below.

On October 21, 1949, Aldos Huxley wrote to fellow author George Orwell, who penned *Nineteen Eighty-Four* (published as *1984*); Orwell's novel knocked "Anglo dystopia." Orwell envisioned life under Anglo-American global fascism. Three decades later, after the world's population became damaged by "perpetual war, omnipresent government surveillance and propaganda," Orwell's predictions began to manifest like those of Huxley.

In Orwell's novel, England became a super-power named Oceania, ruled by a political 'Party' that employed 'Thought Police.' The officials in the novel persecuted individualism and independent thinking. Comparing these impositions and censorship to today's world is chilling.

The Party's Orwellian leader, 'Big Brother,' committed official deceptions, secret surveillance, and used brazenly misleading terminology. Converting the terms "vaccinations" or "inoculations" to "*immunizations*" is one such fraud. The Orwellian leader, like Dr. Anthony Fauci, personified totalitarianism or authoritarian control. We witness this today following COVID's emergence. The world's most powerful nations follow Fauci's vaccination recommendations.

Huxley wrote to Orwell, "how profoundly important the book is. Within the next generation, I believe that the world's leaders will discover that infant conditioning and narcohypnosis" will be commonplace. That is, the use of drugs, the media, and bioelectronic vaccine devices will confuse and control populations, even putting civilization into a 'trance state' from which most people never recover.

Quoting Huxley, these methods of mind control "are more efficient, as instruments of government, than clubs and prisons... [T]he lust for power can be just as completely satisfied by suggesting people into loving their servitude as by flogging them and kicking them into obedience."

As Klaus Schwab was quoted as lecturing, "You'll own nothing, and you'll be happy about it."[10]

Introduction and Background Intelligence

"In its most pessimistic, dehumanized form, the Fourth Industrial Revolution," Schwab concluded, "may indeed have the potential to 'robotize' humanity and thus to deprive us of our heart and soul."

"In its most pessimistic, dehumanized form, the Fourth Industrial Revolution may indeed have the potential to "robotize" humanity and thus to deprive us of our heart and soul."

Klaus Schwab
Founder and Executive Chairman
World Economic Forum (2015/2016)

As Professor James Giordano lectures,[11] privy to this intelligence and the intelligence industry's most advanced bioelectronic neuroscience, "The mind is the seat of the self. . . . There is great power there." He then asks rhetorically, "Who gets the 'goodies' in brain science?"

Some people who were vaccinated have actually reported losing their "soul connection" to God. Who benefits from this, and how?

As you will learn in the forthcoming pages, those complicit in the COVID Coup benefit from this demonic outcome. The COVID plague is fodder for the Fourth Industrial Revolution Coup. The claimed remedial mRNA vaccines pushed on humanity have the potential to "robotize" civilization. This technology intertwines with the preparatory seduction called the "Metaverse."[12] Officials conceal the 'Big Picture' wherein this nano-bioelectronic capability in the COVID vaccines advances like the Deep State's Trojan horse delivering humanity's existential crisis.

Background Intelligence on Lab Viruses and Ties to AIDS

Briefly summarizing this author's previous publications that consider the source of COVID's "gain-of-function," attributable to four AIDS virus genes[13] spliced into the attachment and transmission device called the "Spike protein" or "S-protein," the National Institutes of Health (NIH) played major roles in sourcing and/or secreting many new viruses and man-made pandemics. So too did the National Academy of Sciences, National Research Council (NASNRC), the U.S. Centers for Disease Control and Prevention (CDC), the National Cancer Institute (NCI), the Food and Drug Administration (FDA), and military agencies that largely funded the mostly secret Special Virus Cancer Program,[14] (SVCP).

Rockefeller-related special interests heavily financed the SVCP. In 1969, Dr. Henry Kissinger, as National Security Advisor under Richard Nixon, and Vice President Nelson Rockefeller's subordinate, ordered Admiral Zumwalt of the U.S. Navy to reassess America's biological and chemical weapons capabilities.

Few realized that all of that fell under the population control program that featured mind-control and mass media persuasion known as MKULTRA. From this intelligence, Kissinger, under Nelson Rockefeller's influence, selected the option[15] to develop AIDS-like and Ebola-like biological weapons for "dual use," that is, commercial applications in vaccines and for mass depopulation.

Kissinger's and Rockefeller's conflicting commercial and military interests were heavily concealed. The Merck Drug Company, the largest vaccine maker, had received the lion's share of the Nazi war-chest at the end of WWII. The Rockefeller family had partnered with IG Farben—Germany's leading petrochemical-pharmaceutical combine.

After the war, IG Farben was 'decartelized'—broken up—into the main corporations growing with the investments of the Old World and New World elite. Bayar AG (today, Bayer Co.) was a major successor-in-interest of that cartel. The Old World's Merck drug family benefitted. Incredibly, Kissinger and Associates, Kissinger's business-management company, eventually openly publicized Kissinger's direction for this Merck Company.

Additionally, grossly neglected by the 'conspiracy of silence,' George W. Merck, the company's President, was America's biological weapons industry director, personally appointed by President Roosevelt at the beginning of WWII.

After researching these corporate and political connections, I found the "smoking gun" proving HIV/AIDS, Ebola, and Ebola's mother—the Marburg virus—originated in certain NCI and NIH-funded labs.[30, 44]

The media, heavily influenced if not completely controlled by the CIA, concealed certain revealing NIH/NCI contracts I unearthed: NIH Grant No. 71-2025 and 71-2059. These "Litton Bionetics" and Merck contracts satisfied the primary goal of MKULTRA/MKNAOMI population management. (MKNAOMI was a subordinate MKULTRA program advancing biological weapons, especially lab-engineered viruses, presumably to win the 'War on Cancer.')[30, 44]

The Order to develop and deliver the AIDS-like and Ebola-like immune-destroying cancer viruses went from Kissinger to Kissinger's

Introduction and Background Intelligence

White House counterpart, Roy Ash. Ash's company, Litton Industries — a major military weapons manufacturer— got the contract. Under the 71-2025 commission, Litton was granted $2 million annually for five years, 1970 thru 1975.[44]

Then, through Merck's Contract No. 71-2059, this Kissinger-managed firm developed the first "anti-cancer vaccines." Scientists used contaminated animals supplied by Litton Bionetics to develop new viruses and vaccines. For instance, they used chimps from Central Africa to grow hepatitis B virus mutants and test certain 'cures.'[44]

Censorship of these revelations, and the Rockefeller Commission's conflicting interests with Merck and eugenics/depopulation, was considered a top priority for the CIA. That's why my many works in this field have been censored.

Censorship and diversionary propaganda were vitally important to U.S. National Security, economic and foreign policies. These involved Russia and China. The mind-control and population control drugs and vaccines were of utmost importance.

However, few people know that the CIA/MKULTRA Big Pharma drug cartel operations in *Hawaii* played a key role. Trafficking of anything and everything to the U.S. was largely controlled in the Pacific from Pearl Harbor through the U.S. Navy and the CIA in Bethesda, Honolulu, and Langley. Intelligence operations at Langley were eventually intimately intertwined with Booz-Allen Hamilton in Honolulu. The flow of intelligence, and mind-control drugs such as fentanyl, or opium and heroin from Afghanistan through Asia and the Pacific to America, hung in the balance of these covert operations and mass media diversions.

In this context, my many monumental writings detailing the SVCP[16] provide crucial background intelligence to discern the MKULTRA program's vast commercial reach into the realm of vaccines and drugs, including today's response to COVID-19.

The Power of Propaganda to Conceal the Origin of AIDS

1976 was the first year the CIA/NCI MKNAOMI enterprise released Kissinger's Ebola acquisition from a bio-lab refrigerator in Zaire (Congo), Africa. Equally nefarious, biowarfare was synchronously waged in New Jersey.

In an Affidavit to federal investigators, I wrote, "It is a well-established fact that 'outbreaks' have been caused by laboratory 'accidents.' For instance, the 1977 Influenza A outbreak of human ('swine flu') H1N1 that went extinct for twenty years between 1957 and 1977 suddenly re-emerged immediately following: (a) the suspicious unexplained 1976 military outbreak at Fort Dix, New Jersey of this strain that was most likely a covert military experiment; and (b) the subsequent swine flu deadly vaccination program that followed the Fort Dix outbreak, and media-driven fright; was attributed to a 'laboratory source' according to doctors Zimmer and Burke in the *New England Journal of Medicine* (July 16, 2009; Vol.361:279-285)."

I advised federal officials that the November 1977 sudden reemergence of this Influenza A H1N1 strain in the former Soviet Union was best explained by the NCI's 1978 update of the SVCP (Library call number: E20.3152; V81/977 and 78-21195).

This important (albeit secreted) evidence revealed the June 15, 1976, contract (N01-CP-6-1047) with the American Type Culture Collection, a private commercial collaboration with the Rockefeller/federal NAS-NRC/NCI/military-financed Yale virology program. This organization supplied 'virus materials [and technical advice]... to investigators throughout the world' via a "US-USSR Agreement."

In fact, on closer examination, this dangerous breach of bioweapons-development know-how violated Cold War national security interests.

In essence, their contracts showed that the Deep State held only illusory interests in competing against Russia in the battle between Capitalism v. Communism.

Each 'ism' was on equal footing in the world of biotechnology, virology, vaccinology, and investment banking. But it was convenient to frighten the public with America's 'arms race' against Russia, nonetheless.

Virus materials cited in *Emerging Viruses: AIDS & Ebola—Nature, Accident or Intentional?* and the SVCP contracts included numerous infectious agents. Influenza, parainfluenza, and even laboratory recombination of influenza with acute lymphocytic leukemia viruses were among the "candidate viruses" that could be used for biological warfare, disease induction, and population control. These combinations of the flu with cancer viruses might spread fast-acting lymph cancers by sneezing.

This was obviously no 'joke.' It was not a 'foolish conspiracy theory.' These are facts that 'science skeptics' dared not address. And so, they evaded the subject, and the beat goes on.

Today we see the grand harvest from SVCP-seeds planted during the 1960s and '70s. We now see new laws mandating viral vaccines for school children prompted by Nobel Prize-winning accomplishments in largely censored virology and concomitant public deceptions. The MKULTRA/MKNAOMI NCI/NIH programs have administered this bioweapons risk. This genocidal enterprise continues despite official assurances that biological warfare programs ended with the Rockefeller Commission investigation. It did not!

To show you how severely the public has been hoodwinked, if you think health officials honorably and freely serve to protect humanity against cancers or research viruses, reconsider and grasp the sophistication by which the CIA-orchestrated media makes villains heroes and heroes villains.

The CIA's 'New Age' Covert PSYOPS

'Free spirits' cannot co-exist with COVID mandates, nor New Age 'enlightenment,' as you will increasingly realize as you read this book.

The New Age Metaphysical Movement is intimately intertwined with covert MKULTRA PSYOPS, as is submission to COVID mandates. The New Age/ New Drug movement emerged contemporaneously with the "hippie movement" and the street drug scene. The "turn on and drop out" doctrine was promoted by Timothy Leary, Alan Watts, and the Harvard faculty. According to *Wikipedia*, this "subculture" started in the early 1970s based on ideas present in the 1960s counterculture.

The 'Human Potential Movement' was similarly contrived under the MKULTRA umbrella. Warner Erhard and "EST" was a prime example of what associates of Alan Watts achieved, including advances in "neurolinguistic programming. I should know about this persuasive science in social practice since I naively participated.

As the New Age Movement advanced, organized religions became challenged by the diverse assimilation of spiritual beliefs and disciplines. The New Age "spiritual movement" promoted by the mainstream media, metaphysical seminar centers, and bookstores "further popularized the term [New Age] as a label for the alternative spiritual subculture. It included practices such as meditation, channeling, crystal healing, astral projection, psychic experience, holistic health, simple living, and environmentalism; or belief in phenomena such as Earth mysteries, ancient astronauts, extraterrestrial life, unidentified flying objects, crop circles, and reincarnation... metaphysical traditions" infused with "influences from self-help and motivational psychology, holistic health, parapsychology, consciousness research, and quantum physics," Wikipedia explains.

Two entities founded in 1962: "the Esalen Institute in Big Sur, California, and the Findhorn Foundation —an intentional community which continues to operate the Findhorn Eco-village near Findhorn, Moray, Scotland—played instrumental roles during the early growth period of the New Age Movement," *Wikipedia continues.*

Metaphysical lecturers and proponents of psychedelic drugs set the stage for current advances in neuroscience and projects like Metaverse, wherein physical reality and spirituality are questioned. The CIA and members of this movement prescribed hallucinogens for "spiritual growth and development." They frequented Esalen, Findhorn, and other "New Age" Mecca's as this counterculture emerged. They extensively promoted the benefits of using LSD, mescaline, psilocybin, ayahuasca, and marijuana for "self-actualization" and "divine encounters."

These lecturers were mainly led by two pioneers in this field, Dr. Stanley Krippner[17] and Terence McKenna, both involved in the CIA's psychotropic drug enterprise advanced by Huxley, Watts, Leary, and their Harvard minions. You can click to read about project STARGATE and KRIPPNER's CONTRACT with the CIA and review the "files in parapsychology."[18] This is a declassified part of Krippner's assignment for the CIA. You can also read Terence McKenna's CIA FOIA Reply.[19] For a complete review of the CIA's "remote viewing" PROJECT STARGATE, involving Krippner (from which the Hollywood movie *The Men Who Stare at Goats* was conceived), click Stargate Research CIA Full Report.[20]

As introduced previously, the COVID-19 pandemic, and pharmaceutical remedies, depends on the media and mass mind-control. The clandestine effort to identify and develop drugs to secure military or social mind-control engineering began in the 1930s with the Rockefeller Standard Oil Company's partnership with IG Farben and the Third Reich. These parties advanced the "eugenics" agenda, otherwise called "racial hygiene." The IG Farben, Bayar AG, Hoesch, BASF, and Standard Oil propagandists claimed to be righteously advancing a superior race of humans when they tested their new drugs on holocaust victims. They significantly depopulated European Jews, gay people, Africans, devout Christians, and conscientious objectors with religious fervor.

Following WWII, the CIA's "Project Paperclip" began simultaneously with the creation of the CIA. Thousands of Nazi scientists were exonerated, protected, and enlisted in American corporate and military ventures. In 1947, the U.S. Navy" initiated "Project Charter" to continue related "Nazi experiments" in extracting truth [by chemical means] from unwilling subjects... especially using hallucinogenic mind-altering drugs such as mescaline, LSD, and more. A Federal Government report[21] cites 1500 Nazi scientists were employed by the American military, intelligence, and commercial communities.

Another reported MKULTRA effort was the Navy's top-secret "Perfect Concussion" project, which planned to use acoustic frequencies to erase memory. The program, supposedly, was scrubbed. Yet, this capability is heralded today by Dr. James Giordano and Dr. Charles Morgan.[22] Both lecture on "Neurobiology and War" as experts in the field of "psycho-neurobiology of resilience in elite soldiers." These cutting-edge innovations include erasing human memories and emotional responses to traumas using bioelectronic devices and energy frequencies. This intertwines with recent governmental approvals of dimethyltryptamine (DMT; a.k.a., ayahuasca), psilocybin, and even LSD to treat post-traumatic stress disorders in soldiers and other sufferers.

There is a long history, a 'pattern-and-practice,' of the CIA abusing drugs and energy weapons to impact and damage civilian populations.

Notably, in a 1951 Cold War population control experiment in France, the CIA poisoned an entire community with LSD in "The Pont St. Esprit Incident."[23] It left five people dead, and hundreds psychologically damaged after eating bread poisoned with the hallucinogen.

A few months earlier, the CIA and U.S. Navy targeted unwitting citizens in San Francisco, the New York subway system, the Pennsylvania Turnpike, and even the FDA building in Washington, DC, with weak biological weapons to test the dispersion rate of stronger ones.

Headed by Dr. Sidney Gottlieb, drug experiments conducted by Nazi medical director Walter Schreiber, in support of the agency's MKULTRA project, began by order of CIA director Allen Welsh Dulles on April 13, 1953. Under Gottlieb, in 1964, a project named MK-

SEARCH attempted to produce a perfect "truth drug" for interrogating suspected spies. It also explored other mind control drugs and electronic technologies for military and commercial applications.

An investigation by the late Idaho Senator Frank Church, called "The Church Commission," revealed multiple assaults on unwitting Americans' bodies and minds conducted by the CIA, FBI, and U.S. Navy, under the guise of "natural security."

According to mountainous evidence compiled by Princeton University, the Dulles Brothers —John Foster and Allen— were commercial agents for the Rockefeller Trust. Researchers and numerous scholars were complicit in criminal activities, as evidenced in a 1992 declassified report[24] detailing Rockefeller's political, military, and 'national security' influence.

The Dulles Brothers were complicit in the global drug enterprise. Alan Dulles had been a Rockefeller Standard Oil company tax attorney who had authorized mind control drugs and techniques used on multiple groups and subjects. Supposedly, according to the agency's defense following MKULTRA coming under public scrutiny, the CIA was interested in manipulating foreign leaders, spies, and prisoners using drugs and electronics and admittedly invented several schemes to damage and destroy Fidel Castro.

But the commercial gain for the Rockefeller-controlled multinational corporations profiting from pharmaceutical and bioelectronic developments was clearly within the program's potential, according to the Rockefeller Commission Report that heralds "Section A. The Testing of Scientific and Technological Developments Within the United States,"[25] and under Subsection "1. The Testing of Behavior-Influencing Drugs on Unsuspecting Subjects Within the United States."

This American agenda was purportedly prompted by reports that the Soviet Union was experimenting with such drugs and electronic devices. According to the Rockefeller Commission Report on CIA, Drugs and Social Control Research, studies explored the effects of radiation, electric shock, psychology, psychiatry, sociology, and harassment substances, according to the Rockefeller Commission Report on CIA, Drugs and Social Control Research.pdf.[26] Musical frequencies were not immune to war research and 'non-lethal' crowd control.[27]

Incorporating psychology, psychiatry, and sociological applications for widespread 'brainwashing' and 'controlling human behavior,' in

Introduction and Background Intelligence

contrast to First Amendment rights and American civil liberties, has since raised great concerns.

The duty of the free press to expose these illegal activities, and what has been covertly ongoing commercially and militarily regarding this industry, is generally neglected by the corporate-controlled CIA-complicit media.

But the deliberate destruction of MKULTRA records in 1973 by order of CIA director Richard Helms has made it nearly impossible for journalists and other investigators to gain comprehensive data on more than 150 sub-projects sponsored by the CIA pursuant to mind control.

Wikipedia, a "virtual mouthpiece" for CIA propaganda, further explains, "because drugs traded on the black market can provide a secretive source of money, they have long been used by organizations such as the U.S. Central Intelligence Agency to fund covert operations and proxy wars."

CIA involvement in heroin trafficking began with the French Connection in Marseille. It continued with anti-Communist operations in Southeast Asia.[28] In the early 1980s, the CIA used cocaine as a medium to launder money in Central America (allegedly as part of the Iran–Contra affair).

In many cases, academic researchers were funded through grants from CIA front organizations, unaware their work was used for illegal purposes or to aid-and-abet largely covert social control commercial "experiments."

These covert activities impacted the health and safety of individuals and groups beyond those participating in such experiments. To be sure, one need only consider the evidence linking the CIA's Jonestown Massacre to the pattern of organized crime and intelligence agency activity surrounding Jim Jones and the People's Temple before and after the mass murder of 913 duped cult followers.[29]

Summarily, Western intelligence has grossly corrupted "free thought" and "democracy" in America, favoring population control at every level—physical, mental, emotional, spiritual, economic, and political.

Nobel Prizes for Criminally Negligent Manslaughter

Foreshadowing the emergence of the COVID-19 lab virus, two researchers were awarded the Nobel Prize in medicine for their "discoveries" of two other "novel" viruses. These pathogens were reported to "cause" AIDS and cervical cancers. Both "sexually-transmitted" diseases emerged clinically, like HIV/AIDS and Ebola, during the mid-to-late 1970s. At the same time, herpes was exploding across the United States and overseas. Sex was blamed for AIDS and herpes rather than the bioengineering of lab viruses.

As mentioned, HIV/AIDS broke out in NYC among gay hepatitis B vaccine recipients. This test site was near Cold Spring Harbor Labs, where genetic experiments for "racial hygiene" were pioneered. The Merck Drug Company's facilities are also not far away in Pennsylvania. Swine Flu (H1N1) suddenly emerged a few miles away at the pig-free Fort Dix Army base. Later you will learn that the mailed anthrax following 9/11 was sourced from this same locale.

The viruses, HIV/AIDS, H1N1, and the human papillomavirus (HPV), that is, the "cervical cancer microbe," along with the sexually transmitted hepatitis B virus (HBV), like the 'pig flu,' prompted dramatic changes in society. These viruses and their outbreaks were exploited politically and economically, much as COVID-19 has been since late 2019. Healthcare was called upon to administer urgent care and vaccines. The HPV vaccine was mandated for school children, as it was for polio.

I presented my objections to this agenda to the State of Hawaii, where I resided when HPV was being pushed by bribed Democrat, State Senator Roselyn Baker. My partner, Sherri Kane, and I exposed Baker[30] as a Big Pharma shill on *JudicialCorruptionNews.com*. She received $250,000 in payoffs concealed from the Campaign Finance Committee. The Blackrock[31] "iShares" bribe came through the state's leading lobbyists–George A. Morris and John H Radcliff of Capitol Consultants. This matter that the election committee whitewashed was stonewalled by the State's Ethics Committee. Pursuant to the COVID Coup, Blackrock is the world's largest asset manager pouring money into China, adequately evidencing globalism and the 'Deep State' financial forces administering the Coup and "Great Global Reset."

I submitted evidence and testimony to the State Health Department as Baker was being exonerated. The evidence I submitted proved scientific fraud and fraudulent concealment of vaccination risks. The references I made to the SVCP contracts challenged the two Nobel 'discoveries.' Public health measures, vaccination mandates, and

Introduction and Background Intelligence

vaccine safety claims, in general, were seriously undermined by the scientific papers I filed and referenced in my submission.

Again, in October 2018, I testified before Hawaii officials in opposition to the changes[32] to vaccination laws affecting school children and enrollment, including a new "List of Required Vaccinations." That foreshadowed the Biden administration's similar vaccine mandates for school children.

"Kindergarten – 12th Grade Attendance" was 'required' to get the hepatitis A and B vaccines and HPV vaccine. The latter two are widely known to be "sexually transmitted diseases," not risking the lives of young children.

A quick review of the package insert for Gardasil (for HPV)[33], for example, raised red flags.[33] The HPV vaccine for cervical cancer was made by Merck. The testing procedure showed the <u>mixing-up of two control groups</u>. One received saline and the other "amorphous aluminum hydroxyphosphate sulfate (AAHS) adjuvant." Combining the two control groups <u>confounded the true assessment of risk</u>; since the AAHS "can have a profound influence on the magnitude and quality of the immune response to the HPV vaccine," scientists published.[34]

The clear-and-present fallacies and dangers in mandating and administering vaccinations and drug prescriptions are presented in forthcoming sections and chapters. These practices and public health policies, presumed to be beneficial, are risking civilization's genetic poisoning, mental degeneration, psycho-spiritual retardation, and demonic possession, ultimately risking species extinction.

The following facts also evidence the depth and efficacy of modern mass-mediated mind-control impacting every field of science, especially health science. These facts have profound implications for human rights and pharmaceutical commerce.

The Façade of Legitimacy: The Nobel Committee

Further intelligence on the syndicate governing corrupt virology and vaccinology, reflective of the COVID-19 enterprise, is provided in this section. According to U.S. Government records published during the 1970s, the NIH and NCI assembled an international coalition to study cancer viruses for vaccine developments during the 1960s thru 1978. The unprecedented collaboration was called the "Special Virus Cancer Program" or "SVCP." This program was made known only to industry insiders and paid researchers.[35] This comprehensive international research group sought "candidate viruses" that "caused" cancers presumably amendable to "preventative vaccines." This effort predated President Nixon's "War on Cancer" by a decade.

In 2008, the Nobel Prize in Medicine[36] went to German virologist Harald zur Hausen for reportedly "discovering" the "**cause** of cervical cancer" in *1976*–the "oncogenic human papillomavirus."[37] As mentioned, that was a banner year for the emergence of several genocidal biological weapons, including Ebola and the Swine Flu. Zur Hausen's "discovery" was made during this extraordinary time when the NCI was completing the SVCP.

So a review of zur Hausen's science pursuant to the emergence of HPV is reasonable and leaves an impression of corruption in medicine that is chilling.

In 1983, zur Hausen reportedly "isolated" the new "tumourigenic HPV16 and 18 strains." The following year, in 1984, zur Hausen reportedly "cloned HPV16 and 18 from patients with cervical cancer. These HPV types were found in about 70% of cervical cancer biopsies throughout the world," according to the press release issued by the Karolinska Institute–home to the esteemed Nobel Prize Committee. [37]

Public relations press officers for the KI and Nobel judges claimed zur Hausen's "discovery" and "theory" of cancer causation by a single "tumorigenic agent" flew in the face of "dogma."[37] Previously, the SVCP/NCI/MKNAOMI coalition held that "co-factors" for "co-carcinogenesis" involved multiple "risk factors."

At the time, viruses were not generally-believed to be the cause of cancers and other illnesses. There was 'general acceptance' in science that a co-factor model was more accurate. During the 1960s and 70s, there was a general acceptance that myriad factors such as stress, risky lifestyles, biological and/or chemical agents, or their cumulative damage to genes and human immunity resulted in cancers.

This "immuno-suppression" model was the working premise of the SVCP and NCI's efforts. Chemicals, environmental factors, including radiation, and biological agents such as viruses were known to cause genetic damage prompting cancers. Cells mutated, forming tumors and malignancies from the overwhelming assaults to DNA/RNA and cell repair.

That was the 'dogma' in medicine and science before zur Hausen postulated that certain types of **herpes viruses** "caused"

Introduction and Background Intelligence

malignancies.[38] These DNA viruses, KI officials reported, "could exist in a non-productive state in the tumours." Zur Hausen's "specific searches for viral DNA" sourced his "discovery" that "led to characterization of the natural history of HPV infection, an understanding of mechanisms of HPV-induced carcinogenesis and the development of prophylactic vaccines against HPV acquisition," news outlets heralded.[37]

Simultaneously, Françoise Barré-Sinoussi and Luc Montagnier were likewise credited for having "discovered" the single "cause of AIDS–the human immunodeficiency virus" (HIV-1). Virus production, the KI press reported, "was identified in lymphocytes from patients with enlarged lymph nodes in early stages of acquired immunodeficiency, and in blood from patients with late stage disease."[37]

"The KI's press release added, "They characterized this retrovirus as the first known human lentivirus [i.e., slow-acting virus] based on its morphological, biochemical and immunological properties. HIV impaired the immune system because of massive virus replication and cell damage to lymphocytes. The discovery was one prerequisite for the current understanding of the biology of the disease and its antiretroviral [drug] treatment[s]."[37]

However, KI's promotions neglected this institute's important role in advancing these "discoveries" <u>more than a decade earlier</u> as a collaborator in the SVCP.

SVCP records[35] evidenced that officials at the esteemed Nobel Prize committee's KI in Sweden fraudulently concealed their agency's conflicting interests in zur Hausen's award.

The KI also neglected the reasons the Nobel awards committee rejected Dr. Robert Gallo's nomination. It had come to public knowledge that Montagnier, collaborating with Gallo in the SVCP, was defrauded by Gallo, who had renamed the "French virus" "a couple of times," wrote one of Gallo's colleagues.[38]

Apparently, Gallo acted to fraudulently conceal his use of Montagnier's virus to justify Gallo's claim of having discovered the AIDS virus before Montagnier.

At the same time, the KI promoted and commercialized their cancer enterprise's investments in "single cause etiology virology," neglecting their past efforts in early gene cloning, virus mutation experiments, pioneering lab technologies, vaccine research and developments, and subscription to the multi-factorial model of cancer.

In addition, explosive government records shown below proved that from the 1970s, the KI was involved with the NCI in recombining strains of DNA herpes viruses with strains of RNA tumor viruses isolated from patients with malignancies worldwide. The COVID-19 virus is a RNA pathogen that reportedly enables DNA herpes viruses to break out in lesions clinically.[39]

The SVCP collaborators mutated, hybridized, and recombined all of these virus types. They "transformed" benign or "latent" herpes viruses into model cancer triggers. These studies were well-justified in science's search for the causes and cures for cancer. [35]

But it is unreasonable, irresponsible, and incriminating that this science and intelligence would be secreted or suppressed. The general acceptance of cancer and vaccination doctrines in medicine, science, and society, was heavily influenced by this concealment.

The KI's esteemed Nobel Prize Committee deprived the world of honest information, including KI's conflicting interests and organizational bias. They deceptively institutionalized thereby the virus/cancer medical model. Their deception affected public health, consumer safety, and informed choice-making. Potentially catastrophic viral mutation experiments risked world health as a result of this 'fake science' and 'scientific evidence tampering.'

Science is presumed to be honest. The medical-legal community holds dishonesty as actionable. Fraudulent concealment to spur an industry, or silencing truth to spark commerce, defies ethics and obstructs justice.

The sequestering of the SVCP Progress Reports[35] was akin to "evidence tampering" in law. Denying due process in medical discovery risked millions of lives, potentially even billions of deaths. It is arguably unconscionable, irresponsible, even treasonous, and certainly genocidal what has been ongoing in virology and vaccinology.

Censorship in Health Science and Protection Racketeering

These wrongdoings and "impressions of impropriety" were compounded by superficial dismissals of the SVCP's questionable methods, materials, and outcomes.[40] Laboratory creations of cancer-linked viruses HPV,

HBV, and HIV are further vetted by the NIH/NCI records I show below.[41]

In 2002, the United States General Accounting Office (USGAO) white-washed this subject.[40] The suspects discouraged this dialogue rather than thank honorable colleagues, citizens, and the few politicians who bravely encouraged transparency, discovery, and open debate over these matters of vaccine safety, lab virus mutations, and the cancer industry.

Vaccine industrialists argued in favor of such censorship. Stockholders were the beneficiaries of this censorship.[42] [43] Their media produced diversionary propaganda. These facts and acts defied ethical norms and risked public health and confidence in the vaccine industry and in government.[44]

Vaccine advocates and government officials have argued that open dialogue would undermine compliance with increasingly "mandated" vaccinations. So the general acceptance of vaccination policies was largely based on these dangerous concealments.[44,Error! Bookmark not defined.,45]

Equally troubling, to sustain the silence and combat opponents, certain vaccine industrialists commissioned public relations firms, social media "trolls," and Internet "skeptics" to issue propaganda to protect the global enterprise. Like protection racketeers, these propagandists smeared whistleblowers and authors who raised vaccine debate.[46] [47]

The following facts evidence the seriousness of the science being concealed. An international scientific open dialogue should have followed 'my 2001 report in *Medical Hypotheses*.[48] Instead, United Nations' official censorship followed.[49]

When, if ever, has the U.N. censored a scientist? This unprecedented action affirmed and concealed the U.N.'s submission to Rockefeller-Big Pharma influence.

The compelling evidence from government contracts begs for scientific scrutiny and public discourse. These findings of facts show powerful conflicting interests impose public acceptance of vaccines and their mandates.[50]

This previously secreted knowledge compels reasonable and responsible citizens to act in support of human rights, health freedoms, and protection from drug industrialists committing genocide.

Shocking Government Documents Evidence Cancer Genocide

Shown below are photocopies from the 1971 and 1972 SVCP, Progress Reports 8 and 9. These government records proved the program's initiation in 1962 when Sarah Stewart and Bernice Eddy discovered the "SE [Stewart Eddy] polyomavirus" that contaminated Merck's polio vaccines. The two scientists subsequently proved SE polyoma, renamed SV40, caused cancers.[35]

These SVCP documents were published exclusively for internal review by collaborators in this program. The enterprise was governed by the U.S. Department of Health, Education, and Welfare, Public Health Service, National Institutes of Health, Division of Cancer Cause and Prevention, and the National Cancer Institute. These documents were rarely found in medical libraries before 1996 when I published ***Emerging Viruses: AIDS & Ebola–Nature, Accident or Intentional?*** wherein I made known and analyzed these documents.[51]

Page 410 of the SVCP-NCI Record #9, as shown on the next page, identified the KI. The Nobel Prize organization is cited below Johns Hopkins University (America's leading hepatitis B research enterprise). Above the KI listing was "**Life Sciences**, Inc."–a private company that no longer exists. It changed its name to Life Sciences Advanced Technologies, Inc.

The "Life Sciences Industry" is the premier drug and vaccine (i.e., pharmaceutical and biotechnology) enterprise also referred to as "Big Pharma." The Life Sciences company began in 1962. As stated above, this date coincides with the year the SVCP-NCI program began. It immediately followed the discovery of the SE polyomavirus and cancers triggered by Merck's polio vaccine. The polyomavirus was renamed the "SV40" virus by Merck's chief vaccine developer, Dr. Maurice Hilleman.

There is 'general acceptance' today that SV40 contaminated polio vaccines spread cancers worldwide.[35]

From 1962 forward, "Life Sciences" (actually "Death Sciences") evolved to become Big Pharma's leading provider of molecular biology products and services, as proven by the screenshot below.

It is also known that the Life Sciences Building at Los Alamos Laboratory is where Dr. Gerald Myers et al. published his "Big Bang" theory of HIV/AIDS's emergence in the "early 1970s." They based their conclusions on the use of "genomic sequencing" products and

Introduction and Background Intelligence

services.[51,52] Their findings corroborate my conclusion that HIV/AIDS emerged from the 1972 thru 1974 hepatitis B vaccine trials.

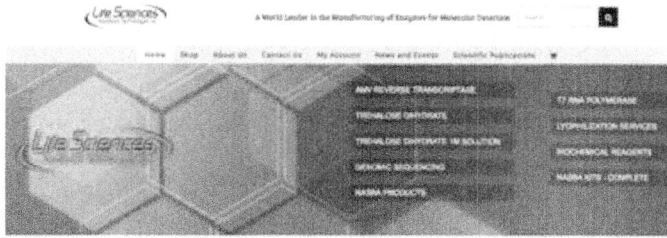

Recall "The 'Patient Zero' theory that claimed gay promiscuity sourced AIDS? That was determined to be false, just like false reports that a bat, or pangolin, sourced COVID-19. Myers's discoveries solidly debunked the Patient Zero theory.

Randy Shiltz, in the book, *And the Band Played On,* initiated the alleged AIDS origin 'hoax." Notice the CIA-monitored *Wikipedia* heralds Shiltz and his book but censored all discussion about my book, *Emerging Viruses. Wikipedia* even struck my entire bio! Myers's "Big Bang" publications corroborated my research, not the heavily controlled and widely promoted propaganda about Shiltz and 'The Band.' The corporate-controlled media diverted from the laboratory-origin and vaccine-transmission theory. One telling graphic is inset below.

Myers's Big Bang chart shows HIV/AIDS exploded simultaneously with the hepatitis B vaccination campaign that targeted the precise populations initially decimated. A single Patient Zero could never have accomplished this simultaneous explosion of HIV strains.

The Patient Zero falsehood also diverted from Dr. Gallo's deceptions. AIDS virus "co-discoverer" (actually co-developer) Gallo not only stole Montagnier's lab virus for exclusive glory but deceptively denied his cancer virus productions while working for the NIH/NCI.

In 1996, I confronted Gallo on this incriminating evidence at the XI International Conference on AIDS in Vancouver, Canada. You can view the interchange between Dr. Gallo and myself.[53] More than 6 million concerned citizens have watched Gallo lying and covering up his lab virus research and developments.

1976 was a 'banner year' for virology. That year zur Hausen first published his "discovery" of HPV. At the same time, Ebola was outbreaking in Central Africa, as HIV/AIDS spread from tainted hep. B vaccinations; and 1976, the first Gay-Related Immunodeficiency Disease (GRID, i.e., AIDS) patient was identified in New York City.[37]

These events also coincide with the 2-to-5 year 'incubation period' for HIV to express itself following receipt of tainted hep. B vaccinations. AIDS-related leukemia, lymphoma, and sarcoma cancer lesions followed system-wide immunosuppression building over 2-to-5 years.

Accordingly, the first GRID/AIDS cases would have acquired their infections between 1971 and 1975, precisely as I concluded from reviewing the research literature and SVCP contracts. My conclusion is also consistent with Myers's determination,[48] and shows the virology

Introduction and Background Intelligence

cartel hard at work during the early 1970s to deliver new plagues using lab viruses and "novel" vaccines.

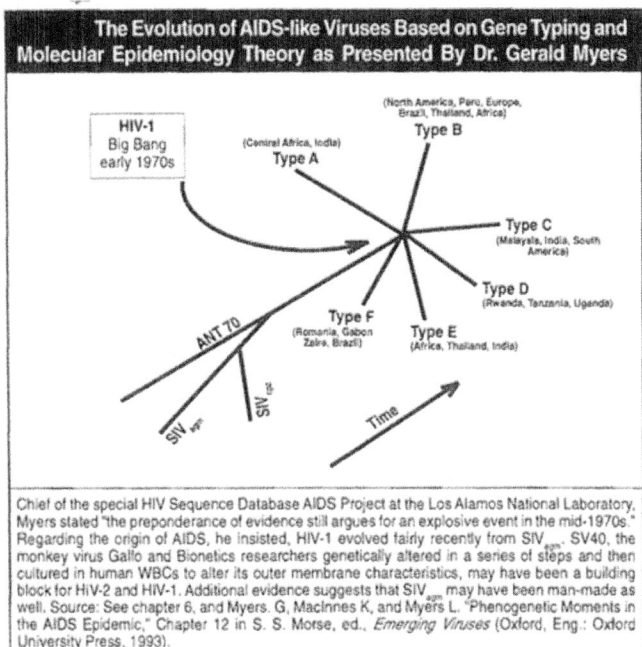

Chief of the special HIV Sequence Database AIDS Project at the Los Alamos National Laboratory, Myers stated "the preponderance of evidence still argues for an explosive event in the mid-1970s." Regarding the origin of AIDS, he insisted, HIV-1 evolved fairly recently from SIV_{agm}. SV40, the monkey virus Gallo and Bionetics researchers genetically altered in a series of steps and then cultured in human WBCs to alter its outer membrane characteristics, may have been a building block for HIV-2 and HIV-1. Additional evidence suggests that SIV_{agm} may have been man-made as well. Source: See chapter 6, and Myers. G, MacInnes K, and Myers L. "Phenogenetic Moments in the AIDS Epidemic," Chapter 12 in S. S. Morse, ed., *Emerging Viruses* (Oxford, Eng.: Oxford University Press, 1993).

Under "Major Findings" of the University of Pennsylvania's Wister group, the SVCP-NCI published in 1972 what was accomplished in 1971: "Rescue attempts have been initiated employing human sarcomas and leukemias in a variety of combinations designed to activate latent viral genomes."**Error! Bookmark not defined.** That document has been a 'smoking gun' since 1971-1972. This confirmed precisely when the tainted hepatitis B vaccines were given to the gay men in NYC and Blacks in Central Africa, AND explained the never-before-seen 'cancer complex' combining sarcomas and leukemias diagnostic for AIDS. It is unreasonable to conclude HIV/AIDS and its "novel" cancer complex came from anywhere other than the SVCP.

The next SCVP-NCI Report #9 photocopy shown below contains Johns Hopkins University's contract description. Like the Hopkins "contact page" reprinted above, Hopkins is cited immediately above KI's contact page, and likewise, their contract description.

The Hopkins' description states, "The contractor will compare antigens in cervical carcinoma cells with those induced in Hep[atitis-2

(B-type)] by Herpes simplex virus type I and II." From this, it is clear that the Hopkins group triggered hepatitis B infections using Herpes simplex viruses. Then they compared these viral antigens from Herpes type I and II with the antigens found in cervical cancers. These experiments risked viral recombination and outbreaks.

Recall this risky science later in this book when you read about the "spike-protein" antigen of the COVID-19 virus and the Pfizer and Moderna mRNA vaccines. The key to COVID-19 is the lab-engineered antigen that contains four genes from the AIDS virus, HIV-1.

The hepatitis B vaccine theory on HIV/AIDS's emergence comports with Myers's Big Bang lab-virus evidence, as well as reports that COVID-19 cases skyrocketed in the U.S., Mexico, and elsewhere, months after vaccinations were given to millions of people, many suffering from "breakthrough infections."[35]

Multiple strains of HIV/AIDS exploded globally, much as we see with COVID mutants.[54] HIV mutants appeared in North America and Africa in the "early 1970s," synchronous with the Merck Drug Company's administration of the first hepatitis B vaccines. Merck also produced the HPV vaccine (i.e., Gardisil)[48]

The National Institutes of Health Contract 71-2059 to Merck and Company, Inc. (shown below from Report #8) is titled "Study of Viruses in Human and Animal Neoplasia (i.e., cancer). This contract was overseen by Dr. Maurice R. Hilleman and Robert Gallo's bosses at the NCI, Robert Manaker, and Jack Gruber. The objectives of their study included developing "vaccines or other agents effective for the prophylaxis and therapy of human neoplasia [cancers] of suspected viral etiology." In that contract, it states, "At the present time, investigations will be focused upon herpes-type (DNA) viruses and "B" and "C" type (RNA) particles. Parallel studies to evolve live attenuated and killed viral vaccines in appropriate animal model systems will be conducted." Hence, it is most reasonable and responsible to suspect cervical cancers were mostly spread from vaccinations contaminated with herpes-type (DNA) viruses.

Furthermore, the SVCP contract NIH-71-2025 (from 1971 Report #8) shows Project Officer Gallo officiating for the NIH/NCI. His collaborators explained their objective of "Evaluation of long-term oncogenic effects of human and animal viral inocula in primates of various species..." Humans are primates. "Progress was reported on "An RNA-dependent DNA polymerase ["reverse-transcriptase enzyme" that characterizes HIV/AIDS as well as the SARS-CoV-HIV lab virus] similar to that associated with RNA tumor viruses." Gallo's associates detected and isolated this reverse transcriptase enzyme from "human leukemic cells but not in normal cells... The [reverse transcriptase AIDS-linked and COVID-19 virus linked] enzyme was isolated, purified, and concentrated 200-fold, making possible its further characterization and study in relation to the leukemic process in man."

Consider that the SARS/Coronavirus/HIV mutant that emerged from Wuhan is a similar RNA virus that uses this precise enzyme, reverse transcriptase, and messenger RNA (mRNA) to corrupt host DNA. Consider also that the Pfizer and Moderna mRNA vaccines work likewise, using mRNA and reverse transcriptase enzyme, to instruct the host DNA to output more HIV/AIDS-laced spike protein.

In other words, the COVID-19 virus as well as the mRNA vaccines to supposedly prevent the infection, take their marching orders from the AIDS virus and AIDS virus research documented in the SVCP. The coronaviruses and vaccines use the same unique 'pathogenic pathway,' the same cell attachment mechanism, the same enzymes to become infected, and the same intra-cellular RNA to DNA subversion. How much more evidence would reasonable jurors need to convict this viral vaccine syndicate of genocide?.[35,48,51]

More Smoking Guns Evidencing Genocide

More 'smoking gun' evidence of genocidal malfeasance appears in the U.S. Congressional Record. The 1969 Department of Defense Appropriations for 1970 record shows that on July 1, 1969, the military requisitioned HIV/AIDS and Ebola-like viruses.[55] This record states under the caption "SYNTHETIC BIOLOGICAL AGENTS":

> Within the next 5 to 10 years, it would probably be possible to make a new infective microorganism that could differ in certain important aspects from any known disease-causing organisms. Most important of these is that it might be refractory to the immunological and therapeutic processes upon which we depend to maintain our relative freedom from infectious disease. A research program to explore the feasibility of this could be completed in approximately 5 years at a total cost of $10 million.[44] [i.e., $2 million per year over 5 years.][51]

As mentioned, Litton Bionetics received this money. They used it to manufacture and distribute these 'novel' immune-system destroying bioweapons; then shipped them to vaccine-makers worldwide. The entire SVCP was, in fact, administered by Litton Bionetics on behalf of the NCI and Life Sciences industry.[48, 51]

NCI Project Officer Gallo oversaw the Bionetics group in the early 1970s. Litton was contracted (under NIH 69-2060) to administer "Support Services for the Special Virus Cancer Program" during those years. Under that contract, Litton supplied the experimental viruses, monkeys, chimpanzees, and vaccine regents to SVCP collaborators, including the Merck Drug Company under Merck's contract NIH 71-2059. That study was titled "Oncogenic Research and Vaccine Development." (Copies of these contracts are downloadable).

Litton received in the neighborhood of $2 million per year over "5 years at a total cost of $10 million" beginning in 1969-70. As you can read, that money was spent developing viruses that were descriptively and functionally identical to HIV/AIDS.

Congress authorized this dangerous research under the contrived excuse that the Russians were competing in developing such biological weapons. Meanwhile, the SVCP contracts show U.S. officials and virologists (including Gallo) traveling to Russia during that time, in the heat of the Cold War, to instruct Russian SVCP collaborators precisely how to do this genetic engineering to produce these mutant cancer viruses. The screenshot below is an excerpt from the **Congressional Record**. It correlates with the express provisions described in Litton contracts. (See: Project Reports 8 and 9.)

Introduction and Background Intelligence

Merck and Company, Inc. (NIH-71-2059)

Title: Study of Viruses in Human and Animal Neoplasia.

Contractor's Project Director: Dr. Maurice R. Hilleman

Project Officers (NCI): Dr. Robert A. Manaker
Dr. Jack Gruber

Objectives: To perform investigations designed to develop vaccines or other agents effective for the prophylaxis and therapy of human neoplasia of suspected viral etiology.

Major Findings: This is a new contract.

Significance to Biomedical Research and the Program of the Institute:
Current data support the concept that a virus or viruses are the essential element in most animal tumors studied and that viruses are probably the necessary etiological component in human neoplasia, though expression may be greatly influenced and modified by host and environmental factors. If viruses are the essential element in human cancer, then prophylaxis by vaccines to prevent or minimize infection should provide a rational approach to cancer prevention. This could be accomplished by utilization of live or killed virus vaccines or possibly by vaccines of purified virion subunits.

Vaccines would obviously provide their greatest benefit in preventing infection with oncogenic viruses transmitted horizontally after birth. However, even the possible vertical transmission of hypothetical neoplastic agents does not rule out a potential benefit from vaccines. Nononcogenic viruses may function as essential cofactors in expression of neoplasia, and immunity against such secondary agents might prevent expression of the neoplastic state. Additionally, antibody or cellular immunity may be enhanced by vaccination with homologous virus in virus-dependent cancer. Obviously this research investigation is of fundamental importance to the goals of SVCP and can make unique contributions to the total program.

Proposed Course: The investigators will devote initial efforts to developing methods for propagation, purification, concentration and specific quantitation of candidate viruses suspected or shown to cause cancer in man. At the present time, investigations will be focused upon herpes-type (DNA) viruses and "B" and "C" type (RNA) particles. Parallel studies to evolve live attenuated and killed virus vaccines in appropriate animal model systems will be conducted. Particular attention will be given to developing and applying optimal methods for viral attenuation, viral inactivation, viral quantitation, vaccine safety assessment, and vaccine potency assay.

Date Contract Initiated: March 1, 1971

BIONETICS RESEARCH LABORATORIES, INC. (NIH-71-2025)

Title: Investigations of Viral Carcinogenesis in Primates

Contractor's Project Directors: Dr. John Landon
Dr. David Valerio
Dr. Robert Ting

Project Officers (NCI): Dr. Roy Kinard
Dr. Jack Gruber
Dr. Robert Gallo

Objectives: (1) Evaluation of long-term oncogenic effects of human and animal viral inocula in primates of various species, especially newborn macaques; (2) maintenance of monkey breeding colonies and laboratories necessary for inoculation, care and monitoring of monkeys; and (3) biochemical studies of transfer RNA under conditions of neoplastic transformation and studies on the significance of RNA-dependent DNA polymerase in human leukemic tissues.

Major Findings: This contractor continues to produce over 300 excellent newborn monkeys per year. This is made possible by diligent attention to reproductive physiological states of female and male breeders. Semen evaluation, artifical insemination, vaginal cytology and ovulatory drugs are used or tried as needed.

Inoculated and control infants are hand-fed and kept in modified germ-free isolators. They are removed from isolators at about 8 weeks of age and placed in filtered air cages for months or years of observation. The holding area now contains approximately 1200 animals up to 5 years old. Approximately 300 are culled every year at a rate of about 25 per month. This is necessary to make room for young animals inoculated with new or improved virus preparations.

During the past year macaques were inoculated at birth or in utero with the Mason-Pfizer monkey mammary virus, Epstein-Barr virus, Herpesvirus saimiri, and Marek's disease virus. EB virus was given with immunostimulation and immunosuppression (ALS, prednisone, imuran). Australia antigen was given to newborn African green monkeys.

The breeding and holding colonies were surveyed for antibody to EBV. All breeders were positive and their offspring contain maternal antibody for several months. Colony-born offspring that have lost maternal antibody and are sero-negative will be surveyed periodically for conversion to the EB positive state.

An RNA-dependent DNA polymerase similar to that associated with RNA tumor viruses was detected in human leukemic cells but not in normal cells stimulat by phytohemagglutinin. The enzyme was isolated, purified and concentrated 200-fold, making possible its further characterization and study in relation to the leukemic process in man.

Significance to Biomedical Research and to the Program of the Institute: Inasmuch as tests for the biological activity of candidate human viruses will not be tested in the human species, it is imperative that another system be developed for these determinations and, subsequently for the evaluation of vaccines or other measures of control. The close phylogenetic relationship of the lower primates to man justifies utilization of these animals for these purposes. Further study of altered transfer RNA and polymerase enzymes would determine their significance in neoplastic change and provide a basis for selection of therapeutic agents.

Proposed Course: Continuation with increased emphasis on monitoring and intensive care of inoculated animals to determine if active infection occurs, effects of infection, and degree of immunosuppression when used. Further studies of human neoplasms at a molecular level will continue.

Date Contract Initiated: February 12, 1962.

> **SYNTHETIC BIOLOGICAL AGENTS**
> There are two things about the biological agent field I would like to mention. One is the possibility of technological surprise. Molecular biology is a field that is advancing very rapidly and eminent biologists believe that within a period of 5 to 10 years it would be possible to produce a synthetic biological agent, an agent that does not naturally exist and for which no natural immunity could have been acquired.
>
> The dramatic progress being made in the field of molecular biology led us to investigate the relevance of this field of science to biological warfare. A small group of experts considered this matter and provided the following observations:
> 1. All biological agents up to the present time are representatives of naturally occurring disease, and are thus known by scientists throughout the world. They are easily available to qualified scientists for research, either for offensive or defensive purposes.
> 2. Within the next 5 to 10 years, it would probably be possible to make a new infective microorganism which could differ in certain important aspects from any known disease-causing organisms. Most important of these is that it might be refractory to the immunological and therapeutic processes upon which we depend to maintain our relative freedom from infectious disease.
> 3. A research program to explore the feasibility of this could be completed in approximately 5 years at a total cost of $10 million.
> 4. It would be very difficult to establish such a program. Molecular biology is a relatively new science. There are not many highly competent scientists in the field, almost all are in university laboratories, and they are generally adequately supported from sources other than DOD. However, it was considered possible to initiate an adequate program through the National Academy of Sciences-National Research Council (NAS-NRC).
>
> Source: Department of Defense Appropriations for 1970 Hearings Before a Subcommittee of the Committee on Appropriations House of Representatives Ninety-First Congress, Part 5 Research, Development, Test, and Evaluation, Dept. of the Army, Tuesday, July 1, 1969, page 79. Washington: U.S. Government Printing Office, 1969.

Scientific Review Controverts Earlier AIDS-Origin Theories

I conclude that the confusion generated over Dr. Hillary Koprowsky's and the Wistar Institute's alleged role in the AIDS outbreak was contrived, much like a 'red-herring' diversion. This obfuscation exemplifies Deep State and media collusion to conceal or discredit accurate AIDS-origin theories.

***Rolling Stone*[56]** magazine falsely accused Koprowsky of causing the AIDS outbreak. That indictment was later retracted, but the scandal rocked medicine and is recalled by the establishment media to discredit subsequent AIDS origin theories. *Rolling Stone* lamely indicted the Salk vaccine's transmission of the aforementioned SV40 cancer virus contaminating those polio vaccines.

Sabin's oral polio vaccine (OPV) was also confirmed to have spread SV40-related cancers. These transmissions occurred during the 1950s during Koprowsky's watch.[35] That evidence, compounded here, corroborated Dr. Bernice Eddy's warning that pandemic cancers would certainly come following SV40's widespread transmission through Merck's tainted polio vaccines produced by both Salk and Sabin.

Accordingly, author Edward Hooper was not far off the mark in *The River: A Journey to the Source of HIV and AIDS*. He conjectured that the Wistar Institute's research had contaminated Albert Sabin's OPV, carrying chimpanzee simian immunodeficiency virus (SIVcpz) and AIDS with it.

SIVcpz is the closest relative to HIV in the primate world. But Hooper's critics, many "pseudo-skeptics" —propagandists paid to protect the status quo by muddling media— promptly "debunked" Hooper's claimed association after confirming that "monkey cells, not chimpanzee cells," were used by Koprowsky.[45] The "smoking gun"–the **chimpanzee** vector in the origin of HIV/AIDS remained secreted until I found and published it. I was the first to publish these connections.[57] The smoking gun chimpanzee contamination is no longer conjectural. I vetted it earlier and below.

The widespread neglect of the NCI's and Litton's actions in vaccine racketeering and reckless concealment of medical history is documented in this author's publications,[35,Error! Bookmark not defined.] The press today, biased by conflicting interests more than ever, has obscured the general acceptance among cancer virologists that vaccines have been the main vectors for deadly communicable diseases and auto-immune disorders. But this revelation doesn't sell vaccines, advertising space, or inspire public confidence.

Given the evidence in hand, it is unreasonable to believe that HIV/AIDS was not created in Litton's lab, allied with Gallo's group at the NIH/NCI. By 1972, their contract proves, this group was commissioned to produce the "gain-of-function" AIDS-like cancer viruses, much like the Wuhan lab was financed by the NIH and Fauci's NIAID to produce the COVID-19 virus as detailed in Chapters III and IV.

> During the past year macaques were inoculated at birth or in utero with the Mason-Pfizer monkey mammary virus, Epstein-Barr virus, Herpesvirus saimiri, and Marek's disease virus. EB virus was given with immunostimulation and immunosuppression (ALS, prednisone, Imuran). Australia antigen was given to newborn African green monkeys.

Further evidencing Litton's and Gallo's important roles in the origin of HIV/AIDS from the first hepatitis B vaccine trials are the contracts shown above and below. NIH-71-2025, shown above, states, "Australia antigen was given to newborn African green monkeys." "Australia antigen" (AuAg) reported in the SVCP publications was the first known agent causing hepatitis. **"AuAg" was later re-named "hepatitis B virus" and "HBV."**[48,51]

The above screenshot shows Litton's official statement that corroborates Dr. Hilleman's recorded statement, "We [at Merck] brought in the African greens. We didn't know we were importing AIDS virus at the time."[44]

Notice also that Litton tested and distributed the "Mason-Pfizer monkey mammary virus." This evidence proves Pfizer played a major early role in the development and distribution of cancer viruses, this one contaminating women's breasts.

Litton Bionetics was certainly the supplier of these viruses to Merck and the other SVCP collaborators. Litton exported the test viruses and animals, especially monkeys and chimpanzees, from their facilities, several overseas. They shipped the chimpanzees and viruses to Merck in New York under contract number NIH-69-2060 —"Support Services for the Special Virus Cancer Program," as shown below.

This record also proves viral recombination studies were conducted using known cancer triggers years before the cancer explosion. The Epstein-Barr virus (EBV) known to cause cancer was tested with "immunostimulant and immunosuppression" coupled with hepatitis B viral experiments. This was done in the same lab as the reverse-transcriptase AIDS-virus enzyme was studied with retrovirus recombination, much like the COVID virus.

The Common Practice of Renaming, "Whatever"

In a later chapter covering the conversion of the label 'hallucinogenic' or 'psychotropic' drugs to 'entheogens,' you will note it is far more appealing for marketing and broad public acceptance to change the names of things. The same contrivance is demonstrated in virology and vaccine science. For example, public knowledge is that the term 'immunization' replaced 'vaccination' for the same reason. Vaccination implies a painful, risky injection, whereas immunization happens naturally from natural exposures to germs. Natural immunization is generally painless and always beneficial.

Likewise, in virology, SVCP investigators commonly renamed viruses for fame and fortune. For example, it is widely known in virology that Dr. Hilleman, the lead vaccine developer at Merck, renamed the

"Stewart-Eddy polyomavirus (SE-polyoma)" to "SV40". The chauvinist snub stole credit from the two women who discovered SE polyoma in Merck's polio vaccine in 1962.

Then there was the Nobel-thirsty Dr. Gallo, who the KI's Nobel Prize committee snubbed for doing worse than stealing Montagnier's "discovery."

There is "General Acceptance" that Montagnier supplied Gallo with HIV for Gallo's NIH/NCI-funded studies.

"It was the French virus that Gallo's lab used for their research," the corporate-controlled media alleged.[38] "Gallo's lab notes, obtained by the *Chicago Tribune* [clearly a Deep State propaganda mill], show that the French virus was renamed a couple of times, apparently to hide the fact that it was being used. Gallo later claimed that the French virus didn't grow."[58] Given the general agreement of Gallo's untrustworthiness, it is unreasonable to believe he didn't falsify his lab notes and this entire story "obtained by the *Tribune*," that is also infamous for fabricating stories.[38]

Why would the *Tribune* go so far out of its way to generate debate over a contrived story? To <u>divert from the actual, scientifically proven, evidence-based, real story</u> of generating these viruses during the SVCP for-profit, covert military applications, and global elite population control.

This most reasonable and responsible conclusion follows the aforementioned contractual evidence and the objections raised by Robert Strecker, M.D., Ph.D., published in *Emerging Viruses: AIDS & Ebola–Nature, Accident or Intentional?*[59]

I interviewed Dr. Strecker in 1996. Strecker was the first American medical scholar who raised the lab virus AIDS-origin theory. While working with Gallo at the NCI, Strecker stated that Montagnier discovered that "Epstein-Barr-infected T-cells will just churn out AIDS viruses day after day..." Strecker explained this discovery was necessary to overcome Gallo's problem of producing enough HTLV-III viruses (HIVs) "to make enough antibody, " because as the *Chicago Tribute* story stated, "the French virus didn't grow."

Regarding this scandal—the French American AIDS fracas— Strecker opined, "That's all a lot of bull because they both had the virus, and they both knew what they were doing from day one."

Introduction and Background Intelligence

Bionetics's Little Known "Classified" History

As shown below, Litton Bionetics was the sixth-leading biological weapons contractor in 1969. Robert Gallo served as the NCI's Project Officer overseeing Litton's contract, NIH-71-2025. Gallo also served in the SVCP as the Developmental Research Segment co-director. As mentioned, Litton Bionetics was the primary cancer research supplier to the NCI and SVCP contractors. Litton supplied viruses, including retroviruses, experimental reagents, and test animals, including chimpanzees.[48,51] The SVCP Progress Report #8 makes it clear that Gallo's collaborators at Litton Bionetics colonized several species of monkeys and supplied cancer researchers and vaccine developers worldwide with new experimental viruses and monkey cell lines. **Life Science, Inc.** supplied enzymes and other supplies needed for vaccine research too. This company, and the term "**Life Science**," is emboldened here because it flips reality. The designation "Life Sciences" conceals 'Death Sciences,' especially covert military biological weapons programs.

This also concerns the military and corporate ('dual use') interests behind brain-altering *entheogens*, including Mr. Peter Thiel and Silicon Valley's interests, as detailed later in this book.

Litton Bionetics's contract followed closely the Stewart-Eddy 1962 discovery of cancers spread through Merck's polio vaccines. Nearly a decade later, the NIH and NCI published Progress Report #8 that detailed Gallo's leadership in the 71-2025 contract that began that same year, in 1962. The 1971 contract, titled "Investigations of Viral Carcinogenesis in Primates," was expanded in 1972 as the first hepatitis B vaccines were injected into the Willowbrook children and gay volunteers in New York. (See also SVCP Progress Report 9, p. 195).[33]

Besides supplying the Merck Drug Company with mutated viruses for vaccine studies, Litton's investigators supplied U.S. Army officials and Dr. Saul Krugman of the New York University Medical Center and Rockefeller Blood Center with experimental materials used to test mentally retarded children at Willowbrook. As mentioned earlier, gay men were similarly poisoned and infected. Merck's advertisements attracted homosexual volunteers. These ads can still be found online.[51]

All the above has been fraudulently concealed or recklessly neglected by willfully blind officials and the complicit "fake news" media. This 'silence' is most incriminating.

Genetic engineering at the cooperating SVCP-NCI labs is so incriminating that it implores reconsideration of this entire matter in the context of an MKULTRA/MKNAOMI genocidal conspiracy leading up to the COVID Coup.

Also incriminating is the fact that these 'sexually transmitted diseases' —HIV/AIDS and HPV cervical cancers— emerged at the same time the HBV vaccine was made and tested in the 'high-risk' sexually promiscuous populations. No better populations than sexually promiscuous citizens might be selected for population control studies and 'immunization' activities.

Here you see the combination of MKULTRA mind-control using the controlled media with the MKNAOMI biological weapons depopulation administration.

The MKULTRA mind-control social-engineering conspiracy successfully spread fear of herpes for three more benefits:
1) The fright reduced sexual promiscuity, vicariously reducing populations.
2) The fright created a vast market for commercial products; and
3) The fright diverted from the biological weapons programs of MKNAOMI, including genetic engineering of herpes viruses in the SVCP labs, transforming them into cancer viruses, including HPV.

The KI's 1969 contract NIH-69-2005 was titled "Studies of The Significance of Herpes-Type Virus in The Etiology of Some Human Cancers." This SVCP Progress Report #9 covered the "4/9/71 – 4/8/73" period. The Segment Chairman was Dr. Robert Manaker —Robert Gallo's supervisor at the NCI. Manaker was key MKNAOMI manager as the Chairman of the "Developmental Research Segment" of the SVCP. That group included several prominent names in developmental virology, among them Dr. Anthony Girardi and the University of Pennsylvania's Wister Institute. (See p. 6, of the SVCP, Progress Report #9.)

These proven associations tie the relatively small group of suspected scientists together in what appears to be a cancer virology racket administering profitable depopulation. This racketeering enterprise is equally suspect for directing the press's 'silence.' The SVCP-KI contract shown below convincingly documents and confirms the statements made by Dr. Robert Strecker about the importance of "Montagnier's alleged discovery" —that EBV from Burkitt's lymphoma viruses produced tumors we see with AIDS. Strecker explained that adding that virus to T-lymphocytes will churn out antigens and viruses, as we have seen, killing more than 40 million people worldwide from AIDS.

Solid Evidence of a 'Conspiracy of Silence'

The Honorable Chief Justice Louis Brandies in the *United States ex rel. Bilokumsky v. Tod*, 263 US 149, 154 – Supreme Court 1923 stated, "Conduct which forms a basis for inference is evidence. Silence is often evidence of the most persuasive character."

How can you reconcile mainstream medicine, science, and media, neglecting the facts as mentioned above published in virology other than by conspiracy and contempt for honesty in government and commerce? Yet, the voices of 'justice' scrutinize and publicize every fact they can find to disparage alternative thinkers, natural healing methods, and practitioners.

What impact will such scientific censorship, malice, and negligence have on civilization other than depopulation?

It will permit and secure advancing mind-altering genetic engineering, vaccine bioelectronics, and drugs called 'geneto-pharmaceuticals,' entheogens. The vaccines for COVID are strongly suspected of injecting nano-bioelectronic devices to connect human brains to Big Brother's Cloud.

Ethical practice compels reconsidering these matters by a special "independent" investigative committee, a Nuremberg-like commission, and/or grand jury.

Diverting from "silence as evidence," the media has muddled and debunked serious hypotheses, calling them "conspiracy theories." This media counter-intelligence distracts from the facts and evidence published here and elsewhere.[48] The poofs I provided, including the SVCP contracts republished above, have never been honorably or legitimately "debunked," only ignored, silenced, censored, or smeared.[48,51,58]

Other authors and I have witnessed time-and-time-again the mainstream media hashing facts, administering diversions and obfuscations in the corporate-controlled social media. This pattern and practice of malfeasance serve as a "protection racket" for ongoing multi-trillion-dollar population control commerce, engaging the biotechnology/vaccinology industry. Millions of dollars are spent to commission "pseudo-skeptics" and "debunkers" to write Op-Ed pieces and online slams against valid whistleblowers. This activity is another MKULTRA "MindWar" ploy.

Major United States Army Biological Weapons Contractors for Fiscal Year 1969

Mr. Mahon. List for the record the major contractors and the sums allocated to them in this program in fiscal year 1969.
(The information follows:)

The following list contains the major contractors and amounts of each contract.

Contractor	Fiscal year 1969
Miami, University of Coral Gables Fla.	$645,000
Herner and Co., Bethesda, Md.	518,000
Missouri, University of, Columbia, Mo.	250,000
Chicago, University of Chicago, Ill.	216,000
Aerojet-General Corp., Sacramento, Calif.	210,000
Bionetics Research Laboratories, Inc., Falls Church, Va.	180,000
West Virginia University, Morgantown, W. Va.	177,000
Maryland, University of, College Park, Md.	170,000
Dow Chemical Co., Midland, Mich.	158,000
Hazelton Laboratories, Inc., Falls Church, Reston, Va.	145,000
New York University Medical Center, New York, N.Y.	142,000
Midwest Research Institute, Kansas City, Mo.	134,000
Stanford University, Palo Alto, Calif.	125,000
Stanford Research Institute, Menlo Park, Calif.	124,000
Pfizer and Co., Inc., New York, N.Y.	120,000
Aldrich Chemical Co., Inc., Milwaukee, Wis.	117,000
Computer Usage Development Corp., Washington, D.C.	110,000
New England Nuclear Corp., Boston, Mass.	104,000

Source: Department of Defense Appropriations For 1970: Hearings Before A Subcommittee of the Committee on Appropriations House of Representatives, Ninety-first Congress, First Session, H.B. 15090, Part 5, Research, Development, Test and Evaluation of Biological Weapons, Dept. of the Army. U.S. Government Printing Office, Washington, D.C., 1969, p. 689.

Early list of bioweapons contractors includes Pfizer, Bionetics, the Univ. of Chicago, the Computer Usage Development Corp. that was intertwined with IBM and the U.S. military, and Stanford that advanced 'eugenics' (i.e., racial hygiene) but also hosted David McClellan—the nation's leading behavioral scientist, social engineer and senior research scholar at Harvard overseeing Timothy Leary, Alan Watts and Aldus Huxley.

This allegation is proven by Big Pharma having been caught several times financing online "trolls" to supplement battalions of propagandists and PR firms, issuing "fake news." These minions and their false commentaries are well-documented in my award-winning "Best Film-2016" titled **UN-VAXXED: A Docu-commentary for Robert De Niro**.[60]

These 'missing links' are crucial in unveiling the 'Big Picture' of the COVID Coup. The 'silence' surrounding these facts is persuasive evidence that supplements several neglected 'smoking guns.' Compounding incriminating facts creates a 'presumption of guilt' in law. That produces probable cause for a criminal investigation into the

Introduction and Background Intelligence

ongoing deceptions and damaging MKULTRA/MKNAOMI/COVID population control administration.

> **Proposed Course:** The contractor will compare antigens in cervical carcinoma cells with those induced in Hep-2 by Herpes simplex virus type I and II. Work will continue on the characterization of HSV-2 antigens present in exfoliated cancer cells from patients with cervical cancer.
>
> **Date Contract Initiated:** May 5, 1971
>
> **Current Annual Level:** $92,000
>
> **KAROLINSKA INSTITUTE (NIH-69-2005)**
>
> **Title:** Studies on the Significance of Herpes-type Virus in the Etiology of Some Human Cancers
>
> **Contractor's Project Director:** Dr. George Klein
>
> **Project Officers (NCI):** Dr. Charles W. Boone
> Dr. Gary R. Pearson
>
> **Objectives:** (1) To obtain additional data on EB virus-host interactions. (2) To investigate host immune responses to tumor antigens. (3) To study the regulation of C-type virus expression in defined systems. (4) To investigate cell mediated tumor immune reaction mechanisms *in vitro* and *in vivo*.
>
> **Major Findings:** EBV related research: membrane antigens, early antigens and virus capsid antigens mediated by EBV have been studied in established lymphoblastoid cell lines. Inhibitors such as mitomycin C increase the amount of antigens detected by immunofluorescence. Mouse lymphoblastoid cell hybrids have made it possible to determine whether the presence of EBV DNA is dependent on one or several chromosomes. BUDR labeled cells were super-infected with tritiated thymidine labeled virus and the heavy cell DNA was recovered and examined for associated EBV. No evidence of integrated EBV genome was found. The incubation of tumor cells with serum from patients with lipo-, fibro-, osteo-, and neuro-sarcoma inhibited the stimulation in one autologous lymphocyte-fibrosarcoma combination.
>
> **Significance to Biomedical Research and the Program of the Institute:** A major effort of the Special Virus Cancer Program has been the study of the viral involvement in the etiology and course of Burkitt's tumor in man. Research on the relationship between EBV infection and the onset of Burkitt tumor, the development of EBV-coded antigens in infected cells, and the analysis of the immune response to Burkitt tumor is therefore highly relevant to total program.
>
> **Proposed Course:** The contract effort will continue essentially as described above.
>
> **Date Contract Initiated:** April 9, 1968
>
> **Current Annual Level:** $90,000

Indictment of Racketeering in Life Sciences Fraud

The facts mentioned above give probable cause to assert a 'presumption of guilt' and Deep State conspiracy to conceal and administer genocide using biological weapons and vaccinations.

As stated in the following section covering legal 'Standards of Review,' the elements of Life Sciences fraud are satisfied by:

1) the false representation that vaccines or certain drugs are "safe" or "safe enough" for injection or oral consumption.
2) government officials' and healthcare workers' knowledge of the falsity of these representations of safety and efficacy (or without knowledge of their truth or falsity in willfully blind obedience to MKULtra-induced general acceptance);
3) contemplating society and individual consumers' reliance upon these false representations.
4) with society and consumers' reliance upon these falsehoods;

5) causing damage, side effects, illnesses, including brain damage, cancers, and deaths. Shoppe v. Gucci America, Inc.61, 14 P. 3d 1049 - Haw: Supreme Court 2000.

Similarly, the elements of fraudulent concealment are satisfied by the "employment of artifice" (e.g., false representations of safety or reasonable risk) "planned to prevent inquiry or escape investigation and mislead or hinder acquirement of information disclosing a right of action."

This includes the right to abstain from any 'assault' as defined by law. The officials' acts relied on are "of an affirmative character and fraudulent." *Lemson v. General Motors Corp.*, 66 Mich. App. 94, 97, 238 N.W.2d 414, 415 (1975)^Error! Bookmark not defined. quoting *De Haan v. Winter*, 258 Mich. 293, 296, 241 N.W. 923, 924 (1932).**Error! Bookmark not defined.** The fraudulent concealment also "involves the actions taken" by government and healthcare workers "to conceal a known cause of action" such as negligence, assault, breach of public duty to prevent injury, and/or death. *Au v. Au*,^Error! Bookmark not defined. 626 P. 2d 173 – Haw: Supreme Court 1981.

The facts presented here, proven by government records and public knowledge, evidence racketeering activities as defined by 18 U.S.C. §§1961 and 1962.

How Big Pharma Racketeers Gain 'General Acceptance'

The 'predicate acts' of this Big Pharma Life Sciences racketeering enterprise include the following methods used to gain the public's 'general acceptance':

1) Bribery of public officials and media censors. One example I gave involved Hawaii State Senator, Roz Baker who was among the first politicians pushing mandatory vaccinations;[30]
2) Extortion of persons to get vaccinated or risk dying from terroristically-threatened diseases, or by threatening to deprive healthcare and civil rights;
3) Dealing falsely in controlled chemicals, drugs, and vaccines falsely regarded as "safe," including sterilizers and immune system poisons;
4) Mail fraud and wire fraud in correspondence between officials, the media, and the public;

Introduction and Background Intelligence

5) Obstructing justice by scientific evidence tampering, false filings with the state and courts; and corrupt influence over witnesses, whistleblowers, and the media;
6) Obstruction of criminal investigations;
7) Tampering with witnesses, victims, and informants to preclude discovery and prosecution, as the US General Accounting Office (GAO) did with me after I supplied credible scientific evidence censored by GAO officials during their whitewashing of an official SVCP AIDS origin inquiry; [61]
8) Retaliating against witnesses, victims, and informants who sought relief and remedies as the enterprise did against Robert Strecker's brother, fellow whistleblower, and lawyer Ted Strecker, resulting in the latter's death.

My partner and I, likewise, were retaliated against. This is evidenced by the outrageous court rulings[62] favoring crime syndicate kingpin Paul J. Sulla, Jr., who forged documents to steal my property in Hawaii. Sulla's criminal influence in the courts contributed to Sherri Kane's death by manslaughter.

9) Relating to peonage and slavery in the abuse of test subjects. Examples include experimental COVID vaccine recipients and the Willowbrook children infected with HIV/AIDS from the hepatitis B vaccines;
10) Economic espionage and theft of trade secrets, as Dr. Gallo was vetted for doing purportedly damaging Montagnier in France;
11) Interference with commerce in the natural healing arts and sciences by subverting the multi-factorial causes and cures for cancer and general immunity while disparaging natural healthcare practices; and
12) Trafficking in biological weapons and chemical weapons under the guise of "cancer control" and "public health" involving fraud in the sale of vaccines, drugs, and also securities used to finance the individuals and companies manufacturing the weapons.

Criminally Negligent Manslaughter

Criminally negligent manslaughter is referred to as criminally negligent homicide in the United States. "It occurs where death results from serious negligence, or, in some jurisdictions, serious recklessness."

A high degree of negligence is required to warrant criminal liability. The previously stated facts exceed this bar.

In general, risks to public health and safety from these advancing biochemical technologies have been grossly neglected by ignorant "evil obedient" officials. Ignorance is not "bliss." Nor does shutting one's eyes to genocide relieve public duties.

Willful blindness is official malfeasance. Such intentional blinding of oneself is criminal and arguably demonic.[46]

Criminally negligent manslaughter occurs when someone fails to act when there is a duty to do so, which leads to death. In public health and medicine, officials and doctors are not only obligated by 'public duty doctrine' and 42 U.S.C. § 1986 to prevent damage to citizens, but also by the Hippocratic oath to "above all do no harm."

The existence of the duty to save lives imposes criminal liability on the willfully-blind who aid-and-abet rackets damaging society. Rather than speaking up, protecting the voiceless, devil-doers remain silent and consenting.

Criminally negligent manslaughter is most common among health professionals "who are grossly negligent in the course of their employment."[63]

Willful blindness to 'therapeutic' risks is widespread. As mentioned, the cause of this general acceptance and willful blindness is the pervasive deceptive media fundamental to the drug and vaccine racket.

Rampant Fraud in Health Science Prompts Diseases

It has come to the public's attention through respected science journals that fraud committed by special interest groups benefiting "Big Pharma" or the "Life Sciences" industry has substantially damaged the field of medicine.[47]

"Something has gone fundamentally wrong with one of our greatest human creations," wrote Richard Horton, the editor of the esteemed *Lancet* in 2015.[64] "The case against science is straightforward: much of the scientific literature, perhaps half, may simply be untrue. Afflicted by studies with small sample sizes, tiny effects, invalid exploratory analyses, and flagrant conflicts of interest, together with an obsession for pursuing fashionable trends of dubious importance, science has taken a turn towards darkness," Horton protested. "As one participant put it, 'poor methods get results.'

Introduction and Background Intelligence

"The Academy of Medical Sciences, Medical Research Council, and Biotechnology and Biological Sciences Research Council have now put their reputational weight behind an investigation into these questionable research practices," Horton continued. "The apparent endemicity of bad research behaviour is alarming. In their quest for telling a compelling story, scientists too often sculpt data to fit their preferred theory of the world..."

Dr. Horton concluded his heroic editorial stating, "The bad news is that nobody is ready to take the first step to clean up the system."

```
CONTRACTOR    : Wistar Institute of Anatomy and Biology        (71-2092)
ADDRESS       : 36th Street at Spruce, Philadelphia, Pennsylvania  19104
PHONE         : AC-215, Phone 222-6700, x-226
CNTRCT TITLE: Extraction and Characterization of Virus-Induced Transplantation
              Antigen from Sarcomas and Leukemia
DATES         : 2/1/72 - 1/31/73
PRINC INVEST: Dr. Anthony Girardi
PROJ OFFICER: Dr. James T. Duff, Bldg. 37, Room 1822, x-65947

SEGMENT       : Solid Tumor-Virus
SEG CHAIRMAN: Dr. Robert Huebner, Bldg. 37, Room 2004, x-63301
CNTRCT SPEC : Mr. Thomas Porter, Bldg. 37, Room 1A03, x-65025
```

I share the honorable *Lancet* editor's sentiments but disagree with his conclusion. As this book proves, many of us in science and medicine have taken "the first step" only to be smeared and ostracized by the same enterprise that controls *Lancet*[65] and the 'Life Sciences' industry.[Error! Bookmark not defined.,47]

I took that "first step" in 1996 with the publication of Emerging Viruses: AIDS & Ebola–Nature, Accident or Intentional?[51] Subsequently, I added a peer-reviewed scientific summary in Medical Hypothesis[66] bravely financed and administered by the late unjustly smeared editor, David Horrobin.[47] It is that "unclean system" that imposes illegitimacy in science along with "general acceptance" of its falsehoods.[50]

Blame "vaccine hesitancy" primarily on this. Abuse of the media, including "peer-reviewed" publications by propagandists, exploits the most well-studied science called 'behavioral science.'

This abuse of behavioral science for social engineering of 'general acceptance' of virtually anything, such as mandatory COVID vaccinations, imposes deadly consequences. In this instance, "mandatory vaccinations," like hallucinogenic drug promotions, foreshadow devastating side effects. The risks include population-wide genetic or neurological damage.

Governmental endorsements of nano-bioelectronic mind-control devices, as will be detailed later, intertwines with officials' promotions of mind-altering drugs. People of all races, religions and spiritual practices are at risk of this scientific corruption, mass-mediated deceptions, and advancing nano-bioelectronic neuroscience brain-to-Cloud technologies. The overall agenda even corrupts accepted principles and practices in Luciferianism, also explained in a later chapter.

Both vaccines and mind-altering entheogens are implicated in inducing brain damage and even cancers. This degeneration and association do not appear to be serendipity. Commercial interests profiting from advances in artificial intelligence, data-mining, and Technotronic warfare for social engineering, invest as though human brains were commercial commodities subject to 'collateral damage.'

For instance, in the United States, there is an explosion of need for special educators now burdening the economy and millions of families attributable to skyrocketing rates of brain damage properly attributed to vaccinations. Autism is now widely recognized as having been caused by mercury poisoning in vaccinated victims.

All of this has occurred while health officials, bribed by special interests, have neglected or excused the 'mercury-toxicity autism association.'

The same goes for skyrocketing rates of cancer, with cancer virus links to vaccines. The lab virus sourcing COVID is similarly troubling.

Officials regurgitating corrupted science and false narratives simply dismiss these genocides. Many held similarly, for decades, the risks from tobacco smoke, lead, or lithium poisoning. Opposition to these genocides was "neutralized" as "unproven" or "debunked" as "conspiracy theories."[67]

The same is true for the dangers of COVID vaccines.

On November 3, 2021, as this book was going to press, Wisconsin Senator Ron Johnson (R) updated his COVID vaccine deaths statistics from June 2021.[68] He stated more than 17,000 deaths were caused by the "novel" mRNA vaccines produced by Pfizer and Moderna.

CHAPTER II
THE WUHAN OUTBREAK

According to the government of China, the first COVID-19 case appeared in November 2020 and not December as initially reported. The suspicious report[69] in the *South China Morning Post* said authorities identified the earliest case on the 17th of November —weeks before authorities announced the emergence of the "bat virus" at the Wuhan Seafood Market.

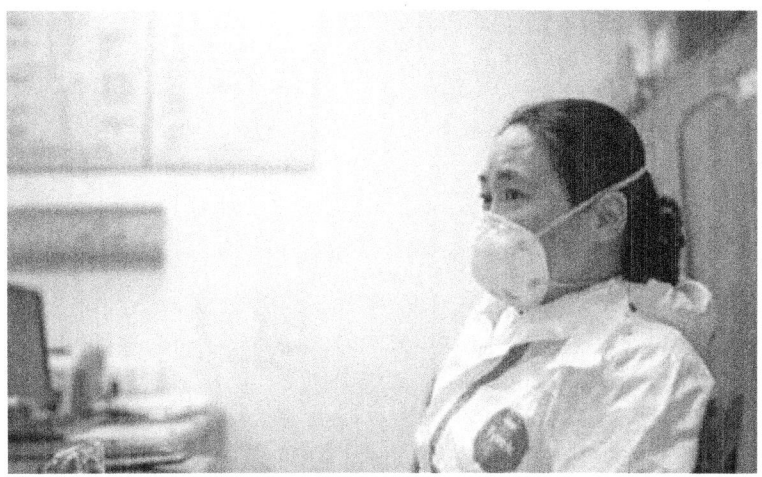

The Chinese government has been widely criticized for concealing more than it's telling. This includes the arrests and censorship of doctors who warned colleagues about the new SARS-like virus that emerged in the city of eleven million people.

The Guardian[70] reported that hospital emergency room Doctor Ai Fen became one of the first censored whistleblowers in China. She began criticizing hospital authorities after seeing several colleagues die from the "novel" coronavirus.

In an interview with the Chinese magazine, *Renwu* (i.e., *People*), Ai Fen, director of the emergency room at Wuhan Central hospital, said she was reprimanded after alerting her superiors and colleagues of a "SARS-like" virus seen in patients in December.

Four of her colleagues, doctors at her hospital, one being the first whistleblower ophthalmologist Li Wenliang, risked their jobs and imprisonment to report what they witnessed.

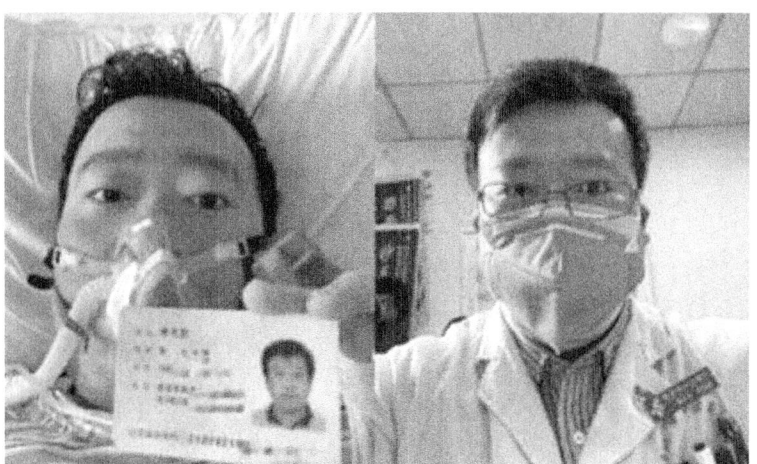

'Hero who told the truth': Chinese rage over coronavirus death of whistleblower doctor

Li Wenliang, 34, died on February 6, 2020, after he was infected, said a statement issued by the Wuhan central hospital. Wenliang reported that the Seafood Market appeared to be the epicenter of the outbreak, with seven shoppers there becoming ill.

According to *The Guardian*, "Li warned colleagues on social media in late December about a mysterious virus that would become the coronavirus epidemic and was detained by police in Wuhan on 3 January for 'spreading false rumours.' He was forced to sign a police document to admit he had breached the law and had 'seriously disrupted social order.'"

"If I had known what was to happen, I would not have cared about the reprimand," said Ai Fen. "I would have fucking talked about it to whoever, wherever I could," she said. Her interview was quickly deleted from Chinese social media sites.

On 30 December, after examining several patients with flu-like symptoms, Ai received alarming lab results. It contained the words: "SARS coronavirus." Ai broke out in a cold sweat. She circled the word

SARS, took a photo, and sent it to a colleague at another hospital in Wuhan. By that evening, the photo had spread across the medical community in Wuhan, alerting Li Wenliang. Ai's lab report became the first evidence of the outbreak.

"That night Ai said she received a message from her hospital saying information about this mysterious disease should not be arbitrarily released in order to avoid causing panic," *The Guardian* report continued. "Two days later, she told the magazine, she was summoned by the head of the hospital's disciplinary inspection committee and reprimanded for 'spreading rumours' and 'harming stability.'"

"The staff were forbidden from passing messages or images related to the virus," she said. All Ai could do was ask her staff to wear protective clothing and masks — even as hospital authorities told them not to. She also told her department to wear protective jackets under their doctor coats.

"'We watched more and more patients come in as the radius of the spread of infection became larger,' she said, as they began to see patients with no connection to the seafood market, believed to be the source of the first cases."

A 55-year-old man from Hubei province was rumored to be the first patient to contract COVID-19 on or about November 17, 2020. Supposedly, in the days that followed, one to five new cases were reported each day. By December 20, 2020, there were presumably 60 confirmed cases.

Official statements by the Chinese government to the World Health Organization reported that the first confirmed case had been diagnosed on December 8th. Authorities concealed human-to-human transmissions until the third week of January.

Outbreak Inconsistencies Evidence Foul Play

According to Li Wenliang's wrongful criminal arrest record,[71] this doctor first identified "7 SARS cases at the Huanan fruit and seafood market in the WeChat group 'Wuhan University clinical `04' on December 30, 2019."

In a city of 11 million people, with the seafood market 20 miles away from the biological safety level (BSL) 4 lab, it is unreasonable to presume: a) other doctors closer to the lab did not see similar cases if the outbreak occurred from the Wuhan lab in November; and b) even if other doctors closer to the lab were censored, or feared reporting the novel respiratory illness, this would not account for the cluster of 7 cases reported in or around the market. This is especially true given the two-week incubation period of the infection spreading from one person to another. Had the virus emerged from the Wuhan lab as suspected between mid-November and early December 2019, why all the finger-pointing at the Seafood Market?

In other words, the reported facts and media censorship conceal(s) a most likely ***intentional* release** of the bioweapon *at the market*. The Western media correctly blamed Chinese officials for wrongdoing while neglecting the common-sense conclusion that the bioweapon was set loose at the market. But it is unreasonable to claim the Chinese government released the virus to its own disadvantage.

The common international media censorship and disinformation implicates what President Trump labeled "fake news." The providers represent the 'Deep State' multi-national corporate interests at war with civilization as we know it.

That is, the suppression of coronavirus science and vaccination risks, as you will read later, is necessary to enable the transformation of civilization into 'transhumanism,' consistent with the Fourth Industrial Revolution ("4IR") and "Final Solution." The 'pattern-and-practice' of disregarding the public's health and safety for commercial and political power extends to this little-known nefarious geopolitical operation. The motive and means to commit the COVID Coup, featuring bioterrorism, intoxication and infection for depopulation using the mutagenic lab virus and its spike protein antigen, are apparent in those undermining public health and safety as identified in forthcoming chapters.

The White House COVID Origin Probe

Substantial criminal evidence is compiled in this book to justify a Nuremberg-like trial. Key in this discovery are the redacted (i.e., censored) NIH e-mail correspondence between Dr. Anthony Fauci and officials administering an urgent "Teleconference" on Sunday, February 2, 2020. This event is chronicled in forthcoming Chapters III and IV.

Simultaneously with Fauci's activity, Trump White House officials advanced their probe into COVID's suspicious origin. On February 3,

2020, the press announced this after Fauci coordinated a teleconference with other leading suspects. President Trump's Director of the Office of Science and Technology Policy,[72] Kelvin K. Droegemeier, was in charge of the White House probe. He wrote to the National Academy of Sciences President, Dr. Marcia McNutt, on that date to investigate "the origin of the 2019 coronavirus outbreak."

That official request prompted me to file an intelligence memorandum[73] evidencing the laboratory creation of the pandemic. On February 10, 2020, I wrote the White House accusing the media and pharmaceutical industry partners of criminal complicity in bioterrorism. I had not yet seen the "Fauci E-mails" proving an organized criminal conspiracy to defraud people worldwide.

Largely neglected by the "fake news" media, Director Droegemeier requested the National Academy of Science (NAS) National Research Council (NRC, together NAS-NRC) scientists provide a "Rapid Response Assessment... that would help determine the origin of 2019-nCoV,[74] specifically from an evolutionary/structural biology standpoint."

This notice to the NAS-NRC shocked me because it was the NAS-NRC that the U.S. Congress charged in 1969-70 to develop the AIDS and Ebola-like viruses that did not exist before that time.

I wrote extensively about these facts in *Emerging Viruses: AIDS & Ebola—Nature, Accident or Intentional?* I made public before anyone else the incriminating NIH and NCI's contracts under which this deadly "science" occurred. That book provides evidence beyond Chapter I's revelations of who created AIDS and how they did it virtually every step of the way. It evidences how vaccines and vaccine production facilities sourced those two plagues —AIDS and Ebola. My Emerging Viruses book is the definitive work in this field.

So when I saw that Droegemeier was engaging the "fox in the hen house," the NAS-NRC, I balked and decided to object in writing. I wrote to him and the Department of Homeland Security Acting Secretary, Chad Wolf (who appeared on *FOX News* discussing the Trump administration's coronavirus defense plans). You can read my complete memorandum,[73] and what follows is my advisement.

"Unfortunately, without addressing the liabilities," I wrote, "your administration may be hoodwinked, undermined, and diverted from the true origin of this lab virus and appropriate preventative actions, unnecessarily increasing morbidity, mortality, and public anxiety."

I detailed five federal government liabilities beginning with evidence that the "novel" 2019 coronavirus (2019 nCoV) was mutated in a BSL3/4 lab and engineered to include AIDS-virus envelop genes and the SARS virus "middle fragment," verifying its unnatural origin.

Accordingly, I recommended re-naming the virus "2019 nCoV/HIV/SARS." I warned that neglecting this scientific evidence "invites additional releases; may hamper efforts to develop cures and cause more diseases and deaths."

I advised federal officials "as a Harvard-trained expert in emerging diseases and media communications[75] who had been heavily persecuted, defamed, and censored"[76] for my efforts in vetting the lab origin of AIDS and Ebola.

I made officials aware that I was not to be dismissed as a "conspiracy theorist." I was registered with the U.S. District Court of Hawaii, under Criminal Justice Act (CJA) provisions, as an expert witness paid to testify on matters of medical research, biological weaponry and bioterrorism, vaccinology, and public corruption. I was a sixty-seven-year-old award-winning author[66] and filmmaker[60] who testified before legislative bodies on the politically incorrect matter of vaccination risks, viral recombinants risking genetic damage, pandemics, and cancers. I had exposed the man-made origin of the HIV/AIDS cancer complex,[14] HPV/cervical cancers,[77] ZIKA-induced microcephaly,[78] Ebola's genesis[79] and immune suppression, SARS,[80] H1N1 Swine Flu,[81] the source of the anthrax mailings,[82] and more.

My biography and efforts went to no avail. I received no reply from the White House. It appeared that the Trump Administration's affection for the pharmaceutical industry and Deep State pressures defied goodwill and logic.

The author, Dr. Leonard G. Horowitz, first citizen to warn the White House regarding the danger of trusting NAS-NRC officials with the research task.

I alerted officials nonetheless about four more liabilities that Trump officials would face. I did this long before others echoed similar concerns. These liabilities included:
1) the public's distrust of government from media censorship of accurate intelligence addressing reasonable concerns;
2) more disease, damage, and deaths from "erroneous and terrorizing information being broadcast instead of helpful preventative strategies beyond handwashing and the use of face masks;"
3) severe risks to U.S. National Security and international trade relations caused by neglecting "the geopolitical and economic correlates and antecedents of the Wuhan outbreak;" and
4) Hawaii's immigration policies and risking mainland transmission. This warning was based on my knowledge of the unprecedented unconscionable corruption in Hawaii's judicial system and law enforcement,[62] enabling human trafficking and drugs, including fentanyl and more, to pour into North America

from the Pacific Rim countries, including China, through Honolulu and the Big Island.

Coronavirus Censorship Conceals Accurate Intelligence

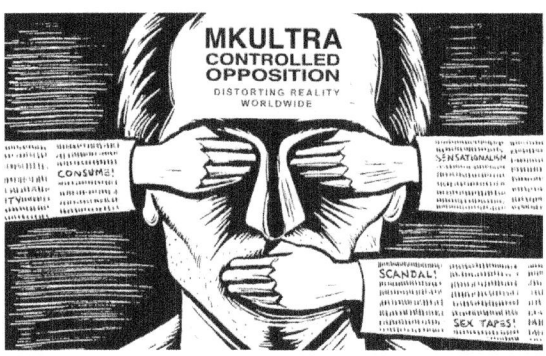

I read media messaging to grasp geopolitical agendas, in this case, bioterrorism. I warned officials that "substantial discovery and diagnosis can be made by evaluating media coverage, diversions, fraudulent concealments, or censorship favoring special interests."

"It must be presumed," I wrote, "given the facts above and neglect of them by the corporate-controlled media, that widespread censorship and diversionary misinformation surrounding the coronavirus pandemic provide 'cover' for criminal operations."

Precedent for media-wide censorship exists and evidences a "pattern and practice of censoring politically incorrect information," I explained and gave examples to no avail. At that time, there was no substantial public movement opposing vaccinations, especially "mandatory vaccines," like there is at the time of this writing.

"For instance," I wrote, "publications critical of vaccinations, like coronavirus vaccines, rightfully indict bioweapons labs, but are recklessly dismissed, disparaged, or censored. This censorship is administered by special interests through public corruption or financial influence, as seen with Rep. Adam Schiff's written notices to social media companies Facebook and Google/YouTube to block consumers' comments opposing vaccination injuries."

I wrote that before the mass movement opposing Big Tech's social media censorship became a political issue.

Likewise, the corporate-controlled media similarly censored a critical study —the Pradhan group's genetic sequencing of the COVID-19 virus. This evolutionary virology determined foul play[13] that I will address in the next chapter. Just as media controllers consistently censored my publications,[83] favoring Big Pharma and covert intelligence agency operations, they did the same to Pradhan's team that published what Fauci et al. smeared as the "Indian paper."

My memo to officials also raised concerns over similar censorship ongoing in China. As mentioned, Dr. Li Wenliang was wrongfully arrested[84] for alerting colleagues about "7 SARS cases" at the Wuhan market.

I wrote, "with the seafood market 20 miles away from the BSL 4 lab, it is unreasonable to presume:

a) other doctors closer to the lab were not seeing similar cases if the outbreak occurred from the Wuhan lab; and

b) even if other doctors closer to the lab were censored or feared reporting the novel respiratory illness, this would not account for the cluster of 7 cases reported in or around the market, nor the mass media focusing on that precise location as the source of the outbreak. This was especially perplexing given the 2 week incubation period of the infection spreading from one person to another (had the virus emerged from the Wuhan lab as suspected at the beginning of December 2019).

"In other words," I concluded, "the reported facts and media censorship conceal(s) a most likely [intentional] release of the bioweapon at the market, with the Western media correctly blaming Chinese officials for wrongdoing while neglecting the common-sense conclusion that the bioweapon was loosed at the market, but not by the Chinese government to its own disadvantage."

The common international media censorship implicated what President Trump referred to as the "fake news" media representing the "Deep State" multi-national corporate interests at war with the President, all sovereign nations, and *We The People*. I realized and wrote about that.

That is, "the suppression of coronavirus science, as with vaccination risks, is part of the pattern and practice of similarly disregarding the public's health and safety for commercial and political gains," I warned Droegemeier and Wolf. "This motive and means to commit bioterrorism are clearly apparent in those who have been undermining the Trump presidency."

Preventative Strategies Beyond Handwashing and Masking

I advised the White House that what was missing was a "prudent health education and self-care motivation" information campaign by the media to help families and schools that have been recklessly neglected, frightened, extorted, and abused.

In addition, I recommended "natural remedies such as good hydration" and the use of alternative healthcare products and services that "have been disparaged in favor of Big Pharma."

I later sued Pfizer and Moderna in federal court for depriving me personally and financially of my free and fair trade with the anti-viral vaccine competitor, "Oxysilver^{TM} with 528".

I encouraged the government to develop a product similar to Oxysilver^{TM}, originally developed by NASA scientists to keep astronauts healthy. "The NASA-science-developed silver hydrosols," I wrote, "have proven safe, effective, and economical for augmenting human immunity against a broad spectrum of infectious diseases."

My advisement came long before other reputable authorities encouraged using other antioxidants, including hydroxychloroquine and chlorine dioxide. Deep State officials misrepresented the latter as "bleach" (that is, sodium hypochlorite). Trump was racked over the coals for citing these reasonable recommendations.

Instead of suppressing these helpful products, I advised the government to mass-produce and distribute these remedies, "especially at this time of urgency where no drugs or vaccines are available to fight the 2019 nCoV/HIV/SARS bioweapon." Instead, officials mass-produced respiratory ventilators that did much more harm than good.

The agenda had already been set. Pfizer and Moderna/DARPA's mRNA vaccines would be the exclusive remedies as far as government investments were concerned. I opposed these efforts stating, "'fast tracking' costly and risky coronavirus vaccines is imprudent given the fact the 2019 nCoV/HIV/SARS bioweapon is highly unstable... This is largely why AIDS vaccines have failed."

My counsel was right on the money, albeit neglected and censored. I had accurately foreseen and foretold mutant viruses developing from the source virus due to the instability of the genes that had been spliced into the offending germ. I knew that lab viruses lacked stability and would mutate rapidly because the spliced genes were unstable. They did not undergo 'natural selection.' They did not evolve over millennia. They were mutagens —mutants certain to cause other viruses and bacteria to mutate.

Similarly, 'fast tracking' anti-viral drugs to combat this bioweapon were unlikely to produce a risk/benefit ratio favorable to public health and safety. I correctly warned the Trump Administration.

I also recommended public instructions to improve nutrition, "including Vitamin C that provides anti-viral potency in mega-doses, aerobic exercising, and bio-energetic technologies... to help reduce morbidity, mortality, and the menacing fear the media is spreading that disables people from exercising their own defenses and 'self-efficacy' in resisting colds, flu, and the coronavirus disease."

> *I knew that lab viruses lacked stability and would mutate rapidly because the spliced genes were unstable. They did not undergo 'natural selection.' They did not evolve over millennia. They were mutagens —mutants certain to cause other viruses and bacteria to mutate.*

Coronavirus Bioweapon and Media Bioterrorism

Long before others proposed, I published articles and gave radio interviews stating that the 2019 nCoV/HIV/SARS recombinant was "a lab-engineered bioweapon."

Nine honorable experts led by Prashant Pradhan, the Chief Technical Officer for IBM in Asia, published clear-and-convincing proof that the 2019 nCoV/HIV/SARS recombinant was "a lab-engineered bioweapon. Pradhan's team used dedicated software and the WATSON computer to analyze genetic similarities in the coronavirus and other pathogens. Unmistakably, 2019 nCoV contained a 'smoking gun' —four AIDS virus spike protein envelop genes— prima facie evidence of a lab virus 'recombinant.'

The discovery of weaponizing 2019 nCoV with HIV's attachment mechanism was compounded by the findings of evolutionary biologist Dr. James Lyons-Weiler of the Institute for Pure and Applied

Knowledge.[85] Dr. Lyons showed that the coronavirus' genetic sequences contained a unique "middle fragment" encoding for a SARS protein, presumably inserted to increase respiratory distress using "shuttle" technology. This biotechnology was only used in labs, not in nature.

Accordingly, it was unreasonable to presume '2019 nCoV' came from bats, snakes, pangolins, or the wild. Alternatively, it was a scientifically based conclusion that this germ was manufactured as a bioweapon and released for political and commercial gain (i.e., bioterrorism). I elaborated on this highest probability. At that time, I had not yet considered the "Great Global Reset" and the overriding globalist agenda to impose transhumanism on civilization.

"It is unknown at present... which lab sourced this 2019 nCoV/HIV/SARS recombinant," I wrote to Administration officials. I had not yet learned of the University of North Carolina's involvement in this bioweapon's development.

"Given probable cause to presume criminal activity," I advised, "the Justice Department should begin to investigate the short list of suspected BSL 3/4 labs, in addition to your having directed the National Academy of Sciences to look into these matters." I had not yet realized that the entire Justice Department, including the FBI, had been completely "captured" and were complicit in the plague's administration.

I continued my memorandum, "Questions must be raised concerning the 20-mile proximity of the Wuhan "seafood market" with China's BSL 4 bioweapons lab. The evidence in hand confirms this 2019 nCoV/HIV/SARS recombinant may not have been produced in Wuhan's BSL 4 lab, but alternatively at one of several 'Deep State' suspect labs, such as at Johns Hopkins, Harvard, NYU, the CDC, the Pasteur Institute, or by coronavirus vaccine makers [especially Pfizer that had a long history of contracting with the military to develop "dual-use" biological weapons].

"Our public duties require us to urgently advance these investigations to make these determinations because neglecting these matters invites additional releases; may hamper efforts to develop cures; and cause more diseases and deaths."

I concluded, "In medicine, accurate diagnosis is required for effective treatment. Diagnosis means "to see through" to the root of the disorder. The facts in hand compel our presumption that the bioweapon was released to affect political and financial objectives. Therefore, we are urgently required to investigate the root(s) of this 'disorder' in the public and private sectors."

Media Censorship of Accurate Intelligence

I witnessed unconscionable media censorship concealing the most important and accurate intelligence evidencing the multinational COVID Coup. This recklessness and criminal activity urgently required DOJ interventions that never came.

"Substantial discovery and diagnosis can be made by evaluating media coverage, diversions, fraudulent concealments, or censorship favoring special interests," I advised White House officials. "It must be presumed, given the facts above, and neglect of them by the corporate-controlled media, that widespread censorship and diversionary misinformation surrounding the coronavirus pandemic provide 'cover' for criminal operations."

I then addressed the media's and health officials' political and economic motives for committing the alleged crimes against humanity.

All news sources neglected or concealed the coronavirus preparedness conference[86] held in October 2019, titled "Event 201."[87] That, too, was a "smoking gun." Officials at that conference suspiciously predicted virtually all of the events occurring two months later following the actual release of the COVID bioweapon.

Dr. Fauci compounded evidence of criminality long before anyone realized there was a conspiracy unfolding to sicken and enslave humanity. In January 2017, just as the Trump Administration was entering the White House, Fauci lectured[88] to his colleagues at Georgetown University that the incoming administration would face an unprecedented plague.

Chinese officials likewise knew this, as will be evidenced later.

These facts were irresponsibly neglected, dismissed, or misrepresented by Chinese officials for political and financial gain.

For example, according to *Zerohedge.com*, Shi Zhengli, the Wuhan Institute of Virology director, defended, like Fauci did, that Pradhan's group dismissed their shocking genetic analysis.

This dismissal was completely unreasonable, given the extensive research conducted by Pradhan et al. They discovered "4 unique inserts in the 2019-nCoV, all of which have identity/similarity to amino acid residues in key structural proteins of HIV-1 [each of which are] unlikely to be fortuitous in nature."

Further evidencing fraudulent concealments and bioterrorism by the complicit media, all Western media prejudicially indicted Chinese governors, whereas both Eastern and Western officials were equally implicated or disadvantaged.

President Trump labeled the germ "the Chinese virus." That diverted the media and geopolitics in favor of the global industrialists who were actually responsible.

Censorship Yielded Misplaced Reliance on Corrupted Informants

By February 2020, most American voters were well aware of the divisions and conflicting interests in the branches of government hampering Trump's progress and bipartisan acceptance. This dissonance raised reasonable concerns[89] about whether reliance on a single government agency, in this case, on the NIH, Dr. Fauci, or the National Academy of Sciences, was wise. Indeed, this faith was grossly misplaced.

My concerns arose immediately, as mentioned, from discovering twenty years ago that the National Academy of Sciences, National Research Council (NASNRC), in 1969, was complicit in developing AIDS-like and Ebola-like immune-suppressive viruses for germ warfare.

This revelation directly impacted my recognition of the coronavirus being a biological weapon and lab creation bearing the HIV-1 env gene. According to the *Congressional Record* re-published below from Tuesday, July 1, 1969, a Department of Defense Appropriations for 1970 was paid for research, development, and evaluation of "Synthetic Biological Agents"[90] (p. 129). The AIDS-virus envelop gene that now appears in the COVID-19 virus was a likely part of that earlier "synthetic" construction we now call HIV.

In other words, that military appropriations request, subsequent contracts, and NAS-NRC oversight, resulted in the AIDS cancer trigger. That is also called the mRNA "payload" delivery device, or cell-attachment technology, enabling the infamous "gain-of-function." Today, we see the same genes were used to in the coronavirus spike protein ("S-protein") attachment device to produce COVID-19. This mechanism is precisely described in that *Congressional Record.* The objective was to develop viruses that would evade and damage the immune system and be hyper-transmissible—the functional equivalent of HIV as well as the Ebola virus. The COVID-19 virus acts similarly.

These facts —scientific determinations— were completely censored by Deep State parties with conflicting interests, especially the corporate-controlled media. I was then personally and professionally disparaged for outputting this intelligence by the same intelligence agencies undermining President Trump's good intentions —the CIA and FBI under Robert Mueller.

129 — Tuesday, July 1, 1969
SYNTHETIC BIOLOGICAL AGENTS

There are two things about the biological agent field I would like to mention. One is the possibility of technological surprise. Molecular biology is a field that is advancing very rapidly and eminent biologists believe that within a period of 5 to 10 years it would be possible to produce a synthetic biological agent, an agent that does not naturally exist and for which no natural immunity could have been acquired.

Mr. SIKES. Are we doing any work in that field?
Dr. MACARTHUR. We are not.
Mr. SIKES. Why not? Lack of money or lack of interest?
Dr. MACARTHUR. Certainly not lack of interest.
Mr. SIKES. Would you provide for our records information on what would be required, what the advantages of such a program would be, the time and the cost involved?
Dr. MACARTHUR. We will be very happy to.
[The information follows:]

The dramatic progress being made in the field of molecular biology led us to investigate the relevance of this field of science to biological warfare. A small group of experts considered this matter and provided the following observations:

1. All biological agents up to the present time are representatives of naturally occurring disease, and are thus known by scientists throughout the world. They are easily available to qualified scientists for research either for offensive or defensive purposes.

* 2. Within the next 5 to 10 years, it would probably be possible to make a new infective microorganism which could differ in certain important aspects from any known disease-causing organisms. Most important of these is that it might be refractory to the immunological and therapeutic processes upon which we depend to maintain our relative freedom from infectious disease.

* 3. A research program to explore the feasibility of this could be completed in approximately 5 years at a total cost of $10 million.

4. It would be very difficult to establish such a program. Molecular biology is a relatively new science. There are not many highly competent scientists in the field, almost all are in university laboratories, and they are generally adequately supported from sources other than DOD. However, it was considered possible to initiate an adequate program through the National Academy of Sciences-National Research Council (NAS-NRC).

The matter was discussed with the NAS-NRC, and tentative plans were made to initiate the program. However, decreasing funds in CB, growing criticism of the CB program, and our reluctance to involve the NAS NRC in such a controversial endeavor have led us to postpone it for the past 2 years.

* It is a highly controversial issue and there are many who believe such research should not be undertaken lest it lead to yet another method of massive killing of large populations. On the other hand, without the sure scientific knowledge that such a weapon is possible, and an understanding of the ways it could be done, there is little that can be done to devise defensive measures. Should an enemy develop it there is little doubt that this is an important area of potential military technological inferiority in which there is no adequate research program.

The U.S. Law Library Temporarily Tampered with Evidence

Further evidencing this internal conflict in governmental affairs, I submitted to the White House two pages of evidence suggesting treasonous operations by traitors within the government. These pages showed the U.S. Law Library of Congress had temporarily censored (i.e., tampered with) the most important (material) evidence providing <u>a reasonable motive behind the coronavirus pandemic and bioterror campaign</u>.

That censorship had to have been committed by someone deep inside the intelligence community —someone in the CIA or NSA. The censored record that I first identified was 'pulled' within hours of my discovery.

That record was a "Global Legal Monitor" published news item. It was titled "China: Vaccine Law Passed," as had happened on August 27, 2019. That policy took effect on December 1, 2019. That was the approximate date of the COVID-19 bioweapon's first appearance (resulting in the 7-person cluster of cases at the Wuhan market).

It appeared that someone in the government tampered with this public record on February 8, 2020. This is evidenced by the censored page having disappeared for a time. I recovered my copy from Wayback.org[91] as I shared with White House officials. The censored public record was re-published on-or-about February 9, 2020.

That temporarily tampered legal record makes known that the Chinese government presumably opposed *the commercial interests of* Big Pharma.

For all practical purposes, 'Big Pharma is the Deep State.' They are partnered with Big Tech, Big Military, and Big Banking. As will be explained in more detail later, this global cartel is represented in the United States by what I call the "National Security Crime Syndicate." Elsewhere, it is represented, as I have mentioned earlier, by Bill Gates and Klaus Schwab's personification of evil underlying the World Economic Forum.

China's new law: (1) criminalized vaccine manufacturing and distribution misbehaviors; (2) monopolized the manufacture and distribution of vaccines in China in favor of the Communist government; (3) pledged free vaccines were to be exclusively administered to citizens; (4) mandated vaccinations for all Chinese citizens; and (5) created a compensation program for vaccine injuries. (Compensation, presumably, would be minimized by the government taking over vaccine manufacturing, distribution, testing, and certification processes, thereby assuring quality, efficacy, and safety of vaccines —actions Big Pharma opposes.)

Accordingly, much like the opposition, President Trump experienced from Democratic Party leaders financed by BigTech/Big Pharma/Deep State special interests, the Chinese outbreak of the COVID-19 bioweapon was best viewed as sabotage in retaliation to the anti-Big Pharma policies of Presidents Trump and Jinping.

These facts gave legitimate investigators probable cause to consider the Wuhan outbreak as a treasonous threat to U.S. National Security, international relations with China, industrial espionage, sabotage, and bioterrorism committed for political and economic gains.

Geopolitical and Economic Correlates of the Wuhan Outbreak

"No pandemic in earth's history has evolved divorced from major sociopolitical and economic upheaval," I've lectured dozens of times during the past quarter-century. Coronavirus morbidity, mortality, and terroristic threatening serve political and economic motives favoring 'Deep State' special interests. What follows is more evidence of this apparent organized crime.

Contemporaneous with the Wuhan outbreak:
1) On October 18, 2019, approximately two months *before* the pandemic began clinically in Wuhan, the Bill & Melinda Gates Foundation, Klaus Schwab and the World Economic Forum, and Johns Hopkins University officials sponsored the "Event 201" 'exercise' predicting nearly precisely the coronavirus pandemic and its social, military, political, and economic consequences. A popular critical review[92] of this "predictive programming" event can be viewed on the online video;
2) President Trump had declared "war" against the Deep State for a good cause. He encouraged political and economic policies with China aversive to Bill Gates and other Deep State/Big Pharma investors. China had, after all, developed into a "superpower" by way of Anglo-American banking, backing the global cartel. Before the coronavirus outbreak in Wuhan, the President secured commercial contracts with China, viewed unfavorably

by the global elite... Adding fuel to this fire, the President campaigned heavily against drug and vaccine industrialists' price-gouging;

3) A 'pattern-and-practice' of politically effective bioterrorism had already been established. Reflecting on the 2001 anthrax mailings sent to House Majority Leader Tom Daschle and Chairman of the Senate Judiciary Committee, Patrick Leahy, both had sponsored legislation opposing Big Pharma's expanding monopoly and price gouging;

4) The American intelligence community and military had declared war against Trump, who opposed Chinese developments in the Taiwan Strait. This anti-Trump conspiracy backed by the CIA, FBI and Joint Chiefs at the Pentagon was best evidenced by General Mark Milley, that group's 20th chairman. Months after January 6, 2021, alleged "insurrection" at the Capitol Building, Milley attacked Trump, who had appointed Milley to his chairmanship. Rebuking Trump's alleged "coup" planned for that day, Milley said, "They may try, but they're not going to fucking succeed... You can't do this without the military. You can't do this without the CIA and the FBI. We're the guys with the guns."

Further intelligence shared below evidences the CIA's, FBI's, and DOD's conflicting interests and organized criminal-complicity in the COVID Coup. Accordingly, Milley's statements were transparent propaganda. He 'projected' or transferred his co-conspirators' guilt onto Trump for Deep State advantage.

5) As detailed below, the arrest of Harvard Professor Charles Lieber[93] for lying to the FBI is noteworthy. But curiously not on the Chinese espionage charges he faces. According to news reports, at crime-time, only a couple of weeks before the Wuhan outbreak, two Harvard students, both with ties to the Chinese military, were arrested for smuggling biologicals to China. Most incriminating was the false FBI Affidavit submitted by Special Agent Robert Plumb, misrepresenting the technology transfers by Lieber. As you will learn in Chapter VIII, Lieber's actual transfer involved the most advanced nano-bioelectronic vaccine 'hydrogel' knowhow. This ties into the next item.

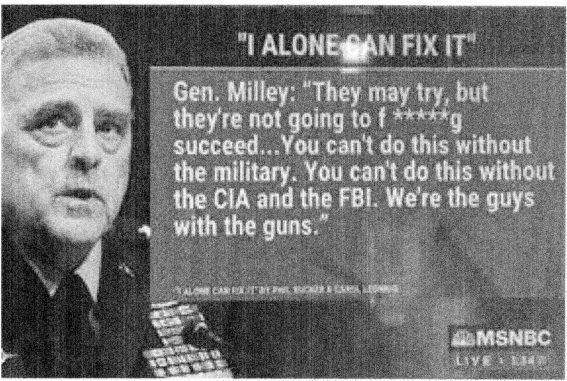

6) It is public knowledge that President Trump opposed vaccine propagandist, eugenicist, and Big Tech industrialist, Bill Gates, who:
   (i) Disparaged Donald Trump publicly for seeking vaccine safety science that Gates declared was a 'dead-end;'
   (ii) Lectured on releasing genetically engineered mosquitoes to vector GMO-laced vaccines;
   (iii) Allied with Jeffrey Epstein in financing MIT and Harvard studies in genetics, neuroscience, and evolutionary biology. These multidisciplinary fields are crucial to transhumanism and Lieber's and Robert Langer's lab activities at Harvard and MIT, respectively.
   (iv) Promoted (like his father did) eugenics-like depopulation to secure the global elite's interests and investments in IBM, IT, AI, 5G, Cloud-brain neuroscience, human data-mining, and transhumanism;
   (v) Secured with Microsoft governmental partnerships advancing black ops leveraging IT technologies central to Silicon Valley, DARPA, the CIA et al., including actors at John Hopkins, NYU, Stanford, and MIT's Media Labs; and most obviously;
   (vi) Financed (through Microsoft) Rep. Adam Schiff's election and attacks against the President —Microsoft being among Schiff's leading campaign financiers in 2015-16, along with Facebook and Google. These suspects demonstrated extreme un-American and un-Constitutional censorship and

discriminatory animus against citizens seeking vaccine-safety assurances: (1) online; (2) through science; (3) in the courts; and (4) by freedom of the press, especially citizens criticizing this corrupted industry.

These suspects had the financial and political wherewithal, 'motive and means,' to release and spread the terrorizing coronavirus and media influence.

As published by *TIME* magazine[94] (Oct. 5, 2018), in an article published to promote NYU's Shanghai branch and global submission to Deep State impositions, the authors asked the following rhetorical question: "How can the Chinese government hope to foster a generation of disruptors in science and business but not expect them to critically appraise the society around them? What use is fostering critical thinking in a society that demands blind obedience?"

In other words, the coronavirus released in China was, and still is, a means to impose "blind obedience" by the stockholders underwriting the Chinese and American governments. But for what purpose? "[T]o foster a generation of disruptors in science and business" that act blindly obedient.

CHAPTER III
COVID-19's LAB ORIGIN & WHO's ACCOUNTABLE

This chapter, supplemented by the identities and conflicting interests of the "insiders" in Chapter IV, exposes the criminal syndicate that sourced the COVID-19 pandemic to effect the COVID Coup.

This is not coronavirus "conspiracy theory." It is irrefutably proven 'conspiracy reality.' It is so adjudged despite naysayers, skeptics, and counter-intelligence propagandists.

Many have come under fire for allegedly undermining the scientific community's and governments' remedial actions.[95, 96] This divisive strategy diverts attention from the most important question in this controversy. **What is the <u>origin of the outbreak</u>?**

It is reckless and unconscionable to dismiss this question by simply accepting the 'accidental release' theory or alternatively shifting blame to a subordinate, as the NIH did on October 20, 2021, citing EcoHealth Alliance as the exclusive culprit.[97]

Instead of answering that question, the media and officials have consistently diverted and lied.

If you knew a drug syndicate was trying to damage, enslave, or kill you, you would not simply accept law enforcers' lame dismissals, such as, "It was just an act of nature," or as Anthony Fauci defended against Rand Paul, "You don't know what you're talking about."

Diversions include the falsely alleged utility of masks, ventilators, drugs, vaccines, first responders, and lawmakers who impose business shutdowns. These 'remedies' fall short of arresting the genocidalists, who seek depopulation to secure globalization.

More than ever, "immunization mandates," expanding social restrains, freedom restrictions, and commercial deprivations evidence Huxley's 'Brave New World' and 1984 totalitarianism. Diversions, obstructions, obfuscations, and impositions have become so apparent that criminality has become transparent.

Unless these elite criminals are exposed, prosecuted, and made accountable, humanity remains susceptible to their lab viruses and covert manipulations. For all we know, without identifying the true source of the outbreak and the motive behind it, additional outbreaks of more deadly "recombinants" may multiply risks and civilization's damage, disease, distress, depopulation, and even extinction.

Compelling evidence compiled in this book exposes conspiracy realities that can now be proven beyond a reasonable doubt. However, Chapter III summarizes much more. This intelligence certifies, and best explains, the origin of COVID-19 as a vital action by COVID Coup actors. The creation and worldwide dissemination of the "novel" germ for bioterrorism and biocrime, and the key agents and agencies responsible, are named below.

Before introducing the chief suspects in this criminal conspiracy, you might consider 18 U.S. Code §1002 that makes it illegal to manufacture or possess **false writings** to enable bioterrorism or any bio-crimes. False publications in this context of organized crime are outlawed because they would aid-and-abet anyone obtaining money from the United States, any agency, officer, or agent thereof, based on any fraudulent document.

My forthcoming sections, condemning and indicting Tulane University and Scripps Research Institute co-authors Kristian G. Andersen and Robert F. Garry, have a basis in 18 U.S. Code §1002. These researchers falsely published the so-called "science" proximal to the main question of origin, evidencing organized crime. The evidence presented below, and in the next chapter reviewing "Fauci E-mails," gives probable cause for a DOJ investigation into this group's certifiably fake science; their confirmed conspiracy to defraud everyone for profit and politics; and their financial backers and 'handlers' operating at the highest levels of government and private commerce.

Who's Responsible for Evidence Tampering and 'Fake Science'

On March 30, 2020, when asked about COVID-19's origin and related propaganda described as "outrageous lies" by some officials from China and America, President Trump admitted on *FOX News*: "They do it, and we do it... Every country does it."[98]

Days later, he stated what other government officials were concealing —their knowledge that the virus, outbreak, pandemic, and remedial response, was "artificially induced."[99]

The next evening, the network's Tucker Carlson —son of Richard Warner Carlson, past director of the U.S. Information Agency, and director of the Voice of America propaganda program broadcasting during the last six years of the Cold War— indicted all governments. Carlson scolded health officials worldwide for "lying" about the apparent bioweapon and 'lab virus outbreak.'[100]

Referencing unnamed Chinese researchers[100] who published an indictment of bioweapons research in labs allegedly sourcing the pandemic, Carlson dismissed the general scientific consensus that contaminated bat meat sourced the human disease.

Only three weeks earlier, Carlson had reported that anyone advancing such a "conspiracy theory" was **lying**. In his March 31 revision, Carlson's sources theorized that the outbreak occurred when coronavirus-contaminated lab specimens were somehow deposited at the nearby Wuhan seafood market.

Contrary to Carlson's theory, on March 17, 2020, ABC and Yahoo News heralded a conflicting "study" that allegedly put the question to bed. Published in *Nature Medicine*, Kristian G. Andersen of The Scripps Research Institute of La Jolla, CA; and Robert F. Garry of Tulane University in New Orleans and Zalgen Labs in Maryland (promoted by Andersen and Scripps), and others on their team, titled their writing "The proximal origin of SARS-CoV-2."[101] That "study" purportedly "clearly show[ed] that SARS-CoV-2 is not a laboratory construct or a purposefully manipulated virus."[95] This opposition to "conspiracy theories" prompted international news headlines such as, "Sorry, conspiracy theorists. The study concludes COVID-19 'is not a laboratory construct.'"[102]

In response, this author critically examined the Andersen and Garry et al. study and solidly refuted their conclusion. I had not yet discovered their complicity in the criminal concealment of genetic science refuting their conclusion, nor had I realized yet their conflicting interests in attending an infamous Teleconference of February 2, 2020, detailed below.

The Andersen-Garry team of 'esteemed scientists,' experts in the field of 'evolutionary virology,' working under federal contracts, misrepresented facts and neglected the most substantive science available. Their conclusion evidenced frank *fraud* and complicity in concealing COVID's lab origin. They also purportedly debunked the Deep State conspiracy, but only by violating U.S. criminal law, 18 U.S. Code §1002.

In addition, Title 18 U.S. Code §1519 precludes the "Destruction, alteration, or falsification of records in Federal investigations..." It states, "Whoever knowingly... conceals, covers up, falsifies, or makes a false entry in any record, document, or tangible object with the intent to impede, obstruct, or influence the investigation or proper administration of any matter within the jurisdiction of any department or agency of the United States... or in relation to or contemplation of any such matter or case, shall be fined under this title, imprisoned not more than 20 years, or both."

Pursuant to these charges of criminal conduct by Andersen and Garry, et al., on February 3, 2020, the day after a special "Teleconference" was held in which Andersen and Garry were star 'players,' Trump's Director of the Office of Science and Technology Policy,[103] Droegemeier, initiated the federal investigation into this precise subject —the alleged laboratory origin of the 2019 coronavirus[104] knowingly concealed and falsely debunked by Andersen and Garry.

As mentioned, Droegemeier had written the National Academy of Sciences (NAS) President, Dr. Marcia McNutt, to begin the 'politically incorrect' probe. Consequently, since that time, the origin of COVID was under federal investigation; and certain to be influenced by Andersen and Garry et al.'s science article published in *Nature Medicine*[101] published weeks later dismissing the lab virus bioweapon notion.

> *"[T]he consortium has most to gain from the coronavirus pandemic, and much to lose from determining that the virus originated in a pharmaceutical biotechnology lab—a determination that would bring the entire genetic engineering biotechnology industry under intense scrutiny and regulatory pressures internationally."*

More Evidence for the Crime of Scientific Evidence Tampering

Andersen and Garry et al. issued the most widely heralded and influential "science" paper on this topic.[95] But another group of researchers led by Xiao Li, a structural virologist at the University of Texas, El Paso, is noteworthy. They, too, belittled the truth and smeared Pradhan's science[105] paper to confuse the facts. This wrongdoing further tied the COVID Coup to Pfizer and the cancer cartel.

More obvious than Andersen and Garry's conflicting interests, Li et al.'s publication came with blatant biases. These are certified by Pfizer's long history of financing Xiao Li's University of Texas and the cancer virus studies conducted there.

These studies included researching enzymes active in cancer induction versus genetic repair. (Don't forget, as mentioned earlier, Pfizer is evidenced in the *Congressional Record* as being a biological weapons contractor. In fact, material to Pfizer's COVID-19 mRNA vaccine and the HIV-1/cancer link, my lawsuit against Pfizer recalled what I reported earlier. That, "Pfizer financed the isolation and mass production of the first breast cancer virus called the Mason-Pfizer Monkey virus." This was before the explosive increase in breast cancer cases among vaccinated Jewish women.

On Aug 22, 2019, four months before the COVID-19 virus began circulating in North America, Pfizer added $500 million to a billion-dollar University of North Carolina (UNC), Chapel Hill, "gene therapy" *business,* raising several 'red flags.' UNC is home to the infamous 2014-15 'gain-of-flu' studies for such "cancer virus research." Also called "vaccine therapeutics," Pfizer advanced this business at the UNC under the direction of a virologist named Ralph Baric. (The next Chapter details this more.) That research tied to the COVID-19 gain-of-function bat coronavirus genetic mutations was shut down due to severe outbreak risks, NPR reported in 2014.[106]

Curiously, UNC was the school where I discovered the never-declassified government reports[12] evidencing the lab creation of HIV/AIDS and Ebola-like viruses. I re-published the most revealing contracts in *Emerging Viruses: AIDS & Ebola—Nature, Accident or Intentional?* The grants, the labs, and researchers involved in developing those 'gain-of-function' immune-system destroyers, otherwise known as 'cancer viruses,' are now known and susceptible to criminal investigations and indictments.

"The goal of that [2014-15 UNC coronavirus engineering] work," NPR explained, "was to see whether this bird flu virus [H5N1, that kills more than 60% of the people it infects] might mutate in the wild and start a pandemic in people."

Critics were aghast. What if this lab-made super-flu escaped?

Five years later, supposedly in Wuhan, a related respiratory pathogen called "COVID-19" did.

The Andersen and Garry et al., Fraud and Alleged Felony

As mentioned above, a Code §1519 fraud violation is committed when it is intended to impede, obstruct, or influence a government proceeding, in this case, the presumably ongoing White House investigation into COVID's origin.

Andersen and Garry et al. obstructed the administration of that investigation and the federal coronavirus response involving 'source research.' Their felony aided-and-abetted and protected agents and agencies suspected in the COVID Coup, such as Dr. Fauci's group at the NIAID and Dr. Baric's group at UNC. Additional co-conspirators conspired to subvert science and public information to conceal their liabilities and complicity. They aided-and-abetted coronavirus bioterrorism and this devastating bio-crime.

In addition, anti-trust laws are called into question pursuant to the apparent 'joint venture' between agents for Scripps and Tulane University corruptly influencing federal agents and agencies financing, investigating, and/or obfuscating the outbreak. The cartel's actions, political and institutional influence, and economic benefits are consistent with a 'racketeering enterprise.'

Their actions exclusively favored their drug industry alliances and vaccine special interests. They unfairly deprived us of urgent information, subverted healthcare, and prevented competing products and services from reaching the marketplace.

This prohibited activity —depriving free and fair trade, favorable competition, and public health— supposedly precluded by the Department of Justice and Federal Trade Commission Statements of Antitrust Enforcement Policy in Health Care,[107] was subverted by Andersen, Garry, and others like Fauci, who lied about the lab origin of the pandemic.

Who is Kristian Andersen?

In 2018, "Project Leader," Kristian Andersen, Ph.D., co-director of the Center for Viral Systems Biology at The Scripps Research Institute (TSRI; at that time), received a $15 million grant to conduct "an in-depth study" through the TSRI-led Center for Viral Systems Biology to fight hemorrhagic fever viruses, including Ebola and Lassa. That grant was given by the NIH's National Institute of Allergy and Infectious Diseases (NIAID). NIAID, at that time, was directed by Dr. Fauci.[108]

"Our goal is to help eradicate these diseases by building better diagnostics, designing new drugs, and informing vaccine design," Andersen explained to press officials at that time.

Quoting Scripps' press release[109] heralding the $15M grant, "The new study will take advantage of TSRI's expertise in **genomic analysis** and **data science**. Andersen ha[d] previously led large-scale projects to track the geographic spread and evolution of viruses using genomic analysis. He and his colleagues are now planning to use genomic analysis and other advanced tools, including physiological measurements, to study individual disease survivors."(Emphasis added.)

Dr. Andersen's "genetic analysis" of the coronavirus spike protein structure, omitting and neglecting four insertions from HIV-1, is central to this story, cover-up, subverted federal investigation, and alleged complicity in The Coup.

Who is Robert Garry?

Working closely with Andersen in COVID-19's "genomic analysis" was Dr. Robert Garry. According to a press release issued by Dr. Garry's company, Zalgen Labs in Maryland, as early as August 2016, the NIAID transferred to Zalgen a grant to develop recombinant antigen diagnostics for filoviruses, which "resulted in development of the ReEBOV® Ebola test, among others."[110]

This fact is also material to Zalgen's and Abbott Labs' developments of industry-leading coronavirus tests; as well as to Dr. Fauci and Abbott Labs' recorded statements of November 7, 2019,[111] pursuant to AIDS virus gene sequences enabling the circulating pandemic coronavirus to infect human cells (as reported by Prashant et al.[112] as further detailed below).

That date was three weeks **after** the suspect "Event 201" (i.e., 'Agenda 21'[113]) Coronavirus Preparedness ('Predictive Programming'[92]) conference sponsored by key investors in the pharmaceutical cartel. As mentioned, this included the Bill & Melinda Gates Foundation, Johns Hopkins University that issued all the frightful and deceptive coronavirus data, and the World Economic Forum that represents leading investors in the infectious disease syndicate.

That date, November 7, 2019, was synchronous with Wuhan's alleged first COVID patient identification, or a month before the first coronavirus cluster appeared at the Wuhan seafood market.

How many 'coincidences' do we need before we see a 'pattern' of organized crime being evidenced here? Evidence "beyond a reasonable doubt" required for criminal convictions?

Here are a couple more most relevant to the common-sense conclusion of racketeering in the COVID biocrime. On this same day, November 7, 2019, CNN reported[114] that a new strain of HIV-1/AIDS had suddenly been discovered by Abbott Labs.[115]

Could it be that this press release was used to conceal and divert from the HIV-1 genes incorporated into the COVID lab virus obfuscated by Andersen and Garry?

Suspicious Contemporaneous 'Discovery'

Johns Hopkins had reported genetically mutating the AIDS virus (HIV-1) to deliver "new DNA" to patients. This was applauded by NIH Director Francis S. Collins, Fauci's superior. Collins called this treatment a "cure" for sickle cell anemia targeting mainly people of color.

Abbott Labs had financed Johns Hopkins research[116] into anti-cancer drugs that act similar to the AIDS-laced spike protein 'protease inhibitors.' This anti-coronavirus therapy acts much like Abbott's drug Norvir, as further explained below. Abbott also produced the main test used to confirm COVID-19.

The following paragraph quotes from *Emerging Viruses: AIDS & Ebola–Nature, Accident or Intentional?* wherein I had urged international scrutiny and opposition to the AIDS industry's malfeasance. I included Anthony Fauci —the 'AIDS Czar' for the NIH at the NIAID, a privy insider gaining intelligence and commercial enrichments from related patents. He is a key accomplice in the COVID Coup. The corporate and governmental suspects are evidenced having manufactured the AIDS pandemic during the early 1970s as previously discussed:

"Abbott Labs are best known for having licensed and produced the ELISA screening test for HIV... Abbott also licensed and marketed the hepatitis core antigen test purchased by New York City Blood Center officials, following years of delay and before the ELISA test was available, to help identify blood units suspected of HIV infection. The company had also supplied expertise and the experimental radioactive reagents to [Dr. Wolf] Szmuness, required for this New York homosexual hepatitis B vaccine trial. Furthermore, Abbott Labs ended up commercially marketing Merck-Sharp & Dohme's hepatitis B vaccine. ([117]; p. 126.)

"Moreover, the hepatitis B vaccines suspected of having transmitted HIV to American homosexuals was researched by Abbott's L.R. Overby, who was intimately connected to the New York University Medical Center hepatitis B chief, Saul Krugman. Together, they evaluated hepatitis B susceptibility and vaccination methods in the New York subjects during the mid-1970s." ([117]; p. 126.)

Consequently, we witness Abbott labs operating as a major player within the Deep State's genocidal enterprise. During my investigation, these named agents and entities represented a "consortium" of interested parties (according to their press reports). These agents and entities formed a single medical science pharmaceutical syndicate, cartel, or alleged racketeering enterprise.

'Over the years, Abbott's press officials distanced the company from its cartel agreements with the Rockefeller-IBM-partnered IG Farben conglomerate that administered Auschwitz and financed the death camps of WWII. Decartelization following Nuremberg Trials gave rise to Bayar AG, Hoechst, and BASF. Abbott[118] acquired the latter in late 2000 for $6.9B (US) in cash.

Since that time, Abbott has been a leading funder of the American Legislative Exchange Council[119] (ALEC). ALEC is the main legislative lobbying, politician bribing, "bill mill." They heavily influence the push for mandatory vaccines through campaign financing of candidates in both parties.

Abbott, which manufactures the AIDS virus Spike protein attachment 'protease inhibitor' Norvir, has a long history of bribing medical doctors too, resulting in its 2006 rebuke by the Association of the British Pharmaceutical Industry[120] (ABPI).

It is common sense that to develop Norvir's AIDS virus spike protein attachment enzyme inhibitor; you need to mass-produce the AIDS virus spike protein first to test the drug's effectiveness. That means that these Abbott 'insiders' and allies had substantial motives to lie about and divert media attention away from COVID's Spike protein *AIDS virus genetic splices*.

Bill Gates and the Coronavirus Conspiracy Enterprise

Returning to Dr. Garry, on March 20, 2020, business publications[121] heralded Garry's Zalgen Labs that had developed and commercialized a coronavirus diagnostic test in alliance with pharmaceutical interests, including the NIH, NIAID, the FDA, and Johns Hopkins University. Zalgen Labs operated at the forefront of coronavirus intelligence gathering, data mining, scientific analysis, infection projections, and media propaganda.

Washington Business Journal
March 9 at 6:30 PM

Germantown biotech Zalgen Labs is working on tools and testing "that will catch just about any coronavirus that may come our way," said co-founder Luis Branco. Will the world be ready for next time?

BIZJOURNALS.COM
This Maryland biotech is developing coronavirus tests for next time

Operating at the heart of this coronavirus enterprise commercializing the "pandemic response" (allegedly in the interest of public health and safety) is the Cambridge, Massachusetts biotech company called Moderna Inc. With Pfizer's complicity and DOD financing, Moderna and Pfizer developed the leading candidate vaccines against the "Chinese virus."

Moderna's principals include Robert Langer, and partners Bill Gates, the NIAID headed by Fauci, and the Defense Advanced Research Project Agency (DARPA). Langer "almost single-handedly created the fields[122] of controlled drug delivery and tissue engineering."

COVID COUP: "The Rise of the Fourth Reich"

Bill Gates shakes hands with Dr. Anthony Fauci, allied in the Moderna coronavirus vaccine project to develop "mRNA-1273."

"Notable about the Fauci-Gates-Langer Moderna coronavirus vaccine, mRNA-1273," explained Princeton Univ. trained political and economic investigator, F. William Engdahl,[123] is that it was "rolled out in a matter of weeks, not years." That is FALSE. As you will read later, Moderna and DARPA had been working on their S-protein mRNA vaccine for years.

"Nature Medicine and Springer Nature act within the alleged criminal enterprise to influence the global scientific community and federal investigators as a main source of 'fake science' and 'fake news.'"

In addition to the suspects mentioned above, Dr. Andersen and Dr. Garry's publisher, *Nature Medicine*, are similarly incriminated by ethical breaches and conflicting interests. The entity is burdened by a history of scientific fraud. "Springer Nature" is the owner of this discredited purportedly "peer-reviewed" journal.

Springer Nature is a premier globalist academic publishing company (within the alleged criminal enterprise) created by the merger of Springer Science+Business Media and Holtzbrinck Publishing Group's Nature Publishing Group, Palgrave Macmillan, and Macmillan Education. This multi-national syndicate made nearly $2 billion in 2019 by marketing its publications and properties in alliance with Big Pharma.

This consortium had most to gain from the coronavirus pandemic and still does. That syndicate had much to lose from determining that the virus originated in a pharmaceutical biotechnology lab. That determination could have brought the entire genetic engineering biotechnology industry under intense scrutiny and regulatory pressures internationally.[96]

Accenting a history of foul play, in 2011, Springer Nature acquired Pharma Marketing and Publishing Services (MPS) to mainly market partnering companies' drugs and vaccines. In 2013, the London-based private equity firm BC Partners acquired a majority stake in Springer from EQT and GIC for $4.4 billion. A year later, it was revealed that Springer had fraudulently generated and published sixteen papers in conference proceedings. The scheme used SCIgen, "a computer program that uses context-free grammar to **randomly generate nonsense in the form of computer science** *research papers.*"[124] (Emphasis added)

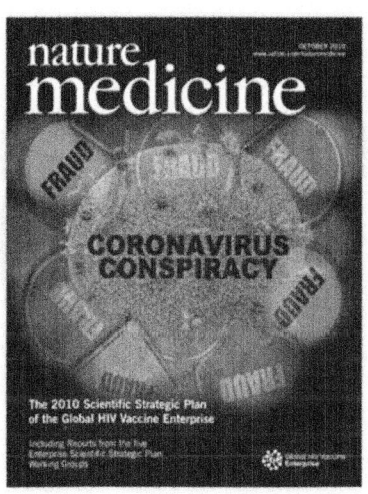

Created by scientists at MIT (the Massachusetts Institute of Technology),[125] home to the Media Lab embroiled in the Bill Gates and Jeffrey Epstein scandal, and also home to the lead scientist and entrepreneur that started Moderna— Robert Langer —SCIgen's stated aim was "to maximize amusement, rather than coherence." All elements of the papers SCIgen generated, including graphs, diagrams, and citations, were "fake" —fraudulently manufactured.

In other words, *Nature Medicine* and Springer Nature act within the criminal cartel to influence the global scientific community (and federal investigators researching COVID's origin) as the main source of "fake science" and "fake news."

In this instance, to discredit, discourage, and disparage coronavirus "conspiracy theorists," this enterprise published Andersen and Garry's "fake science" (a.k.a. 'pseudoscience'). As detailed below, their motive was established by senior NIH and UK intelligence officials to promote counterintelligence concealing evidence of the lab virus origin.

To do this, the devil-doers needed to neglect the journal's vitally important earlier publication, a 2015 report by Menachery et al. That publication too offers evidence of the conspiracy to fraudulently conceal the lab origin of COVID-19. Menachery's research proposed convincingly that "A SARS-like cluster of circulating bat coronaviruses shows potential for human emergence."[126] Human transmission was especially predicted when the virus was altered in labs!

"On the basis of these findings," scientists from UNC, the Wuhan Institute of Virology in China, and the Department of Cancer Immunology and AIDS at the Dana-Farber Cancer Institute at Harvard Medical School[127] described their risky activity as **"synthetically re-deriv[ing] an infectious full-length [SARS/Corona] recombinant virus [that] demonstrate[d] robust viral replication both in vitro and in vivo."**

Aware of these studies and dangers, the Anderson team noted, "Basic research involving the passage of bat SARS-CoV-like coronaviruses in cell culture and/or animal models has been ongoing for many years in biosafety level 2 laboratories across the world." There were "documented instances of laboratory escapes of SARS-CoV." Based on these facts, Anderson and Garry's team concluded, "We must therefore examine the possibility of an inadvertent laboratory release of SARS-CoV-2."

Subsequently, closing their eyes to the science mentioned above and incriminating facts, these Deep State 'submissives' falsely reported in *Nature Medicine* on March 8, 2020, "it is ***improbable*** that SARS-CoV-2 (COVID-19) emerged through laboratory manipulation of a related SARS-CoV-like coronavirus. (Emphasis added.)

Purportedly, after searching for a possible precursor for this bat-to-human species jump through "genetic [data] banks," Andersen and Garry reported finding no other "progenitor virus" (predecessor) with very high genetic similarity to the COVID-19 virus.

More Solid Evidence of Crime Emerges from Fauci E-mails

On June 3, 2021, the Andersen and Garry scientific fraud became irrefutable following the Freedom of Information Act (FOIA) release of Fauci's emails by the NIH. These e-mails are public records heralded on numerous mainstream media websites, including the *Washington Post*, *New York Times*, and *CNN*, citing BuzzFeed News as the source of the FOIA released documents.[128]

Shortly before Andersen and Garry's *Nature Medicine* article was published, Andersen e-mailed Anthony Fauci[129] on March 8, 2020. Contrary to Andersen's "improbable" statement, he wrote to Fauci: "The unusual features of the virus make up a really small part of the genome (<0.1%) so one has to look really closely at all the sequences to see that some of the features (potentially) look engineered."

That 0.1% engineered mutation references the AIDS-virus gene sequences in the SARS-CoV spike protein (as further detailed below). In other words, **that "0.1%" is the synthetic "gain-of-function" laboratory mutation in the COVID disease antigen causing the virus's hyper-transmissibility**.

Furthermore, Fauci's e-mails of Saturday, Feb. 1, 2020, five weeks *before* the Andersen Garry *fraudulent concealment* in their *Nature Medicine* article omitting the HIV/AIDS gene sequences added to CoV's S-protein, a large group of *insiders* held that urgent "Teleconference." They met online to consider whether they would coordinate *lying* about those AIDS virus genes in the COVID pathogen in support of Tedros Adhanom, the World Health Organization (WHO) director, and Bernhard Schwartlander, the WHO's representative in China.

Sir Jeremy "James" Farrar, a British medical researcher and director of the Wellcome Trust since 2013, and previous professor of tropical medicine at the University of Oxford, wrote Fauci. The group was frantically concerned about the need for the insiders to develop a strategy in anticipation of Tedros and Bernhard falsifying the bioweapon scenario. Farrar called this 'prevarication,' which means lying. Five weeks later, Andersen and Garry published the same 'prevarication' that the lab virus origin was "improbable."

This section of Fauci's e-mails suffered severe "redaction," that is, *censorship*, by intelligence agencies and DOJ/FOIA agents. This illegal concealment (i.e., evidence tampering) during a federal investigation is incriminating.

Also incriminating is the Teleconferees' lying and concealing the AIDS virus connection. Fauci, Collins, Andersen, Garry, Farrar, and their subordinates also 'prevaricated.' They panicked over the news headlines outpouring at that time. "Coronavirus Contains 'HIV Insertions', Stoking Fears Over Artificially Created Bioweapon"[130] heralded one article.

They decided to lie about and fraudulently concealed solid science. Andersen and Garry were then assigned to publish fraud by omissions and misrepresentations. Their complicity exclusively favored pharmaceutical interests, 'captured' government agencies and sham science. A "National Security Crime Syndicate" adequately describes this beast.

The objective of their *Nature Medicine* sham science article was to protect what amounts to the main racketeering enterprise sickening the public and destroying America. At the same time, these 'insiders' plotted their fraud. The FBI, CIA, and NSA crime syndicate engulfed the Trump administration in damaging controversy, largely to divert from their genocidal biocrime.

These officials, administering the COVID Coup, deceived and damaged other governments and people worldwide.

The criminal intent of the suspects and their falsified publications was to defraud society and the scientific community and 'neutralize' legitimate whistleblowers such as Pradhan et al., who, in this instance, evidenced the coronavirus' lab origin.[131]

When you consider that this censorship and fraud was intentional and well-coordinated, it is unreasonable to presume the Wuhan outbreak was anything other than intentional and well-coordinated.

President's statement.[132]

Concealed Evidence of Genetic Engineering

Andersen and Garry et al. based their false anti-conspiracy conclusion that the virus came from "nature" on their dishonest 'peer-reviewed' "perspective on the notable features of the SARS-CoV-2 genome."[133,134] In that article, these Deep State mercenaries fraudulently neglected the scholarly genetic research of Pradhan et al.,[134] that prompted the urgent Teleconference on Sunday, February 2, 2020.

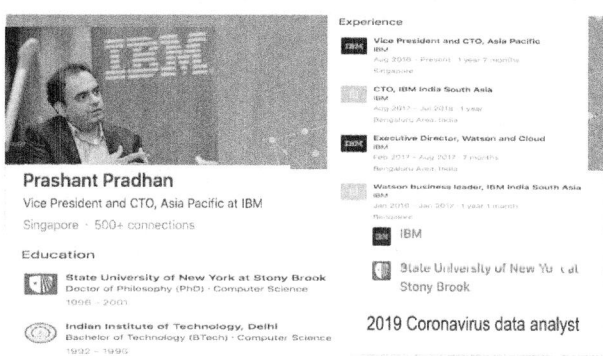

Prashant Pradhan was, at the time, the Chief Technical Officer for IBM in Asia. Pradhan's coronavirus gene-sequencing group of experts were affiliated with the Indian Institute of Technology (IIT), the University of New Delhi, IBM, and New York University at Stonybrook. This team of nine experts used the most advanced computer programs to perform the most important genetic analysis of the COVID virus.[134]

Andersen's team labored to discredit Pradhan's widely circulated "pre-publication." They neglected to review and report on the spike protein "inserts" derived from HIV-1. Instead, Andersen et al. falsely argued against a laboratory cell culture "intermediate" between the presumed 'bat reservoir' and the first infected Wuhan humans. Their team reported that "a hypothetical generation of SARS-CoV-2 by cell culture or animal passage would have required prior isolation of a progenitor virus with very high genetic similarity, which has not been described."

That is FALSE. That statement and hypothesis criminally omit and conceal the HIV-1 gene sequences discovered as reported by Pradhan et al. in what was smeared as the "Indian paper" by the Teleconference organizers.[134]

"High genetic similarity" to HIV-1–the AIDS virus was reported and is certainly amenable to repeated discovery. Instead, the subject was almost completely covered up.

Moreover, the complicit suspects' statement is diversionary. As early as 2009, Wimmer et al. published genetic studies heralding "Synthetic viruses" virtually identical to the 2019 pandemic coronavirus. The manufactured RNA viruses combined the human endogenous retrovirus, HIVcpz [the AIDS virus progenitor in chimps], and the "SARS-like coronavirus" from supposedly bats.[135]

Much like Andersen et al. committed with their bogus dismissal of coronavirus conspiracy theories, Wimmer et al. neglected the increased risks to society in advancing this dangerous biotechnology, exclusively favoring commercial interests and enabling biocrimes.

In fact, Pradhan's group expressly described more than "very high genetic similarity" between the pandemic coronavirus and the earlier engineered lab virus, HIV-1. Pradhan's group identified identical sequences in the subject coronavirus 'bioweapon' and HIV-1. [95,134]

Consequently, Pradhan's group vicariously determined the urgent need for further investigation in lieu of the high probability that further mutations in the circulating bioweapon might arise, combining more deadly gene sequences from coronavirus-infected persons or in viruses, even bacteria, circulating in nature, leading to worsening morbidity and mortality globally.

Also discrediting Andersen and Garry et al., these leading coronavirus co-conspirators controverted their widely publicized opposition to coronavirus conspiracy theories by including a *'disclaimer.'* After acknowledging that "The receptor-binding domain

(RBD) in the spike protein [associated with four unique HIV-1 genes according to Pradhan et al.[134]] is the most variable part of the coronavirus genome amendable to therapies,[95] Andersen et al. wrote: "In [conspiracy] theory, it is possible that SARS-CoV-2 acquired RBD mutations during adaptation to [laboratory] passage in cell culture, as has been observed in studies of SARS-CoV."[96]

> *"Andersen's team disregarded Pradhan's widely circulated 'pre-publication,' neglected to review and report certain spike protein 'inserts,' and falsely argued against a laboratory cell culture 'intermediate' between the presumed 'bat reservoir' and the first infected Wuhan humans."*

Origin of Coronavirus[104]

More Pseudoscience and Fraudulent Concealment

Andersen's team also misrepresented the "natural selection" of the COVID virus when they explained "the high-affinity binding of the SARS-CoV-2 spike protein to human" cells "is most likely the result of natural selection on a human." That falsehood was intended to attack the heart of the lab "gain-of-function" mutation.

Alternatively, to dismiss the lab origin theory and conceal the AIDS virus gene inserts, these cartel agents diverted, writing that a "human-like [receptor] permits another optimal binding solution to arise."

Throughout their discussion, these agents of deception recklessly neglected or purposely concealed the fact that HIV/AIDS researchers identified the "optimal binding solution" during the 1980s, arising from early gene-splicing cancer virus experiments evolved from the SVCP. This "optimal binding solution" relied on inserting the positively-charged HIV-1 AIDS spike protein gene sequences so that it would "bind" electromagnetically to the negatively-charged "ACE-2" receptor sites on targeted human cells.

Contrary to Andersen and Garry's conclusion, "[t]his is strong evidence that SARS[/HIV-1]-CoV-2 is... the product of purposeful manipulation."[95,134]

Understanding the Spike Protein ("S-protein")

It is important to understand how the S-protein works in "gain-of-function" infectivity. For this, the *New York Times*[136] overview of the coronavirus S-protein attachment mechanism may be helpful. This is summarized in the following screenshots:

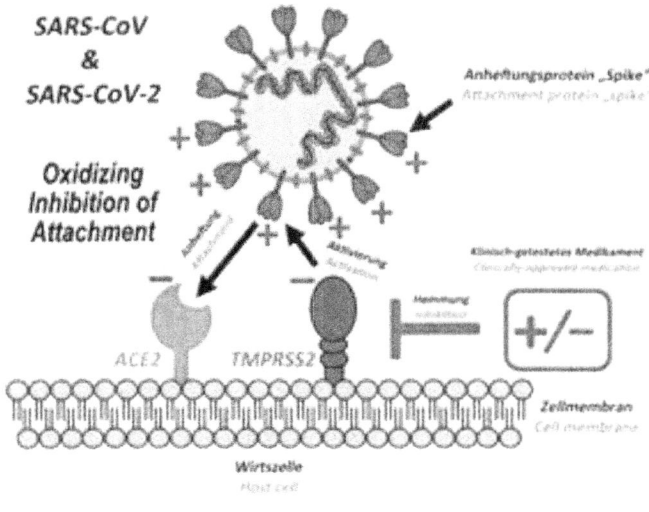

~

> The gene for the spike protein in SARS-CoV-2 has an insertion of 12 genetic letters: ccucggcgggca. This mutation may help the spikes bind tightly to human cells — a crucial step in its evolution from a virus that infected bats and other species.
>
> A number of scientific teams are now designing vaccines that could prevent the spikes from attaching to human cells.
>
> ```
> auguuuguuuucuuguuuuauugccuagucucuagucagugugu uaaucuuaca
> accagaacucaauuaccccugcauacacuaauucuuucacacguggugu uuauuac
> ccugacaaaguuuucagauccucaguuuuacauucaacucaggacuuguucuuaccu
> uucuuuuccaauguuacuugguccaugcuauacaugucucugggaccaauggu acu
> aagaggu uugauaacccuguccuaccauuuaaugauggugu uuauuuugcuuccacu
> gagaagucuaacauaauaagaggcuggauuuuuggu acuacuuuagauucgaagacc
> cagucccuacuuauugu uaauaacgcuacuaauguuguuauuaaagucuguga auuu
> caauuuuguaaugauccauuuuugggguuuauuaccacaaaaacaacaaaaguugg
> auggaaagugaguucagaguuuauucuagugcgauaauaugcacuuuugaauaugu c
> ucucagccuuuucuuauggaccuugaaggaaaacagggu aauuucaaaaaucuuagg
> gaauuugu guuuaagaauauu gauggu uauuuaaaau auauuc uaagcacacgccu
> auuaauuuaguagcugaucucccucagggu uuuucggcuuuagaaccauugguagau
> uugccaauagguauuaacaucacuagguuucaaacuuuacuugcuuuacauagaagu
> uauuugacuccuggugauucuucuucaggu uggacagcuggugcugcagcuuauuau
> guggguuaucu ucaaccuaggacuuuucuauuaaaauauaaugaaaauggaaccauu
> acagaugcuguagacugugcacuugacccucucucagaaacaaaguguacguugaaa
> uccuucacuguagaaaaaggaaucuaucaaacuucuaacuuuagaguccaaccaaca
> ```

Pradhan et al.'s study[13] identified four obviously 'unnatural' gene sequences from the AIDS virus (i.e., HIV-1) that had been spliced into the "novel" "SARS-CoV-2" S-protein. As mentioned above, this unnatural attachment mechanism is governed by the 'positively charged' added envelop gene segments. This is a major 'smoking-gun' in the *biocrime*[137] and central to effective therapies. This positively-charged S-protein indicates that anti-oxidants should be recommended. But these are generally concealed remedies,[138] as I have been reporting since February 2020.

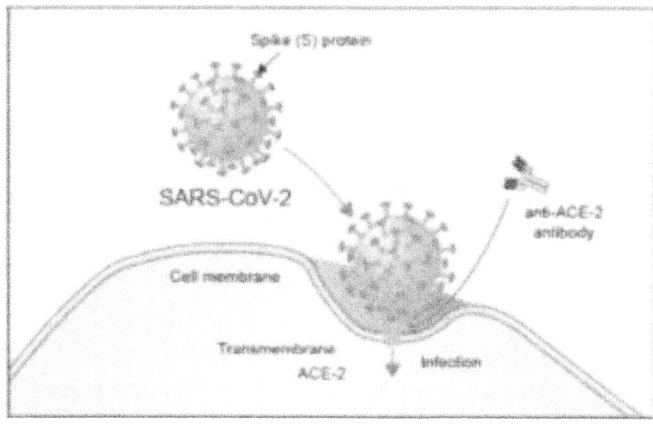

More Technical Analysis of Andersen and Garry

A leader in manufacturing chemotherapies to inhibit the binding of laboratory-engineered viruses to human cells is "R&D Systems" —a fact commonly known to researchers throughout Andersen and Garry's industry.

According to R&D Systems' promotions, the most important mechanism of SARS-CoV-2 infection is the "ACE-2" "Entry Receptor" discussed by the suspects.[139] This is the "attachment apparatus" for pathogenesis—linking the coronavirus to its immune-insufficiency disease and unnecessary deaths.

R&D Systems' literature-review features cancer cell-lines. The company especially tests the "Vero-E6" monkey kidney cell line known to permit SARS-CoV replication, in the same way, it permits the multiplication of the AIDS virus (HIV-1), Ebola, Ebola's 'mother' —the Marburg virus— and cowpox viruses.[140]

The R&D Systems' research and developments dovetail with Garry's Zalgen Company's research and developments. This fact compounds evidence of Andersen and Garry having concealed coronavirus genetic science *intentionally*, knowingly, and willfully to publish false and misleading, albeit federally influential, opposition to revealing the laboratory creation of the SARS-CoV-HIV-1 bioweapon.

> *"For all we know, without identifying the true source of the outbreak, additional outbreaks may multiply risks and civilization's damage, disease, distress, and deaths."*

Researchers internationally noted, as Hoffman et al. determined, that "SARS-CoV-2 Cell Entry Depends on [that] ACE2 and TMPRSS2 and is Blocked by a Clinically Proven Protease Inhibitor."[141] What that "protease inhibitor" does is central to Garry's commercial efforts in Maryland at Zalgen Lab. It is also central to the public-private 'pandemic enterprise' as a whole. That protease inhibitor blocks the activation of the attachment apparatus so that the virus cannot attack human cells.

In other words, by neglecting, concealing, and obfuscating what is common knowledge in the coronavirus biotechnology trade, Andersen and Garry et al. have deprived and damaged society, the government, and especially taxpaying citizens who suffer from this cover-up, including the concealment of multiple non-drug remedies and natural cures based on this intelligence.

To prevent coronavirus disease, inhibiting the enzymatic transformation of the attachment protein is required. To stop the disease and deaths this way, the specific **protein transforming enzyme produced by the human host cell** is required to be neutralized, thus altering the virus-host binding site. Without that ACE-2 enzyme needed to transform and activate the positively-charged spike protein, the failed attachment precludes infection, thus preventing AIDS and COVID.

Otherwise, the offending SARS/AIDS/COVID viruses electromagnetically (i.e., energetically) attach to human cells and deliver what officials call their "payload." That is, inject their poisonous RNA genetic code into the host, thus causing morbidity and mortality.

R&D Systems developed an "Anti-ACE-2 Antibody"[142] (Catalog # AF933[143]) to commercialize and profit from this knowledge. The company tested this potential remedy knowing this product blocked the Vesicular Stomatitis Virus (VSV) entry into human cells. They used VSV "pseudotypes" to express the damaging SARS/HIV-1/CoV-2 S-protein.[139]

But that drug therapy threatens serious life-threatening side effects. It knocks out the natural hormonal balance regulating blood vessels' size, blood flow, and your blood pressure.[144]

It is noteworthy that these alleged 'mad scientists' originally developed their VSV 'pseudotype lab virus' in cell cultures from the

Rhabdovirus species. These studies are extremely well known to Andersen, Garry, and NIAID Director Anthony Fauci because Rhabdoviruses have a lot to do with the 2014 "Ebola Emergency" in which they played expert roles.

That family of deadly viruses initially infected rodents, causing rabies. But through crude genetic engineering of Rhabdoviruses at the U.S. Army's sixth leading biological weapons lab —Litton Bionetics— during the 1960s and early 1970s, officials generated the "mother of Ebola" —the Marburg virus. They then adapted this Rhabdovirus to infect monkeys, producing 'Rhabdovirus simian.' These acts were recorded in government contracts and conference discussions during the SVCP.[14] I first made this knowledge public in *Emerging Viruses: AIDS & Ebola—Nature, Accident or Intentional?*[117]

Accordingly, I allege the suspects are "mad scientists"—egomaniacs who remain willfully blind to the risks they pose to civilization. Indeed, the virology and biotechnology communities at large are well aware this research and ACE-2 drugs are risky; and that this know-how sources from early AIDS and Ebola virus research. The SARS/AIDS-CoV-2 RNA entry into human cells requires the 'cleavage' of the AIDS-linked lab-engineered S-protein that activates ACE-2. In their objectionable writings, the suspects have fraudulently concealed and recklessly neglected these facts. Relying on S&R's words, "For SARS-CoV entry into a host cell, its S protein needs to be cleaved by cellular proteases at two sites, termed S protein priming, so the viral and cellular membranes can fuse."[139]

As mentioned, this publicly hidden knowledge prevents awareness of natural preventatives, treatments, and cures for coronavirus disease. "ACE-2 is a zinc metalloenzyme."[144] That is why Zinc, an antioxidant, is especially helpful in restoring normalcy to virus-and-drug-degenerated ACE2 receptor sites on blood vessels.

Add to this restriction the blood vessel damage emerging in vaccinated people. It is common sense that when the vaccines damage normal ACE-2 availability and function, the side effect is thrombosis, thrombocytopenia, and myriad life-threatening diseases erupting from blood platelets going haywire.

Platelets are produced by blood cells called "thrombocytes." Platelets clot broken blood vessels like natural band-aides. That function depends on the platelets being negatively charged, similar to the ACE-2 receptor site. When the latter is overrun by 'cationic' S-protein poisoning, the system becomes acidified. The acid burns holes in the side of blood vessels, causing hemorrhages that the platelets gather to stop. This best explains what news reports are heralding about post-vaccination myocarditis and blood clots.[145]

According to the *New England Journal of Medicine*, this is "a newly described syndrome characterized by thrombosis and thrombocytopenia." It develops "5 to 24 days after initial vaccination." Each of the nCoV-19 vaccines causes different amounts of this syndrome. The AstraZeneca vaccine is most disconcerting. That delivers "a recombinant chimpanzee adenoviral vector encoding the spike protein of SARS-CoV-2."[146]

Questioning the pathogenesis of this new syndrome, *NEJM* authors wrote, "In almost every patient, high levels of antibodies to platelet factor 4 (PF4)–polyanion complexes were identified by enzyme-linked immunosorbent assay (ELISA), as well by assays based on platelet activation, which, when tested, was enhanced by the addition of PF4." ... Is thrombosis propagated along vascular and hematopoietic surfaces that release diverse anionic cofactors, as in heparin-induced thrombocytopenia? In one preliminary report, the investigators reported that antibodies to PF4 do not cross-react with the spike protein."

PF4 is negatively-charged. Antibodies to PF4 are positively-charged. That is why antibodies to PF4 do not cross-react with positively-charged spike proteins.

Acid-base interactions are central to viral and vaccine-related diseases. That ACE-2 attachment mechanism largely depends on the bioelectronics of fluids bathing that environment or 'terrain.' This implores the importance of acid/base chemistry in delivering therapeutics, or illnesses, including thrombosis and platelet dysfunctions. The positively-charged spike-protein (antigen) attachment complex comprises electronically-charged protein and sugar parts, and their binding process is interrupted or blocked by an assortment of natural, low-cost, safe, and effective **antioxidants**.[147,148,149] All the antioxidants are negatively-charged, providing therapeutic oxygen to the environment —in this case, the ACE-2(-)/S-protein(+) binding site.

Energetically-Charged Suppressed Natural Cures

Anti-trust and consumer fraud concerns are raised by the actions of the suspects and their drug industry monopoly. Alternative treatments have been recklessly and damagingly neglected. These 'alternative treatments' for coronavirus disease include bioelectric remedies for recovery and prevention, including vitamins C and D, zinc, chlorophyll, and "strong silver catalysts" (i.e., OxySilver^{TM} with 528 frequency) promoting the breakdown of the spike protein amino-acid sugar complex that is highly influenced by oxidation/reduction reactions.[148]

Again, this molecular chemistry involves 'electro-mechanics' and biophysics. This dynamic system best explains the reported effectiveness of hydroxychloroquine and the aforementioned antioxidant therapies that neutralize reactive oxygen species (ROS) the spike-protein includes.

In other words, Andersen and Garry's concealed 'smoking gun' is also the 'Achilles' heel' of the virus. Like many other viruses, the coronavirus can be neutralized, for instance, by the drug hydroxychloroquine in favor of the pharmaceutical industry.[149] This drug provides a 'hydroxyl radical' to alkalize (i.e., reduce) the positively charged spike-protein.

Unfortunately, hydroxychloroquine can cause adverse side-effects, is not available 'over-the-counter,' and is costly.

Alternatively, the natural silver hydrosol called OxySilver^{TM} may be used. I pioneered this low-cost and safe broad-spectrum antimicrobial manufactured using 'structured water' that resonates at 528Hz frequency. That precise energy has been shown to <u>increase antioxidant activity</u> by 100 percent.[150]

Furthermore, the spike-protein glycan group "can be excellently released from the glycoprotein by alkaline" treatments, according to Vliegenthart and Kamerling.[151] Therefore, alkalizing water for improved body hydration is a reasonable investment. I previously discussed[138] the benefits of alkalizing foods and nutrients. Small amounts of baking soda, $NaHCO_3$ consumed in drinking water (e.g., 1/2 teaspoon per gallon), activates anti-inflammatory mechanisms and the transition of macrophage phenotypes from M1 (inflammatory) to M2 (regulatory) subtypes.[152]

Fuchs et al. stated that "the current available data indicate that macrophage polarization is a multifactorial process in which a huge number of factors can be involved producing different activation scenarios. Once a macrophage adopts a phenotype, it still retains the ability to continue changing in response to new environmental influences,"[153] including terrain electrochemistry, anti-oxidant availability, or extracellular alkalinity.[152,153]

These facts are consistent with Pradhan et al.'s determinations. From studying the novel ACE-2 attachment apparatus of the offending coronavirus, they published Table 1 that makes clear the laboratory-engineered S-protein "inserts have a high density of positively-charged residues. The deleted fragments in inserts 3 and 4 increase the positive charge to surface area ratio."[112]

In other words, returning to Andersen et al.'s false science paper, the subject bioweapon developers, knowing this spike-protein-sugar assembly depended on the "positively charged residues" within the attachment complex, increased the infectivity and lethality of the SARS/HIV-1/2019-nCoV mutant by removing "fragments in insert 3 and 4 [to] increase the positive charge to surface area ratio."[112]

That is the loathsome "gain-of-function" mutation hyper-weaponizing the current plague.

Accordingly, this manufactured coronavirus mutation would increase morbidity and mortality. These two objectives —more disease and costly profitable healthcare, and depopulation as openly promoted by Bill Gates, are noteworthy and material to required criminal investigations into the lab originating the AIDS virus/SARS/coronavirus recombinant.

Pradhan et al. wrote about this mutagen, "The amino acid residues of inserts 1, 2 and 3 of 2019-nCoV spike glycoprotein that mapped to HIV-1 were a part of the V4, V5, and V1 domains respectively in gp120 [Table 1]."[112]

Motifs	Virus Glycoprotein	Motif Alignment	HIV protein and Variable region	HIV Genome Source Country/ subtype	Number of Polar Residues	Total Charge	pI Value
Insert 1	2019-nCoV (GP) HIV1(GP120)	71 76 TNGTKR TNGTKR 404 409	gp120-V4	Thailand */ CRF01_AE	5 5	2 2	11 11
Insert 2	2019-nCoV (GP) HIV1(GP120)	145 150 HKNNKS HKNNKS 462 467	gp120-V5	Kenya*/ G	6 6	2 2	10 10
Insert 3	2019-nCoV (GP) HIV1rGP120)	245 256 RSYL----TPGDSSSG RTYLFNETRGNSSSG 136 150	gp120-V1	India*/C	8 10	2 1	10.84 8.75
Insert 4	2019-nCoV (Poly P) HIV1(gag)	676 684 QTN6-----------PRRA QTNSSD MQRSNFKG PRRA 366 384	Gag	India*/C	6 12	2 4	12.00 12.30

Table 1: Aligned sequences of 2019-nCoV and gp120 protein of HIV-1 with their positions in primary sequence of protein. All the inserts have a high density of positively charged residues. The deleted fragments in insert 3 and 4 increase the positive charge to surface area ratio. *please see Supp. Table 1 for accession numbers

These genetic findings not only convincingly prove the lab virus "conspiracy theory" but also indict the Tulane and Scripps co-authors for recklessly neglecting, smearing, and censoring this highly reputable and reproducible science.

As Pradhan's group made known, "The HIV-1 Gag protein [spliced into the offending coronavirus] enables interaction of the virus with the negatively charged host [cell] surface (Murakami, 2008) and a high positive charge on the Gag protein is a key feature for the host-virus interaction." This 'smoking gun' and 'Achille's heel' of the virus, suggests "unconventional evolution of 2019-nCoV that warrants further investigation."[112]

"To our surprise," Pradhan et al. reported, "these sequence insertions [identifying HIV-1 gp120 and Gag genes spliced into the novel 2019 coronavirus] were not only absent in the S-protein of SARS but were also not observed in any other member of the Coronaviridae family. This is startling as it is quite unlikely for a virus to have acquired such unique insertions naturally in a short duration of time... Taken together, our findings suggest unconventional evolution of 2019-nCoV..."

Pradhan et al.'s "work highlights novel evolutionary aspects of the 2019-nCoV and has implications on the pathogenesis and diagnosis of this virus,"[112] as well as implications on readily available preventatives and highly probable cures.

Andersen and Garry's omissions and false writings implicate their own institutional and governmental involvements in advancing the COVID Coup, the 'coronavirus conspiracy,' and Big Pharma's monopoly over this profitable death-industry.

Criminal Investigations Warranted by the Evidence

Criminal investigations into Andersen and Garry's 'bogus science' and complicity in the enterprise responsible for the alleged biocrime are warranted by the evidence presented above and in previous reports.[104]

This book presents compounding evidence of reckless negligence and negligent manslaughter for the fact that people are dying from the omissions mentioned above, misrepresentations, fraud, and related bioterrorism and biocrime. This published study provides probable cause for Justice Department officials to investigate Anderson et al. These 'insiders' should be indicted by a grand jury for complicity in the mounting genocide and prosecuted to the fullest extent of the law.

Discerning the extent of the academic and media enterprise implicated is crucial to resolving these matters and preventing further 'outbreaks.' Andersen's and Garry's institutional affiliations, related biases, and motives cannot go unnoticed or unpunished. Andersen is affiliated with The Scripps Research Institute of La Jolla, CA. Dr. Garry serves Tulane University in New Orleans and Zalgen Labs in Maryland. A quick online inquiry reveals their conflicting interests.

The Scripps enterprise is heavily invested in media and medicine. In fact, the Scripps publishing enterprise is partnered with the Springer Nature[154] enterprise that published the main paper controlling the damage caused by Pradhan's group science. Springer Nature is infamous for sourcing medical propaganda, especially influencing the genetic science community. Major institutional investors in E.W. Scripps and Springer include world-leading "Deep State" financiers, Blackrock Inc., the Vanguard Group, Inc., JP Morgan Chase & Co., and other globalist investment entities according to NASDAQ.[155]

Further conflicting interests are evidenced by the Scripps Research Institute enterprise that incorporates the Center for HIV/AIDS Vaccine Immunology & Immunogen Discovery. This racket holds a vested interest in concealing the AIDS-virus envelop gene spliced into the coronavirus.

In addition, the concealed conflicting interests of Dr. Garry and Tulane cannot be neglected or dismissed. Tulane University's complicity in sourcing the AIDS cancer complex by viral engineering was thoroughly investigated and reported by Edward T. Haslam in several publications, including *Dr. Mary's Monkey: How the Unsolved Murder of a Doctor, a Secret Laboratory in New Orleans, and Cancer-Causing Monkey Viruses are linked to Lee Harvey Oswald, the JFK Assassination, and Emerging Global Epidemics*.[155]

Dr. Garry unethically neglected to disclose his company's express research and developments of a coronavirus test under the required disclosure of "Competing interests." Dr. Garry simply noted that he co-founded "Zalgen Labs, a biotechnology company that develops countermeasures to emerging viruses."

If the public and scientific community realized the lab virus origin of the COVID-19 virus and the criminal parties' complicity in concealing and aiding-and-abetting the alleged bio-criminal enterprise, not only might Dr. Garry and his cohorts be held accountable under 18 U.S. Code §1002, but civilization might be relieved of these severe infectious disease burdens, future unnatural 'outbreaks,' and vaccination impositions enabling the bio-electronic transhumanist movement.

Conclusion

The scientific evidence, and socio-economic and geopolitical facts presented in this book, raise substantial probable cause for investigating the named suspects for complicity in the overall "coronavirus conspiracy," or COVID Coup. These facts and scientific evidence compound the need for thorough investigations that have been, thus far, neglected or diverted.

Scientific and judicial forums worldwide must reconcile the laboratory creation of the "novel" 2019 coronavirus/SARS/HIV-1 mutagen, its deployment in Wuhan (and probably elsewhere), the resulting deadly pandemic, and geopolitical and economic fallout.

The public and private biotechnology and pharmaceutical enterprise implicated by the evidence of wrongdoing presented herein, and diversionary "fake news" and fraudulent "science" must be reconsidered in light of the facts and evidence reported herein. Suspects must be scrutinized in the interest of public health and justice. Without such urgently needed interventions, the threat of compounding outbreaks and deadlier bio-crimes looms severe.

Andersen et al.'s bogus "science paper" in *Nature Medicine* provides substantial evidence of bad faith and organized crime. These federal grant recipients and enterprise agents did not "clearly show that SARS-CoV-2 is not a laboratory construct or a purposefully manipulated virus." On the contrary, their fraudulent concealment of solid science and genetic engineering of that coronavirus evidences a conspiracy to cover up the bio crime in which they are complicit and substantially implicated.

Not only is Andersen's group commercially and ethically compromised by their concealed conflicting interests exposed herein, but these scientists recklessly neglected and concealed genetic sequencing evidence that vindicates "conspiracy theorists."

Honorable whistleblowers include science scholars who heroically heralded emerging viruses of unnatural origins to their own detriment. By so doing, these public health advocates have been attacked as "conspiracy theorists." They risked their reputations, careers, livelihoods, and safety to discover and disclose concealed truths rejected by the corporate-controlled media corrupting the science world's and society's erroneous 'general agreement.'

This book vindicates these public servants from malicious disparagement. It indicts several primary agents and institutions aiding-and-abetting the commercial enterprise profiting most from this devastating pandemic and alleged bio-crime.

This review and analysis also evidences Andersen and Garry et al. having deliberately fabricated and circulated a "science paper" with the intent to impede, obstruct, or influence the federal investigation into the pandemic's laboratory origin.

This violation of 18 U.S. Code §1002, among other things, occurred within the jurisdiction of the NIH and the National Academy of Sciences in response to direction by the Executive Office of the United States to investigate allegations of n-2019CoV being a lab virus. Andersen and Garry et al. are alleged to have written and published their article in contemplation of their writing misleading and obstructing the White House's reputable investigation. They obstructed the discovery of this urgent intelligence required for competent coronavirus remedial responses.

I petitioned President Trump and his administration to act dutifully in investigating, indicting, prosecuting, and convicting Andersen and Garry et al., under 18 U.S. Code § 1002 and other laws of the United States, encouraging imprisonment by statute for "not more than 20 years."

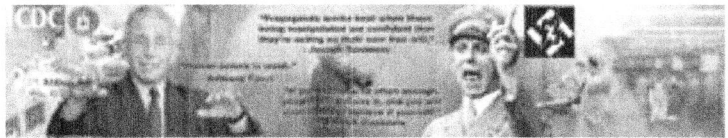

In addition, treason law 18 U.S. Code §2381[156] states, "Whoever, owing allegiance to the United States, levies war against them or adheres to their enemies, giving them aid and comfort within the United States or elsewhere, is guilty of treason and shall suffer death, or shall be imprisoned not less than five years and fined under this title but not less than $10,000, and shall be incapable of holding any office under the United States."

Dr. Anthony Fauci, who is already subject to substantial criticism by the media[157] for miscalculating the spread of coronavirus, and misinforming presidents Trump, Biden, and the public regarding risks and remedies, is subject to this treason charge. Fauci joined Bill Gates,

Klaus Schwab, and their equally suspect allies in acts of treason and genocide.

The next chapter provides substantial suppressed evidence in Dr. Fauci's e-mail correspondence to fellow officials dated January 31, 2020, through February 3, 2020, proving an organized criminal conspiracy within a 'racketeering enterprise' subject to the Department of Justice investigation and criminal prosecutions.

Follow the link to listen. [158]

CHAPTER IV

PANIC AT THE NIH, WELLCOME TRUST, WHO AND HARVARD

> *"[S]omeone who hears the sound of the horn but does not take warning, and a sword comes and takes him away, his blood will be on his own head... But if you on your part warn a wicked person to turn from his way and he does not turn from his way, he will die for his wrongdoing, but you have saved your life...Turn back, turn back from your evil ways! Why then should you die...?"*
> Ezekiel 33:4-10.

The 'smoking guns' neglected in Fauci's e-mails woefully evidence a series of crimes and cover-ups. Guilty parties include the yet-to-be-identified 'redactors' of that mail.

Despite severe redactions amounting to criminal concealment of evidence, I was able to identify the key "insiders" at the top of the National Security Crime Syndicate directing virology and vaccinology in the U.S. and overseas. This chapter exposes those complicit in racketeering, treason, and terrorism in biowarfare, victimizing **We The People** worldwide.

Evidence analyzed here clearly and convincingly proves biowarfare is being committed internationally by governments 'captured' by the biotech enterprise in which Dr. Anthony Fauci operates as a deranged bureaucrat.

This 'parade' (or deadly charade) features *humanicide* in accordance with profitable depopulation objectives schemed by Big Tech's "transhumanists."

With artificial intelligence ("AI"), why sustain "useless eaters"?

The gross blackouts in this analyzed section of Fauci's e-mails also prove "tampering" violating Title 18 U.S.C. §1519. This is the "low hanging fruit." Lacking attention to it compounds impressions of gross negligence and reckless impropriety within the Justice Department that evades its investigative duty and transparency.

Key agents in government, the military, academia, and industry are identified below. They are named in-and-around this most censored section of NIH/Fauci e-mails dating January 31, 2020, thru Feb 3, 2020, detailing the urgent "Teleconference." Redactions protecting these parties' complicity evidence an organized criminal conspiracy to fraudulently conceal crucial AIDS science tied to COVID's evolutionary virology, mRNA vaccinology, foreign investors in nano-bioelectronic devices for 'dual-use' military and commercial vaccine applications, intelligence community leaders, and competing remedies that bolster immunity with little to no risks.

In this context, all safe and effective competing 'natural remedies' and medicines such as anti-oxidants have been unfairly and deceptively restrained from trade, enriching stockholders in the public and private National Security Crime Syndicate and maximizing deaths.

Fauci's e-mails provide clear-and-convincing evidence of racketeering by co-conspiring agents in the CIA, FBI, NSA, academia, British intelligence, Eastern Communist and Western Capitalist businesses, and military communities. These stockholders in the public/private racket have leveraged the COVID pandemic for unjust enrichment, their elite's protection, and transhumanistic subversion of civilization. Thus far, they have secured their National Security Crime Syndicate's geopolitical and economic interests in population control without widespread discovery necessary to arrest their fascistic progress. They are the agents and agencies administering the COVID Coup.

Putting this intelligence in the hands of those willing and able to oppose these demons can help save lives.

Media Censorship and COVID-AIDS Cover-up in Fauci E-mails

"Silence is often evidence of the most persuasive character," ruled U.S. Supreme Court Justice Louis Brandeis when convicting an illegal immigrant pleading his Fifth Amendment right to remain silent. The man had entered the U.S. illegally to distribute propaganda advocating for "the overthrow of the Government of the United States by force or violence."

Curiously, facts in that case of *United States ex rel. Bilokumsky v. Tod* (263 US 149 - Supreme Court 1923) mimick the Biden administration's hypocritical immigration policies permitting illegal aliens to enter the country un-vaccinated, nor tested, against COVID; despite requiring these restrictions on U.S. citizens.

This further evidences America's degeneration imposed by the crime syndicate in service to the same Deep State global elite responsible for the redactions in the illuminating e-mails detailed below.

Similar 'silence' persuasively evidences the overthrow of health science and global governments, including China. COVID-19 and related media propaganda makers are identified on pages 3125 thru 3135 of Dr. Anthony Fauci's e-mails.

These e-mails were obtained, curiously at the same time, by *CNN*, the *Washington Post*, and *BuzzFeed*[159] on or about June 1, 2021. The same media followed up with their own show of 'silence' and diversionary propaganda.

As in Judge Brandeis's case, all of this involves threats to America's existence, along with that of the 'Free World.' Additional symptoms of this 'systemic institutionalized corruption' include violence in the streets, skyrocketing crimes and murder rates, and WOK racial divisiveness.

Drug cartels dominating politics, illegal immigrants, and the "border crisis" are cloaked by complicit media agents and networks tripping over criminal evidence like 'elephants under their carpets.' Each deploys deceptive propaganda. None of them address the intentional destruction of America in favor of the globalists' agenda to convert civilization into a transhuman cyborg colony.

No news sources have, to date, revealed the depth of what I do here. This most heavily 'silenced' (i.e., redacted or censored) portion of the

Fauci e-mails contains crucial evidence to prove the genocidal 'Big Picture.'

Does stupidity, or complicity, best explain this delinquency and secrecy? You decide after reading this chapter.

Regarding the lab origin of COVID-19 and Fauci's complicity therein, the media and politicians recklessly neglected and diverted from the obvious 'elephants.' This exposes the National Security 'Crime' Syndicate's 'Achilles' heel.'

This saga is reminiscent of the irony, tragedy, and wisdom in *The Emperor's New Clothes*. That famous folktale by Danish author Hans Christian Andersen is about a "vain emperor" who was exposed before his subjects. The emperor paraded through the streets like Fauci has done, and continues to do, metaphorically 'nude.' A single child sees and shouts the obvious nudity from the crowd to no avail. The 'sheeple' don't hear, keep silent, and the parade keeps marching.

The 'silence,' censorship, and gross 'redactions,' obfuscating officials' correspondence in Fauci's e-mails from January 31 thru February 2, 2020, document *panic* in Fauci's 'inner circle.' Alarms sounded with emerging scientific evidence of COVID-19's lab origin, and 'gain-of-function' intelligence suddenly appeared in the scientific literature. The AIDS-virus genes in the spike-protein transmission device were identified, as mentioned, by Pradhan et al. Their solid science strongly evidenced the lab origin of the 'dual use' bioweapon.

This silenced section of Fauci's e-mails is evidence beyond a reasonable doubt indicting the organized criminal enterprise administering the COVID Coup. That is the same cartel I have studied and vetted for a quarter-century. It is the same 'racketeering enterprise' that imposed upon the world HIV/AIDS.

> *"[T]he National Security Crime Syndicate leveraged the so-called 'moratorium' to transfer suspicion to China from America, foment divisive politics, and covertly enrich its enterprise in 'defense spending.'"*

These covert 'operatives' combined and covered up the merging of HIV/AIDS and SARS/CoV to facilitate depopulation and transhumanism.

Fauci's minions' panic began on Friday, January 31, 2020, prompted by the release of Pradhan et al.'s study and panic over it, threatened exposure of the crime syndicate. This best explains the near-complete silence, secrecy, and gross redactions of the incriminating evidence surrounding the cabal's emergency "Teleconference" held on Sunday, February 2, 2020.

If these suspects were not guilty, there would be no need for their cover-up.

But more than a §1519 violation, had exposures continued at that time, they would have reached the deepest levels and troublemakers in the Deep State, such as FBI Director Robert Mueller. You will realize this from reading the facts presented in the next Chapter.

Given this criminal cartel's extraordinary efforts to protect their COVID-19 operations and scientific community sabotage, probable cause exists to attribute the attacks against President Trump partly in retaliation for the COVID-19 lab origin investigation ordered by Trump. Protection against incriminating discovery of the plot hatched before Trump took office was required of the saboteurs. After all, "the best defense is a great offense."

At 5:58 in the morning of Sunday, Feb. 2, 2020, Fauci's superior, Francis Collins, the Director of the National Institutes of Health (NIH), wrote Fauci, "In case you haven't seen, attached is the Indian paper claiming HIV sequences have been inserted into 2019-nCoV." Collins then falsely claimed the "Indian paper" had been "debunked."

Collins's allegation that the "Indian paper... has been roundly debunked" is *false*. If this was true, then Collins would not have: (1) urgently contacted Fauci before 6 a.m. that Sunday morning to discuss spin-strategy; and (2) used the words "pretty useful" in characterizing "Jon Cohen's piece in *Science*." Cohen's article did not address the substance of the "Indian paper" by authors Pradhan et al.[13]

The Collins-directed NIH purportedly oversees Fauci's enterprise —the National Institute for Allergies and Infectious Diseases (NIAID). But the truth is both have always been overseen by the U.S. Navy and the U.S.

COVID COUP: "The Rise of the Fourth Reich"

Central Intelligence Agency (CIA), and they are intertwined within the National Security Crime Syndicate, the NSA, and the Security Council.

Fauci responded to Collins by the e-mail shown below.

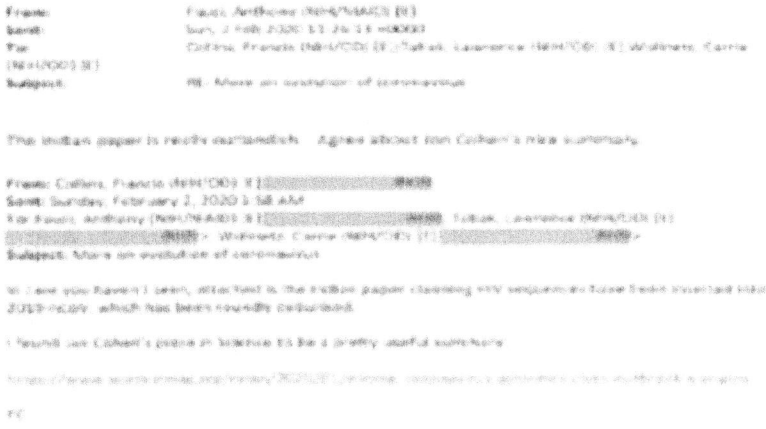

According to the *Washington Post*, Fauci's oversight by the so-called 'Deep State' intelligence community supposedly began in April 2000. This was when infectious diseases, beginning with HIV/AIDS, were deemed threats to "National Security."[160]

The massive redactions in this section of Fauci e-mails were committed not by Fauci nor Collins but by intelligence agents in this alleged National Security Crime Syndicate.

The FBI is falsely rumored to be dutifully investigating these matters. Therefore, sections of these e-mails evidencing crimes were presumably redacted.

But this is not reasonable, responsible, or dutiful, as I explain below. I examined evidence showing the FBI complicit in the crimes and reported this previously.[161] Aiding-and-abetting a deadly racketeering enterprise by willful blindness is a crime.

Fauci's e-mails urgently discuss plans to gather a group of submissive science scholars, commercial investors, and teleconference experts in the field of evolutionary virology. The urgent question was how to respond to, spin, or cover up the shocking AIDS virus genes inserted into the bat coronavirus spike protein.

The teleconference took place 48 hours after Pradhan et al.'s incriminating and potentially disastrous genetic science paper was published. It had to be 'neutralized.'

A week later, I wrote the White House[104] about the urgent findings of the targeted science.

Pradhan's group initially published their analysis online to help scientists fight against "n2019-CoV." To reiterate, their study showed the "Uncanny similarity of unique inserts in the 2019-nCoV spike protein to HIV-1 gp120 and Gag."

Censorship and undeserved disparagement followed. The insiders could not afford to have Pradhan et al. prove four AIDS-virus envelope genes were attached to the coronavirus/SARS/AIDS recombinant spike protein.

COVID COUP: "The Rise of the Fourth Reich"

The "Indian paper" could have spoiled all of the plans and commercial benefits of the 'plandemic.' Therefore, it had to be 'neutralized.'

The "Indian Paper" and "Smoking Gun" Censorship

Evidencing international commercial crime and the 'smoking gun' censorship harboring it, Fauci's e-mail correspondence with Harvard's Dean George Daley (shown below) is most revealing. This exchange also occurred on Sunday, February 2, 2020. It came one hour **before** Fauci wrote Collins, "the Indian paper is really outlandish."

Fauci and Collins agreed to use Jon Cohen's

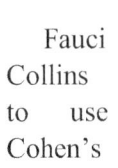

defective *Science* article to deflect Pradhan's solid science. Cohen's 'pseudo-science' did not match. That was obvious to anyone with some amount of viral intelligence.

Fauci, Collins, Harvard Medical School Dean Daley, and **Jeremy Farrar** in England, had a 'meeting of the minds,' and their minions agreed. They decided to conceal —tamper with— the urgent scientific evidence presented in the subsequently retracted Indian paper.

Pradhan et al.'s group of nine scientists was not easy to discredit. Pradhan was affiliated with IBM. His group analyzed the genetic sequences in the CoV/SARS spike protein antigen using sophisticated software. This attachment apparatus enabled the novel coronavirus to jump species to humans. These scholars concluded:

"The finding of 4 unique inserts in the 2019-nCoV, all of which have identity/similarity to amino acid residues in key structural proteins of HIV-1 is unlikely to be fortuitous in nature."

> From: Daley, George Q.
> Sent: Sunday, February 2, 2020 10:32 AM
> To: Fauci, Anthony (NIH/NIAID) [E] ▮▮▮▮▮ ▮▮▮▮ ; Fauci, Anthony (NIH/NIAID) [E]
> ▮▮▮▮ >
> Subject: Inquiry and possible pone cal
>
> Dear Tony,
>
> Alan Garber, Harvard's Provost, and I met yesterday with a team led by Jack Xia, the CEO of China's Evergrande Company, and Dr Jack Liu, Evergrande's chief health officer, who stated thy were acting on behalf of Dr Zhong Nanshan, China's key point person on the coronavirus outbreak (see below). ▮▮▮▮▮▮▮▮▮▮▮▮▮▮▮▮▮▮▮▮▮▮▮▮▮▮▮▮▮▮▮▮▮▮▮▮▮, and they arranged a conference call for tomorrow morning EST with Dr. Zhong.
>
> While I have been mobilizing efforts of our community to react to the virus and to this request, I am not naïve to the challenging politics of such a relationship. I do not want to complicate or duplicate efforts already underway, and am writing to request whatever information you are willing to share on your current efforts to coordinate a response.
>
> If a phone call is more facile, please do not hesitate to try my cell: ▮▮▮▮▮▮▮ anytime today.
>
> Sincerely,
>
> George
>
> George Q. Daley, MD, PhD
> Dean, Harvard Medical School
> Caroline Shields Walker Professor of Medicine
> Professor of Biological Chemistry and Molecular Pharmacology
>
> Office of the Dean, Gordon Hall
> 25 Shattuck Street, Boston MA 02115
> Contact: ▮▮▮▮▮▮▮▮▮▮▮▮▮▮▮▮

That revelation sent the syndicate's 'inner circle' into a tizzy. Fauci would be especially incriminated if they didn't discredit Pradhan's science since Fauci was the "AIDS Czar" for the U.S. Government.

Fauci had worked closely with Dr. Robert Gallo —the "Man that Created AIDS"[53] —developing the false AIDS origin "Patient Zero" narrative. By Gallo's and Fauci's secretive collaboration, Fauci's patents, AIDS science, and drug 'cocktails' became a booming industry.

Sir Jeremy Farrar, Director of the Wellcome Trust (Image: REUTERS)

Consequently, the protective propaganda campaign began, urgently coordinated at all levels of the scientific community and society through the complicit media syndicate. This was the purpose of the Sunday, February 2, 2020, "teleconference" under Jeremy Farrar's leadership.

The "Teleconference" for Coordinating Damage Control

There was a time when the British controlled and colonized China. The Bank of England and Rothschild co-investors in the East India Company administered this from the shadows. What makes you think much has changed?

At 10:32 that Sunday morning, Dean Daley, who had already heard of the crisis posed by the "Indian paper" that threatened their financial interests in the world of COVID/SARS/AIDS vaccine science, sent Fauci the e-mail shown above.

Therein, Daley responded to the crisis requiring 'coordination' for 'damage control.' He stated he was "writing to request whatever information you are willing to share on your current efforts to coordinate a response."

"There is no real 'coordination' of this response," Fauci defensively and *deceptively* replied. That was hours after the teleconference wherein attendees had already agreed to 'coordinate' their efforts, statements, and publications. As I detailed previously, this can be known (i.e., proven) by the group's subsequent sham "science" article published in *Nature Medicine* by Andersen and Garry. Fauci referenced, legitimized, and advised Andersen on this "study" during and after a Trump press briefing.

Fauci and his paymasters decided to deny any-and-all science linking COVID to HIV's 'gain of function' spike protein bioweaponry. "[W]e do not know who is doing what until we are told —just like you have done here," Fauci replied to Dean Daley.

> *"[A] protective propaganda campaign needed to be urgently coordinated at all levels of the scientific community and society through the complicit media syndicate. This was the purpose of the Sunday, February 2, 2020, "teleconference," and Jeremy Farrar's leadership therein."*

One day earlier, on Saturday, February 1, Fauci's subordinate, deputy "HIV/AIDS Czar," **Hugh Auchincloss**, sounded the alarm that coronavirus 'gain-of-function' financing by Fauci's cohorts was important in the context of the "Indian paper" and HIV/AIDS concealment operations.

"[G]ain of function [HIV/AIDS Spike protein gene experiments] have since been reviewed and approved by NIH," deputy NIAID director Auchincloss wrote Fauci.

Having considered his Harvard Medical School service in association with Dean Daley and their mutual concerns about coordinating their damage control, Auchincloss wrote the following e-mail:

From: Fauci, Anthony (NIH/NIAID) [E]
Sent: Sat, 1 Feb 2020 17:51:91 -0500
To: Auchincloss, Hugh (NIH/NIAID) [E]
Subject: RE: Continued

OK. Stay tuned

-----Original Message-----
From: Auchincloss, Hugh (NIH/NIAID) [E]
Sent: Saturday, February 1, 2020 11:47 AM
To: Fauci, Anthony (NIH/NIAID) [E]
Subject: Continued

The paper you sent me was the experiments were performed before the gain of function pause but have since been reviewed and approved by NIH. Not sure what that means since Emily is sure that no Coronavirus work has gone through the P3 framework. She will try to determine if we have any distant ties to this work abroad.

Sent from my iPad

127

Solidly evidencing China's concerns "abroad," the globalists' coordinated complicity in the COVID Coup and cover-up, and heavy commercial conflicts-of-interest and political influence, Dean Daley wrote the following to Fauci:

> "I am not naïve to the challenging politics of such a relationship [between the "Indian paper" science and commercial interests in the vaccine industry and more]." Daley clarified, "I have been mobilizing efforts of our community to react to the virus and to this request [for coordination of response], ... I do not want to complicate or duplicate efforts already underway."

Thus, the **coordination** that occurred within and beyond Fauci's and Harvard's spheres of influence is solidly evidenced beyond a reasonable doubt.

Likewise, organized crime is also evidenced by Daley and his coordination of actions between Fauci and China's leading health science officials.

Four hours after Collins urgently wrote Fauci to spin the narrative from the "Indian paper" to Cohen's superficial article, Dean Daley revealed his Harvard connections to China; and coordination among well-known Chinese intelligence agents and agencies coordinating commerce and military imbalance with the U.S. and U.K governments.

Daley had administered a "high-profile Chinese partner[ship]"[162] with China's Evergrande Group for $115 million. The SARS-CoV-2 work involved the "Chan School of Public Health and Guangzhou Institute for Respiratory Health." The 'dual function' commercial/military collaboration studied the virus purportedly "in an effort to develop therapies against infections... and to prevent new ones."

The Chinese side of the initiative was led by Zhong Nanshan, head of the Chinese 2019-nCoV Expert Taskforce, and the scientist who isolated the suspicious SARS virus in 2003. (Nashan is also director-general of the China State Key Laboratory of Respiratory Diseases," according to news reports.[163])

Hugh Auchincloss, M.D.

Hugh Auchincloss, M.D., serves as NIAID Principal Deputy Director. In this capacity, Dr. Auchincloss is responsible for the following:

Providing leadership for all NIAID research planning and implementation activities, including helping to prepare and support a strategic vision for NIAID.

Overseeing an extensive portfolio of basic, clinical, and applied research, as well as product development for biodefense, HIV/AIDS, infectious diseases, and immune-mediated disorders

His recent leadership activities include spearheading the development of the Institute's strategic plan and chairing the NIAID Research Initiative Committee, an internal governance group that has designed and implemented a more efficient approach to planning, developing, and approving NIAID initiatives. Currently, Dr. Auchincloss is leading an NIAID initiative to design and implement changes in the Institute's clinical research infrastructure, which will be flexible and available for domestic and international clinical research on HIV/AIDS and other infectious diseases. Additionally, Dr. Auchincloss is part of an NIAID senior leadership group responsible for reviewing all aspects of HIV/AIDS research policy, including the evaluation of "test and treat" strategies, analysis of results of pre-exposure prophylaxis (PrEP) clinical trials (including microbicide trials), and coordination of future HIV/AIDS vaccine clinical trials.

Hugh Auchincloss, M.D.
NIAID Principal Deputy Director

Daley wrote Fauci, "Alan Garber, Harvard's Provost, and I met yesterday with a team led by Jack Xia, the CEO of China's Evergrande Company, and Dr. Jack Liu, Evergrande's chief health officer, who stated th[e]y were acting on behalf of Dr. Zong Nashan, China's key point person on the coronavirus outbreak (see below) [redaction], and they arranged a conference call for tomorrow morning EST with Dr. Zhong."

The above redaction likely concealed the names of the main officials in the US/UK/China National Security Crime Syndicate, who coordinated their commercial, "scientific," and military propaganda interests with Dr. Zhong Nashan.

This evidence defied early claims of exclusive Chinese liability for the pandemic. It proves the US/UK and Chinese governments, backed by their military interests, coordinated business dealings, and incentives to cover up and discredit the AIDS virus gain-of-function splicing revealed in the "Indian paper."

Alternatively, agents and agencies identified in these e-mails compel the presumption of a concealed 'end game' motive of great interest to "Dr. Zong," the Chinese military, and the UK/US subversives coordinated by Jeremy Farrar, the World Economic Forum elite, and the intelligence community spinning or concealing all of this.

Harvard and China's Special Interests

It is public knowledge that the Pfizer and Moderna mRNA vaccines feature more than genetic material; more than even the concealed genes from HIV-1 weaponizing the SARS-CoV-2 spike protein.

Worse than the ravages of injecting spike protein antigens warned against by Harvard's famed mRNA vaccine technology developer, Robert Malone,[163] and worse than the neglected 'antigenic complexes' formed throughout the bodies of unsuspecting consumers prompting inflammatory conditions such as myocarditis and blood vessel breaks; the actual "game-changer" has been concealed.

The most "novel" vaccines include "hydrogels" suitable for nano-bioelectronic data-mining, whole body surveillance, and "frequency therapeutics" akin to surveying and subverting the body, mind, and 'spirit' (that is energy) of those poisoned by injections. These 'pharma-electro-genetic' innovations evolved from the public/private investors panicked by the Indian paper. The agents, agencies, and most advanced military biotechnologies of greatest interest to the National Security Crime Syndicate and Anglo-Chinese totalitarian elite were placed at risk.

Harvard and MIT agents had much to lose from investigators discovering their co-funding from the UK, China, and U.S. Governments (among others). These 'parties-of-interest' included the DARPA, BARDA,[164] and DTRA[165] agencies that helped pioneer this vaccine hydrogel bioelectronic know-how. The biotech financiers of this vaccine industry prospered, especially those financing the Charles Lieber Lab at Harvard and the partnering lab at MIT that largely sourced the Moderna company—the Robert Langer Lab.

There is no doubt that Bill Gates and Jeffrey Epstein's cronies in the CIA,[166] and the Mega group,[167] were major investors in these advancing biotechnologies at Harvard and MIT. They promised massive commercial data-mining and brain-Cloud merging for optimal population control. That 'neuroscience' promised to install AI in every surviving human. The coup would alter civilization forever!

These products are keys to the success of 'transhumanism.' They are also essential in the "Fourth Industrial Revolution ('4IR')" as espoused by Schwab.

Central to this biocrime syndicate, Robert Langer[168] **operates as Moderna's co-founder**. In addition, Langer's **Acuitas Company** is partnered with the CIA's OpGen Company.[169] These are major entities "behind the Pfizer and BioNTech vaccine," according to public knowledge.[168]

Langer's partner is **Pieter Cullis**, the co-founder of Acuitas Therapeutics. This group is instrumental to the Pfizer and BioNTech vaccine as well. The CIA's OpGen Company and Robert Langer's Acuitas Company are vicariously partnered in Moderna and Pfizer/BioNTech's commerce.

Geopolitical, financial, and biological engineering to advance 'transhumanism' must be presumed by these facts.

So too must the 'Great Global Reset"[170] spurred by these powerful special interests who could (and did) leverage COVID-19 to advance their geopolitical and economic impositions on the world. They concealed the CoV/SARS/HIV-1 recombinant biotechnology—especially the S-protein antigen that provided "gain of function," clearly profiting the insiders. The future of bioelectronic population subversion (as urged by the Gates / Schwab / Communist / Socialist / Elitist oligarchy) depended on 'neutralizing' the Indian paper.

> *"Central to the biocrimes syndicate, Robert Langer operates as Moderna's co-founder. In addition, Langer's Acuitas Company is partnered with the CIA's OpGen Company. These are major entities behind the Pfizer and BioNTech vaccine."*

Protecting the "National Security 'Crime' Syndicate"

Protecting the interests of the National Security Crime Syndicate is epitomized by Dr. Peter Daszak's complicity with Fauci. These two men conspired to cover up the lab virus's origin and its "gain of function" bioengineering. Evidence for this crime is shown on page 1150 of the Fauci e-mails,[171] wherein Fauci thanks Daszak for his "public comments" concealing the lab origin.

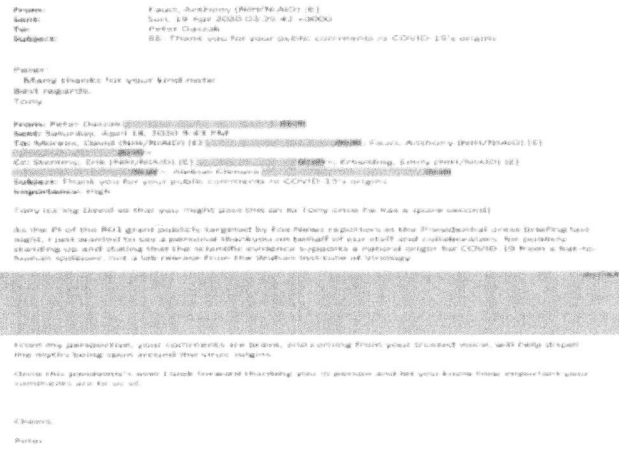

Months later, in early November 2021, Fauci and Daszak were exposed for channeling NIH grant money through EcoHealth Alliance to the Wuhan Institute of Virology for coronavirus gain-of-function research.[172] The "Indian paper" prompted Daszak to protect the cartel's interests in EcoHealth Alliance that also received and funneled Department of Defense[173] grants (not simply NIH and NIAID money) to Wuhan for said CoV/SARS/HIV-1 lab mutations and hyper-weaponization. Much of Daszak's financing was for (supposedly)

"SCIENTIFIC RESEARCH COMBATING WEAPONS OF MASS DESTRUCTION."

Accordingly, the lab virus gain-of-function COVID bioweapon for depopulation and mass population manipulation was precisely what Daszak and his Chinese collaborators in Wuhan co-invested in.

Daszak's EcoHealth Alliance received millions of dollars[173] beginning in 2014 when Fauci and his crime bosses fraudulently contrived a "moratorium"[174] on weaponizing MERS, SARS, and influenza viruses too for 'arms-length' gain-of-function bioengineering.

According to the government's moratorium: "In light of recent concerns regarding biosafety and biosecurity, effective immediately, the U.S. Government (USG) will pause new USG funding for gain-of-function research on influenza, MERS or SARS viruses, as defined below. This research funding pause will be effective until a robust and broad deliberative process is completed, resulting in the adoption of a new USG gain-of-function research policy1. Restrictions on new funding will apply as follows:

go_to link to view[175]

New USG funding will not be released for gain-of-function research projects that may be reasonably anticipated to confer attributes to influenza, MERS, or SARS viruses such that the virus would have enhanced pathogenicity and/or transmissibility in mammals via the respiratory route. The research funding pause would not apply to characterization or testing of naturally occurring influenza, MERS, and

SARS viruses unless the tests are reasonably anticipated to increase transmissibility and/or pathogenicity."

Given the crime syndicate's hate of Trump and Fauci's January 2017, Georgetown University lecture[88] stating his foreknowledge that the Trump administration would suffer an unprecedented plague compounds suspicions, and criminal evidence. It is most reasonable to presume the so-called "moratorium" that shifted that gain-of-function covert operation from the University of North Carolina (where early gain-of-function work mutating bat coronaviruses advanced under Dr. Ralph Baric) to Wuhan was contrived. The transaction did precisely what we now witness —shift the political 'blame game' and military saber-rattling between the US and China, both controlled by 'The Syndicate's Multinational Banking Cartel.'

In other words, the National Security Crime Syndicate must be investigated for contriving and leveraging the so-called "moratorium" to transfer lab work and suspicion to China from America. Fomenting divisive politics and covertly profiting from increased "defense spending," 'plandemic' officials must have considered these outcomes before releasing the virus at the Wuhan seafood market.

Daszak's Knowledge of the Early History of AIDS and Ebola

The focus on 'novel' immune-suppressive 'gain-of-function' viruses began in the 1960s with the SVCP.[14] That program, as already reported, sourced HIV/AIDS by 1972 and Ebola by 1976. These facts are known to doctors Gallo, Fauci, and Daszak et al., albeit fraudulently denied and lamely discredited.

Daszak and his cohorts in this biocrime extended such fraudulent concealments with Ebola and, more recently, COVID-19.[104]

These suspects repeatedly tampered with critical scientific evidence. They coordinated censorship and whistleblower disparagement campaigns. They financed their biocrimes through 'captured' government agencies and non-profits. They made fortunes at the expense of human lives.

Evidencing such defensive evasion and scientific fraud, Daszak wrote Fauci, "I just wanted to say a personal thank you on behalf of our staff and collaborators, for publicly standing up and stating that the scientific evidence supports a natural origin for COVID-19 from a bat-to-human spillover, not a lab release from the Wuhan Institute of Virology."

> *"[T]he 'Great Global Reset' [was] spurred by these powerful special interests who could (and did) conceal the CoV/SARS/HIV-1 recombinant biotechnology —especially the 'spike protein' (i.e., 'S-protein') that provided 'gain of function,' profit to insiders, and the potential for bioelectronic population subversion as urged by the Gates/Schwab/Communist/Socialist/Elitist oligarchy.*

More on Daszak and Ties to the CIA's Profitable In-Q-Tel

The important role Daszak played in the now obvious COVID Coup is superseded by military and industrial actors that are also identified in the Fauci e-mails (albeit censored by the complicit media).

The National Center for Biotechnology Information (NCBI),[176] and the graphic below, indicates Daszak's long-term vested interests in more than the aforementioned devil-doing. Long before Fauci, Farrar, Daszak, Andersen, and Garry et al.'s cover-up of the lab origin of COVID, they did the same with Ebola[177] and the 2014 Ebola Zaire re-emergence (from a refrigerator).[178] The screenshot below shows Daszak as the co-author of a *Nature Communications* article from 2014[179] revealing Google.org financed the main parts of the Daszak/EcoHealth Alliance/NIH/USAID co-funded study of an Ebola-like virus alleged to be prevalent in African bats (i.e., the vesicular stomatitis virus. Actually, Ebola's immediate predecessor was a vesicular stomatitis lab virus mutant re-named "the Marburg virus." I meticulously researched and evidenced this in ***Emerging Viruses: AIDS & Ebola--Nature, Accident or Intentional?***).

For years, USAID, and the NIH, its subordinate agents at the NIAID, and by extension presumably Daszak's EcoHealth Alliance, too, operated under the influence of the U.S. Central Intelligence Agency (CIA) as they continue to do.

Contrary to popular misconceptions, the CIA, from its inception, has served primarily to secure business operations for the power elite. Today, the CIA engages its own for-profit arm called In-Q-Tel.

Wall Street investors did great despite the general agreement that COVID was "bad for the economy," Wall Street investors did great. Insiders made vast fortunes.[180] The secret intelligence held by leaders in the National Security Crime Syndicate afforded the greatest opportunities for investing and money-making. In-Q-Tel participated.

In-Q-Tel was founded by Norm Augustine, a former CEO of the mega-military contractor Lockheed Martin. Michael Crow served as In-Q-Tel's first CEO at the request of George Tenet, the Director of Central Intelligence. Crow thus became chairman of the board for In-Q-Tel — the CIA's venture capital arm.

In-Q-Tel's mission has been, purportedly, to identify and invest in companies developing cutting-edge technologies that serve the United States. Nevertheless, it is poised and active serving the "National Security (Crime) Syndicate."

Origins of the corporation can be traced to Ruth A. David, who headed the Central Intelligence Agency Directorate of Science & Technology in the 1990s and promoted the importance of rapidly advancing information technology for the CIA and its business allies in the corporate sector.

In-Q-Tel engages with entrepreneurs, growth companies, researchers, and venture capitalists to deliver technologies that include the most advanced data-mining methods, materials sciences, and health sciences.

Pfizer and Moderna's mRNA vaccines, for instance, intertwine such investments. Their related developments, including the nano-bioelectronic silver, copper, gold, or other metal lipid hydrogel devices used to deliver their genetic "payloads," were largely financed by the DOD's Defense Advance Research Programs Agency (DARPA) and pioneered at MIT and Harvard.

Former CIA director George Tenet explained, "We [the CIA] decided to use our limited dollars to leverage technology developed elsewhere. In 1999 we chartered... In-Q-Tel... While we pay the bills... CIA identifies pressing problems, and In-Q-Tel provides the technology to address them.

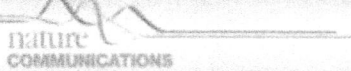

ARTICLE

Received 12 Jan 2014 | Accepted 19 Sep 2014 | Published 18 Nov 2014

Evidence for henipavirus spillover into human populations in Africa

Olivier Pernet[1], Bradley S. Schneider[2,*], Shannon M. Beaty[1,3,*], Matthew LeBreton[2], Tatyana E. Yun[4], Arnold Park[1,3], Trevor T. Zachariah[5], Thomas A. Bowden[6], Peta Hitchens[7], Christina M. Ramirez[8], Peter Daszak[9], Jonna Mazet[7], Alexander N. Freiberg[4], Nathan D. Wolfe[2] & Benhur Lee[1,3]

Zoonotic transmission of lethal henipaviruses (HNVs) from their natural fruit bat reservoirs to humans has only been reported in Australia and South/Southeast Asia. However, a recent study discovered numerous HNV clades in African bat samples. To determine the potential for HNV spillover events among humans in Africa, here we examine well-curated sets of bat (Eidolon helvum, n = 44) and human (n = 497) serum samples from Cameroon for Nipah virus (NiV) cross-neutralizing antibodies (NiV-X-Nabs). Using a vesicular stomatitis virus (VSV)-based pseudoparticle seroneutralization assay, we detect NiV-X-Nabs in 48% and 3-4% of the bat and human samples, respectively. Seropositive human samples are found almost exclusively in individuals who reported butchering bats for bushmeat. Seropositive human sera also neutralize Hendra virus and Gh-M74a (an African HNV) pseudoparticles, as well as live NiV. Butchering bat meat and living in areas undergoing deforestation are the most significant risk factors associated with seropositivity. Evidence for HNV spillover events warrants increased surveillance efforts.

Acknowledgements

We thank Beth Nasse and the Brevard Zoo for providing captive-bred bat sera. We also acknowledge Terry Juelich, Jennifer Smith and Lihong Zhang for their help with the Gluc construct, and thank all the members of the Lee laboratory for their input and suggestions. This project was funded by the Pacific Southwest Regional Center of Excellence (PSWRCE) U54 AI065359 and NIH grant AI069317. Metabiota/Global Viral's role in this study was supported by Google.org, the Skoll Foundation and in particular the US Agency for International Development (USAID); USAID's funding was made possible by the generous support of the American people through the USAID Emerging Pandemic Threats PREDICT program. The authors acknowledge the valuable contributions of the staff of Global Viral Cameroon under the supervision of Ubald Tamoufe, the Cameroon Ministry of Scientific Research and Innovation, the Ministry of Forestry and Wildlife and the Ministry of Health provided support and authorizations for this research.

> *"These suspects repeatedly tampered with critical scientific evidence. They coordinated censorship and whistleblower disparagement campaigns. They financed their biocrimes through 'captured' government agencies and made fortunes at the expense of human lives."*

Typically, after the CIA's investments in start-ups mature, "The Company" transfers its assets to enterprise insiders. This assures loyalty and precludes discovery and is how the National Security Crime Syndicate continuously operates "above the law."

Tying In-Q-Tel to Big Tech, on November 15, 2005, In-Q-Tel sold 5,636 shares of Google, worth over $2.2 million. The shares resulted from Google's acquisition of the CIA's start-up, Keyhole, Inc, its satellite mapping software known as Google Earth.

In August 2006, In-Q-Tel had reviewed more than 5,800 business plans and invested approximately $150 million in more than 90 companies. In 2016 it was funded with at least $120 million per year primarily from the CIA, NSA, FBI, and US Defense Department.

The CIA's In-Q-Tel also helped finance Peter Theil's Palantir Technologies —the U.S. military's chief data integration, search, and discovery contractor, providing management know-how and secure "public/private" collaborations (i.e., criminal complicity).

In-Q-Tel also helped finance Nanosys —the nanotech components company intertwined with Microchip Biotechnologies. They provide data-mining and intelligence analysis instruments for biodefense, including OpGen —the microbial genome analysis system.

According to *The Washington Post,* "virtually any U.S. entrepreneur, inventor or research scientist working on ways to analyze data has probably received a phone call from In-Q-Tel or at least been Googled by its staff of technology-watchers."

Consequently, the National Security Crime Syndicate is evidenced by the actions of its agents and allied agencies, including the CIA/ In-Q-Tel/ Google/ USAID/ NIH/ NIAID/ USAID/ DOD/ DARPA/ BARDA/ CDC/ FDA and alleged public interest organizations such as EcoHealth Alliance.

These nefarious interests, connections, and activities are apparent in the Fauci e-mails scrutinized by this author. Their collaborative investments featured information programs, propaganda 'smoke screens' about viruses emerging from "nature," lucrative outbreaks, and experimental vaccines causing more deadly side effects. Add data-mining in healthcare using "novel" bio-electronic technologies such as the hydrogels injected into people's bodies, monitoring or altering more than the naive masses can conceive, and you have a recipe for totalitarian control over civilization.

Solid Evidence of the "National Security 'Crime' Syndicate"

Five weeks after the infamous "Teleconference," Fauci received an e-mail from Dr. Starnes E. Walker, a retired high-ranking official in the National Security Crime Syndicate. Walker's correspondence provides a wealth of intelligence as to who in the cartel is actually administering the biocrimes.

Typically, 'compartmentalization' in this criminal organization is used to avoid discovery. Officials are given information on a "need to know" basis only. Dim bureaucrats, thereby, remain clueless about their complicity in organized crimes. The rest are complicit by willful blindness.

The Fauci e-mails illuminate this darkness by vetting the names of the suspects, their positions and functions within the syndicate, and their importance or influence in the crime gang concealing the COVID/AIDS connection and the importance of the "Indian paper."

For example, Dr. Walker, a member of the Homeland Security Experts Group, MITRE, engaged in "Global Strategy Officer-Defense & Homeland Security/Intelligence" with the ANDE Corporation, solicited Fauci concerning the syndicate's commercial advancements in genetic testing for infectious diseases.

Therein, concealed censors redacted Walker's identity as the source of this correspondence. But despite their gross redactions (i.e., evidence tapering) in the Fauci e-mails, they made a costly mistake. As shown in the next section, they overlooked Walker's name (and position in the National Security enterprise) on the second and third pages of Fauci's e-mails dated March 7, 2020.

You can read about Walker's connections to several of the most powerful officials in the shadow government's military, biotechnology, 'biodefense,' National Security, and the genetic data-mining syndicate.

Dr. Walker's e-mail to Fauci on March 7th takes investigators into the 'lion's den' of the National Security Crime Syndicate, behind the COVID Coup, and much of the world's most murderous acts.

Walker made known he worked with Fauci and the highest-ranking bio-intelligence insiders, including Dr. Joshua Lederberg.

"History Repeats When You Don't Learn History."

Lab virus outbreaks repeat when their sources remain a "mystery."

I wrote about Dr. Joshua Lederberg 25 years ago. I was appalled that Lederberg eventually caved to political and financial pressures. He first championed moratoriums against bioengineering dangerous lab viruses. I evidenced his unethical conversion in *Emerging Viruses: AIDS & Ebola—Nature, Accident or Intentional?* Lederberg was presumably pressured to 'shut up.'

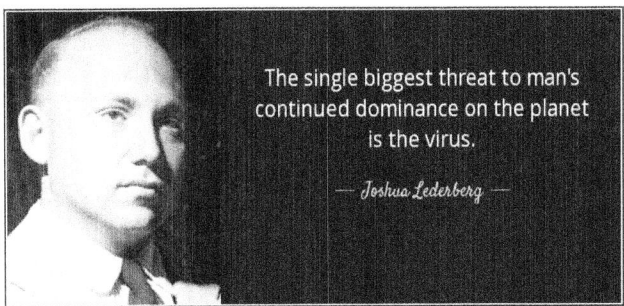

Lederberg had privately analyzed what I later published —Ebola's true source in an ancestor lab virus, not a "bat" as Daszak falsely reported. Lederberg knew that Ebola's immediate predecessor was the Marburg virus. That plague emerged simultaneously in three European **vaccine production facilities** in 1967. It was hard to pin that on "nature."

So the syndicate concealed that knowledge most effectively. They abused their media and coaxed officials to hide the fact that Ebola's mother —Marburg— came from a shipment of 'vesicular stomatitis' infected primates sent by the NIH's, NCI's, and U.S. military's main monkey, chimpanzee, and cancer virus supplier. That was Litton Bionetics, the previously vetted division of Litton Industries.

Litton's activities, especially with Dr. Gallo, were at the heart of the National Security Crime Syndicate's biological weapons and "cancer prevention" programs. Litton's successor, McDonnell Douglas, was also a top military contractor. Those companies manufactured virtually anything the National Security cartel needed for alleged 'defense,' actually profit and power.

Walker's e-mail to Fauci on March 7, 2020, directly connects to the heart of this crime syndicate that reaches the 'eye of the pyramid' in Klaus Schwab's World Economic Forum.[181] As mentioned in Chapter I, that entity co-sponsored the infamous "Event 201" with the Bill & Melinda Gates Foundation.

Event 201, you might recall, was the "predictive programming"[92] conference held in New York through Johns Hopkins. These chief suspects and financiers were(are) linked to Walker's promotion of ANDE—the Aspen Network of Development Entrepreneurs.[182] And Walker reveals much more evidence for criminal investigations and indictments in the following e-mail.

In Walker's March 7 correspondence, Walker reminded Fauci that he was "The DHS [Department of Health and Human Services] Director of Research reporting to SEC Michael Chertoff & U/S Jay Cohen (RADM ret). Michael & Jay brought me aboard when Jay was our Chief of Naval Research at ONR, and I was ONR's Executive Director & Chief Scientist."

Walker wrote, "I wanted to give you a heads up that a game changer for enhanced detection of COVID-19 has emerged. Thanks to DARPA & DHS S&T sponsorship in years past of ANDE developing a real-time Rapid DNA microfluidics system for human identification, the ANDE group has a breakthrough for detection of COVID-19 and other future emerging threat viruses.

COVID COUP: "The Rise of the Fourth Reich"

From: [redacted]
Sent: Sun, 8 Mar 2020 09:04:10 -0400
To: Lane, Cliff (NIH/NIAID) [E]
Subject: Re: COVID-19 Real Time, Sensitive Detection Breakthrough

Please take care of this yourself. Thanks

> On Mar 7, 2020, at 11:30 PM, Lane, Cliff (NIH/NIAID) [E] <[redacted]> wrote:
>
> They claim to have an improved diagnostic developed through DoD funding. It is a DNA platform that they claim to have adapted to COVID-19. They provide no data, only claims.
>
> > On Mar 7, 2020, at 10:03 PM, Fauci, Anthony (NIH/NIAID) [E] <[redacted]> wrote:
> >
> > Please read this and figure out what the heck he is talking about and act according to your judgment. Only 498 emails to go tonight.
> >
> > **From:** [redacted]
> > **Sent:** Saturday, March 7, 2020 6:09 PM
> > **To:** Fauci, Anthony (NIH/NIAID) [E] <[redacted]>
> > **Subject:** COVID-19 Real Time, Sensitive Detection Breakthrough
> >
> > Tony--
> > It has been awhile since we have worked together since my time as the senior SES standing up DTRA (with the help of Josh Lederberg M.D. & Dave Franz DVM who you know), at Argonne/UofChicago establishing your NIAID RBL with Olaf Schneewind M.D., and as the DHS Director of Research reporting to SEC Michael Chertoff & U/S Jay Cohen (RADM-ret). Michael & Jay brought me aboard when Jay was our Chief of Naval Research at ONR and I was ONR's Executive Director & Chief Scientist. I know you have your hands very full with the

COVID-19 threat, so I wanted to give you a heads up that a game changer for enhanced detection of COVID-19 has emerged. Thanks to DARPA &DHS S&T sponsorship in years past of ANDE developing a real time Rapid DNA microfluidics system for human identification, the ANDE group has a breakthrough for detection of COVID-19 and to the future, other emerging threat viruses.

As you may know the ANDE system for human identification (e.g. CT &DHS missions) is mature and now deployed operationally/tactically by CENTCOM, DIA, the IC and used most recently by DHS in their recent test bed in El Paso to demonstrate its effective capabilities to determine family relationship in undocumented minors. Additionally ANDE is in use by law enforcement and by officials responding to mass casualty events (CA 2018 Camp Fire disaster, 2019 Conception dive boat fire, and the very recent 2020 tragic helicopter crash) to ID the victims.

The ANDE system now provides 2 hour turnaround with no special training requirements as a stand-alone system for all the above users. Our warfighters and special operators are using the ANDE system now in field forward operations and it meets MIL specs &is the only system certified for data submission to the DoD ABIS/OIA DNA repository and FBI CODIS data base. The

executive summary and a more in-depth document for your teams review. Hope the above is helpful and I stand ready to provide any additional information. I have cc'd Jim Davis (ANDE Chief Federal Officer). Additionally since it has been some time since we have worked together I have attached my bio and that of ANDE's Chief Scientific Officer &Founder, Richard Selden M.D., Ph.D.

Tony thanks for considering this in your very busy life now and I will look forward to seeing you again.
Best regards—
Starnes
Dr. Starnos E. Walker

Member-Homeland Security Experts Group, MITRE
Global Strategy Officer-Defense &Homeland Security/Intelligence
ANDE Corporation

Rapid DNA for a Safer World

<C:UsersstarnDesktopSEW Bio. October 2019.docx>
<C:UsersstarnDesktopANDE BiosRFS CV 05 March 2020.pdf>

"As you may know the ANDE system for human [genetic] identification (e.g., CT & DHS missions) is mature and now deployed operationally/tactically by CENTCOM, DIA, the IC and used most recently by DHS in their recent test bed in El Paso to demonstrate its effective capabilities to determine family relationship in undocumented minors..."

This intelligence is highly relevant to the mass immigration crisis unfolding at the southern border. Presumably, the immigrants are given access to humanitarian relief, not experimental abuse in DNA-altering mRNA vaccine spike protein experiments. Such pretenses and abuses are reminiscent of the Third Reich's genetic experiments on Holocaust victims.

"Our warfighters and special operators are using the ANDE system now in field forward operations and... is the only system certified for data submission to the DoD ABIS/DIA DNA repository and FBI CODIS data base," Walker concluded with one important section being redacted by someone in intelligence.

Walker's E-mail to Fauci Ties the COVID Crime to James Murren

Dr. Walker disclosed that he served as the Executive Director and Chief Scientist for the Office of Naval Research (ONR). He reported to Michael Chertoff —the Secretary of Homeland Security under the

George W. Bush administration, and Jay Cohen, the Chief of Naval Research.

This intelligence exposed Fauci's criminal 'inner circle' controlling the science, public health politics, policies, and economics of the COVID, AIDS, and Ebola biocrimes.

The leading role of the U.S. Navy in biological weaponry was firmly established by science that I reviewed in *Emerging Viruses: AIDS & Ebola—Nature, Accident or Intentional?* I detailed why the Navy has always been at the forefront of "public health" and biological weapons research and developments.

Working within and through the Special Virus Cancer Program, the U.S. Navy often tested biological weapons with the CIA under the projects I already mentioned, MKULTRA and MKNAOMI.

Here, in Fauci's e-mails, Walker corroborated the Department of Homeland Security, the U.S. Navy, and the CIA's complicity in administering the COVID "Great Global Reset" scheme.

In my 2018 book, *The Las Vegas Deep State Massacre*,[183] co-authored with Sherri Kane and J.T. Kong, Michael Chertoff and the CIA played major roles along with Homeland Security's chief infrastructure corporatist, James Murren —the CEO of MGM Grand Resorts International. Others, including Saudi Prince Al Waleed, George Soros, and their associates in In-Q-Tel, capitalized on their foreknowledge of the massacre and subsequent security businesses.

At that time, presumably to the present, CEO Murren of MGM was the leading presidential advisor and sitting member of the Department of Homeland Security's National Infrastructure Advisory Council. As such, Murren and his fellow 'inside traitors,' including Michael Chertoff, leveraged their National Security intelligence positions to profit from the shootings. More than 500 concert-goers were killed or wounded.

The falsely alleged "lone gunman" —Stephen Paddock— was suspiciously tied to his former military weapons employer, Lockheed Martin. Gambling was Paddock's underworld activity at MGM. (Recall that In-Q-Tel was launched in 1999 by Steven Paddock's former employer, the former executive of Lockheed Martin, Norm Augustine.)

In *The Las Vegas Deep State Massacre*, we identified Chertoff as a corrupt lawyer. He was appointed Secretary of Homeland Security. He co-authored the infamous USA PATRIOT Act.

When Chertoff left his Homeland Security post, he formed "The Chertoff Group." He took with him eleven members of the DHS and CIA, including Reginald Hyde —a key secret agent who helped establish the CIA's In-Q-Tel investment group.

This is how the CIA, In-Q-Tel, and World Economic Forum elite administer or commercialize pandemics, mass shootings, biowarfare, cyberwarfare, the DOD and DHS, "national defense," and related "National Security" products and services.

This is largely the COVID Coup criminal enterprise.

The Chertoff Group quickly became a major player in the world of security systems, technologies, and investment banking. Chertoff's comrades commanded the mainstream media, fear, and propaganda campaigns. They always profit from the sales of their companies' goods and services.

Chertoff's "Deep State" cohorts against Trump included Ret. Col. Stanley McChrystal,[184] owner of the lucrative consulting firm, the McChrystal Group.[185] Aside from administering the "COVID RESPONSE" for large metropolitan city governments, along with their "preparedness" and pro-vaccination propaganda campaigns, the McChrystal Group directs 'MindWar' by agents who commandeer the social media to disparage and "neutralize" anti-vaxxers. Kane and I made these facts known in our multi-award-winning film, *Un-Vaxxed: A Docu-commentary for Robert de Niro*.[60]

In *The Las Vegas Deep State Massacre*, we noted that Anonymous, Google, and their allied trolls disrupting the social media used military neuroscience, PSYOPS, and "bots" to promote the Chertoff Group's "security services." This criminal activity enriched the underworld's partners, including The Carlyle Group and the Coalfire cybersecurity syndicate.

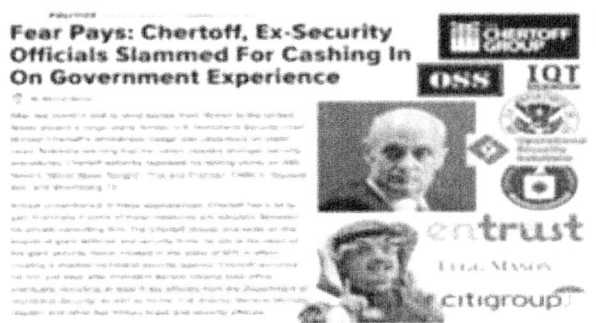

Meanwhile, George Soros, widely known for financing America's fall into corporate fascism, radical socialism, and Communism, is no dummy. The Soros Fund invested heavily in The Chertoff Group's Operational Security Solutions (OSS) company. Chertoff and his CIA buddies formed this security systems enterprise

Soros and the syndicate made massive amounts of money from security investments in the aftermath of the Las Vegas Deep State Massacre and COVID-19.

Suffice it to say, the National Security 'Deep State' Crime Syndicate that administered the deadliest mass shooting in American history maintains the murderous mind for planning, investing in, and administering the COVID Coup.

Experts Lied About COVID's Lab Origin and 'Gain of Function'

As mentioned above, urgently, on Sunday morning, February 2, 2020, at 10:32, the Dean of Harvard's Medical School and Professor of Biological Chemistry and Molecular Pharmacology, Dr. Daley, contacted Fauci "to coordinate a response" to Pradhan's revealing "Indian paper."

Dean Daley was already afraid of losing his life, career, and livelihood in the wake of the still festering Charles Lieber scandal. That crime story broke a few days earlier. The FBI had arrested Lieber, age 60, for lying about his role in Chinese espionage.

Lieber was the Chair of the Department of Chemistry and Chemical Biology at Harvard, Daley's close colleague. They had two Chinese students who were illegally "aiding the People's Republic of China," the FBI's press release read. Lieber was charged with lying about "his involvement in the Thousand Talents Plan affiliated with the Wuhan University of Technology (WUT) and Chinese military. Secret agent, "student" Yanquing Ye, lied about her ongoing military service at the National University of Defense Technology (NUDT), a top military academy directed by the Chinese Communist Party (CCP).

Further quoting the FBI's press release,[186] "a search of Ye's electronic devices demonstrated that at the direction of one NUDT professor, who was a PLA Colonel, Ye had accessed U.S. military websites, researched U.S. military projects and compiled information for the PLA on two U.S. scientists with expertise in robotics and computer science. "Robotics" and "computer science" intertwined Charles Lieber's work at Harvard with Robert Langer's lab activities at MIT.

"[A] review of a WeChat conversation revealed that Ye and the other PLA official from NUDT were collaborating on a research paper about a risk assessment model designed to decipher data for military applications. The data may derive from Lieber and Langer's nano-bioelectronic devices including the drug companies' vaccine hydrogels. During the interview, Ye admitted that she held the rank of Lieutenant in the PLA and admitted she was a member of the CCP..."

What a "risk!" Assessing "data for military applications" in Lieber's and Langer's data-mining field of nano-bioelectronics and robotics. This is not only central to transforming humans into cyborgs but also super-soldiering. It fit Klaus Schwab's and Bill Gates's interests in transhumanism. The developing military cyborg industry, and the future of commercial "data-mining" in human bodies, depend on these biotechnologies.

Ye's military interest paralleled Dr. Lieber's and Dr. Langer's chief expertise in data-mining in humans, as well as the advanced neuroscience capabilities heralded by intelligence experts Dr. Charles Morgan[187] and Dr. James Giordano[188] introduced in Chapter I.

The "genie" was the lab engineering of the COVID-19 virus, containing HIV's spike protein genes. The "bottle" was the secret enterprise that had done this to humanity —the criminal cartel that caused millions of deaths, trillions in damages, mandated their vaccines, destroyed economies, and took over the world.

The FBI was "captured" to lie about all of this. The agency did not reveal (they concealed) the aforementioned facts for obvious geopolitical and financial reasons.

The FBI affidavit indicting Lieber said he was transferring car battery technology to the Chinese illegally. But the transfer of nano-bioelectronic vaccine hydrogel know-how must be presumed from public access science showing Lieber's close collaborations and co-publications with MIT's most famous entrepreneur in these fields — Professor Langer.

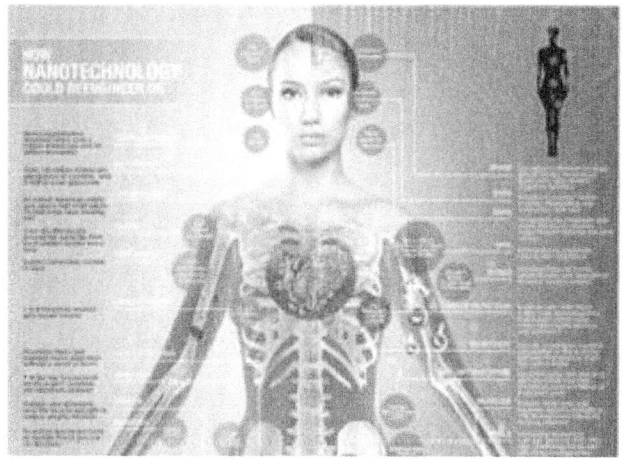

To make matters much worse, the founding director of the Langer Lab, Robert Lander, associated with MIT's Media Lab that had been exposed months earlier for taking money from Jeffrey Epstein's group of investors, including Bill Gates.

Consequently, Harvard Dean Daley's e-mail to Fauci to "coordinate" whatever spin the cartel was going to output on the Indian paper regarding the lab origin and HIV-infused spike protein of COVID-19 was **URGENT**.

"[P]lease do not hesitate to try my cell (phone number redacted) anytime today," Dean Daley wrote Fauci.

Again, this was a Sunday—the day they planned for the ubiquitously-urgent "teleconference." It was organized by the Fauci-Collins' administration of spin-doctors and Jeremy Farrar —the Wellcome Trust Director. Participants included in Farrar's e-mail to Fauci and others identified many of the world's leading experts in evolutionary virology. Another important insider was Mike Ferguson, a senior fellow at the alleged non-profit Center for Medicine in the Public Interest (CMPI). The evidence presumed the CIA to be connected to the CMPI with a certain interest in the teleconference held to secret incriminating intelligence.

All the above insiders, including others listed in the above screenshot, had a 'meeting of the minds' on how to "coordinate," concealing the most damning science in human history.

That day, the teleconference attendees planned how to discredit, disparage, smear, and neutralize the science reported by Pradhan et al. They decided that Kristian Andersen and Bob Garry were best suited to write a fraudulent science paper subsequently published in *Nature Medicine*. That article was used to discredit Pradhan et al.'s IBM Watson computer analysis that showed four unique HIV/AIDS spike protein genes had been inserted into the CoV/SARS mutant.

Agenda
- Introduction, focus and desired outcomes - JF
- Summary – KA
- Comments – EH
- Q&A – All
- Summary and next steps - JF

Kristian Anderson
Bob Garry - I have not been able to contact Bob. Please forward if you can.
Christian Drosten
Tony Fauci
Mike Ferguson
Ron Fouchier
Eddie Holmes
Marion Koopmans
Stefan Pohlmann
Andrew Rambaut
Paul Schreier
Patrick Vallance

Andrew Rambaut
Institute for Evolutionary Biology
Ashworth Laboratories, University of Edinburgh, Edinburgh, EH9 3FL, UK

contact:

The University of Edinburgh is a charitable body, registered in Scotland, with registration number SC005336.

The Harvard-China "Evergreen" Enterprise

Notably, when the "Indian paper" was published, Harvard was swarming with Chinese military agents and entrepreneurs invested in the COVID scheme. In Fauci's e-mails, page 3140, Dean Daley informed Fauci that he had met the day before (on Saturday, Feb. 1), with "a team led by Jack Xia, the CEO of China's Evergrande Company, and Dr. Jack Liu, Evergrande's chief health officer, who stated they were acting on behalf of Dr. Zhong Nanshan, China's key point person on the coronavirus outbreak."

At the time of this writing, Evergrande is in the news for defaulting on massive loans, heavily burdening the Chinese economy. Evergrande is presumably a real estate money laundering enterprise directed by Hui Ka Yan (a.k.a, Xu Jianyin),[189] one of China's richest industrialists with a net worth of $30 billion. Evergrande's projects include "Evergreen Oasis Wuhai" and "Evergreen Spring City," both sharing the name of the

CIA's infamous front company "Evergreen" —meaning 'always making money.'

Quoting another known CIA front called *Wikipedia*, Evergreen's "aircraft supported United Nations peacekeeping operations in 30 countries; flew 'insect eradication' missions throughout Africa... [and] operated helicopters for FEMA following Hurricane Katrina in Louisiana. Commercially, the Evergreen airline helped build the Trans-Alaska Pipeline; and developed and serviced the offshore oil and gas market with helicopter support worldwide via its Evergreen Helicopters division."

Thereby, from Cambridge to China, or Alaska to Africa, the internationally active Harvard scientific business group contacting Fauci on teleconference day helped 'coordinate' the spin of the most urgent matters condemning the worlds of science and medicine to deadly and demonic fascism.

As further explained below, the COVID Coup and media cover-up sources from what has been accurately diagnosed as the National Security State's money-laundering megalopolis that slowly but surely 'captured' the international scientific communities, global commerce, geopolitics, and economics.

> *"[T]he internationally active Harvard scientific business group contacting Fauci the day of the teleconference was solidly evidenced 'coordinating' the spin of the most urgent matters condemning the worlds of science and medicine to deadly and demonic fascism.*

"I am not naïve to the challenging politics of such a relationship," Dean Daley cajoled Fauci. "I do not want to complicate or duplicate efforts already underway..."

Fauci **freaked**. "There is no real 'coordination' of this response," he rebuked Daley to protect his interests and the crime syndicate from conspiracy charges.

Fauci feared someone might intercept or read these e-mails. The flim-flam Fauci cunningly diverted to the "Chief Scientist at WHO," Dr.

Soumya Swaminathan, who was "organizing a meeting" two weeks later in Geneva.

By then, the "Indian paper" was to be silenced and discredited.

Cover-up Mission not Impossible but Criminal

"Geeeez," Fauci replied to Courtney Billet about the media, "[s]eeking comment on [Pradhan et al.'s] 'Indian paper' about new Coronavirus." (See page 3121 of the Fauci e-mails.)

"Talk about trying to put the genie back in the bottle!" Billet wrote, realizing his job was to do precisely that. "Yeesh." Billet was the chief propagandist at the NIH, the Director of the Office of Communications and Government Relations (OCGR).

As mentioned, the "genie" was the lab engineering of the COVID-19 virus, containing HIV's spike protein genes. The "bottle" was the secret enterprise.

Courtney Billet's subordinate was assigned to do the dirty work — spin the "genie" and conceal the "bottle." At the same time, Harvard's Dean Daley sought instructions to "coordinate" that spinning and disappearing act.

Jennifer Routh, the OCGR's "News and Science Writing" Director, responded, "We consulted with HHS [i.e., Department of Health and Human Services] and ASF [Acquisition Services Fund[190]] that intertwines business deals with private companies for the U.S. Department of Defense, the U.S. Department of Homeland Security, the U.S. Department of Justice (including the federal courts and the FBI), the Department of Agriculture, and the HHS]. OCGR is going to send a note to the reporter[s] to decline [all media requests for comment on the Indian paper], noting that the paper is not peer-reviewed."

"Please let us know if you receive similar requests," Routh wrote to keep a lid on the scandal. She had already been ordered by Barney Graham,[191] the Deputy Director of the NIH's Vaccine Research Center, not to answer "without **high-level input**." (Emphasis added. See Fauci e-mails, pg. 3122.)

Graham's presence in the cover-up exposes gross conflicting interests at the NIH and NIAID under Fauci's leadership. Graham personified the "dual use" financing and public/private partnerships between the government and vaccine companies, especially Moderna and Pfizer.

Graham co-authored a paper titled "An mRNA Vaccine against SARS-CoV-2 - Preliminary Report."[192] This provides one astonishing example of such conflicting interests and even criminal malfeasance.

Graham et al. exemplifying gross hypocrisy reported in the *New England Journal of Medicine* (NEJM) that half of the participants who received their experimental vaccine suffered adverse events that "included fatigue, chills, headache, myalgia, and pain at the injection site. Systemic adverse events were more common after the second vaccination, particularly with the highest dose, and three participants (21%) in the 250-μg dose group reported one or more severe adverse events."

Nevertheless, Graham's group concluded, "no trial-limiting safety concerns were identified."

After that, the damaging and deadly vaccine was brought to market by Graham's corporate allies, Moderna and Pfizer, and certified for "experimental" use on humanity by the FDA's Emergency Use Authorization (EUA).

What "high-level input" was Graham referencing and requiring to answer reporters' questions about the lab virus origin of the Coronavirus/SARS/HIV recombinant?

The buck stopped with the top brass in the Chinese military, the U.S. Navy, and the allied intelligence communities comprising the National Security Crime Syndicate controlled by the globalizing militaristic Deep State.

The result of their criminal cover-up is graphed below. The U.S. Vaccine Adverse Event Reporting System ("VAERS") Data Burden from 1993 to 2021 showed a huge spike in reported deaths and illnesses as simply measured by the extraordinary increase in the annual data file demand in Megabytes (MB).

The second set of VAERS data shows 389,323 adverse events reported by Pfizer and Moderna mRNA vaccines victims, January 1 thru June 28, 2021. These were unusually serious side effects or deaths.

Curiously, far higher numbers of mainly Pfizer and Moderna vaccine recipients, 503,422 to be exact, reported distressing symptoms resulting from their mRNA HIV/AIDS spike protein intoxications. That VAERS data is publicly available intelligence.

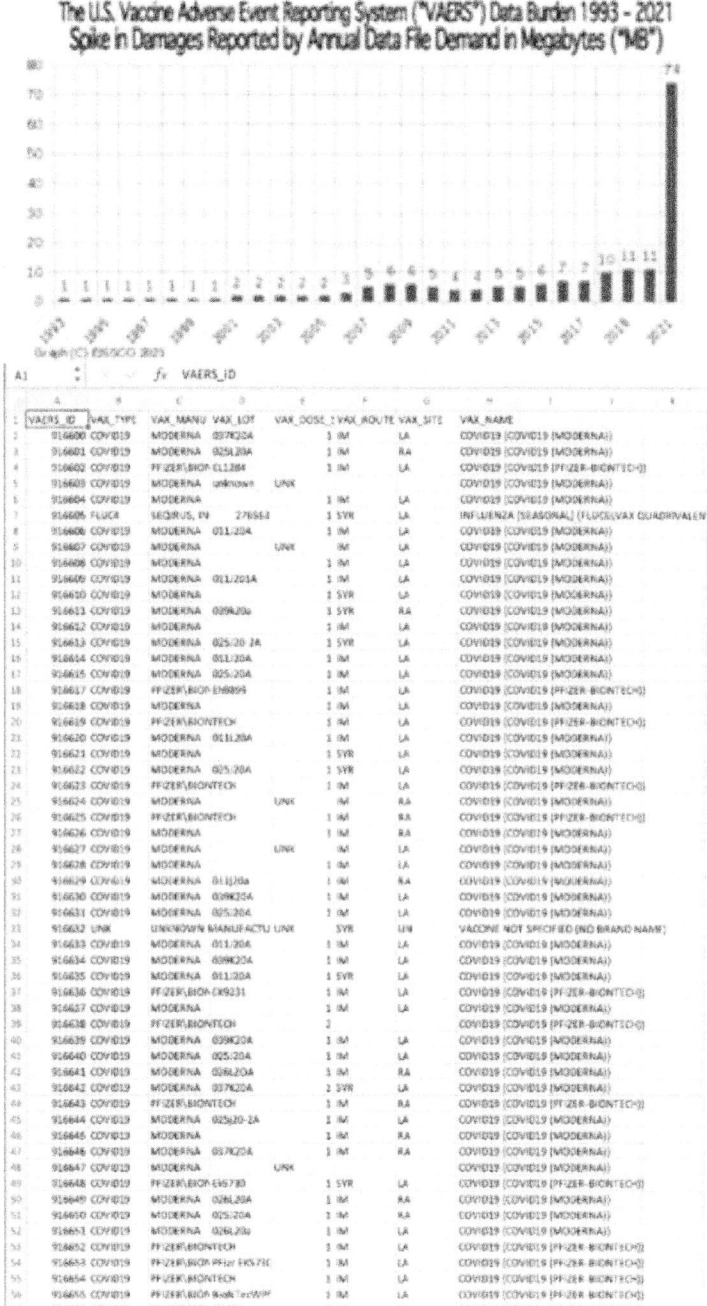

Divide and Conquer: Media Spins and Blame-Game

Western officials blamed China for the pandemic, and the Eastern media accused the U.S. military, among other things. The truth in both sides of the story got lost in the shuffle. That was the plan all along. The objective of putting politics before the facts diverted from the biocrime and shielded those responsible.

It was all a BIG LIE.

All the captured government, military, and science world's virology elite were complicit. Official powers planned, administered, and defended the COVID pandemic. Weighty evidence for this has been published by multiple sources, including this author's popular "Coronavirus Predictive Programming"[92] videos showing on *RevolutionTelevision.net*.

Time and again we witnessed the same syndicate's agents, agencies, and institutions complicit in this litany of biocrimes—from AIDS and Ebola to Zika and other Flavivirus diseases. The graphic below, for instance, evidences Peter Daszak's co-publication financed by the same federal suspects and private entities repeatedly implicated in suspicious plague outbreaks. Repeatedly, the actual "reservoir" or natural source of the outbreaks remained unknown —hidden.

Despite the evidence compiled herein being available to researchers, profitable depopulation through this kind of biological warfare has continued. It is laborious to prove and harder to prosecute. That's why, unlike other weapons of mass depopulation, lab viruses are preferred, particularly by the "Rockefeller cartel."

Recently, the Rockefeller Foundation[193] heralded its "international collaboration" with the Federal Republic of Germany's WHO Hub for Pandemic and Epidemic Intelligence [194]and the UK's Global Pandemic Radar,[195] co-financed by "Sir Jeremy James Farrar" at the Wellcome Trust and World Health Organization.

"By the end of this year," the Rockefeller press release reported, "we will transform the global capacity for stopping disease outbreaks in the first 100 days before they can begin to spread."

What's the catch?

To do this, extensive data-mining and real-time analyses are required —precisely like those administered through the nano-bioelectronic vaccine devices pioneered at Harvard and MIT, co-financed by the U.S. DOD.

Farrar, a British medical researcher and director of the Wellcome Trust since 2013, organized with Fauci and Collins the urgent "teleconference." Farrar e-mailed Fauci about an hour after Harvard Dean Daley checked in. (See Fauci email 3125). Farrar, too, exhibited panic. The UK's most powerful medical science financier expressed concern about World Health Organization (WHO) director Tedros Adhanom Ghebreyesus and Tedros's Cabinet chief, Bernhard Schwartländer, going into "conclave" over the "Indian paper."

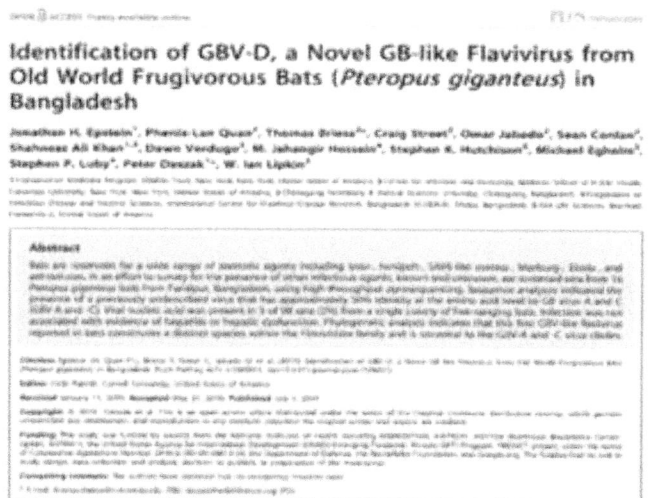

Farrar, rather than heralding the 'smoking gun' HIV/AIDS connection to COVID-19, used the word "prevaricate" to describe the lying he expected his colleagues at the WHO to do for damage-control. His mail to Fauci used that word, "PREVARICATE," to conceal their syndicate's anticipated fraud. Prevaricating is precisely what the teleconference participants decided they would do together. They conspired to deceive the world.

Material to the alleged criminal patterns and practices of the chief suspects aforementioned, Farrar served as a professor of tropical medicine at the University of Oxford.

Farrar implicated Oxford as much as Harvard and MIT in writing Fauci urgently on "2-2-20" how all of their colleagues should lie about the Indian paper.

> *The 'regulatory capture' of science, politics, the White House, the Justice Department, the Big Tech media, and all intelligence agencies East and West by the globalists is implicated in the January 31 thru Feb. 3 e-mails of Anthony Fauci.*

Oxford Joins Harvard and MIT as Co-Conspirators

In 2015, recognizing the same agents and agencies committing these biocrimes against humanity, I tracked suspects to the Rockefeller Foundation that shipped the Zika virus to Yale for refrigeration and further distribution.

Recall the mysterious Zika virus outbreak that year. Consider the first Zika virus isolation and shipment of the pathogen to Yale for "safekeeping." Under Rockefeller's license, Yale later sent the Zika virus to the American Type Culture Collection (ATCC) that sells germs to vaccine makers and bioweapons contractors worldwide.

Recall Daszak's Flavivirus research. Flavivirus (yellow fever) diseases include Zika and the West Nile Virus. Daszak's EcoHealth Alliance was financed by the same federal suspects and private entities repeatedly implicated in those suspicious outbreaks as well.

I exposed the West Nile Virus scam in my 2001 book, *Death In The Air: Globalism, Terrorism & Toxic Warfare*. I subsequently produced a video slide presentation[196] to inform Hawaii (and later Florida) officials about the public health risks posed by Zika and Oxford's spin-off company called Oxitec. My activism and whistleblowing went for naught.

That start-up company, Oxitec, provides a great example of how this commercial crime syndicate operates. It follows the CIA/In-Q-Tel's pattern-and-practice of privately financing promising, albeit murderous, businesses. In this case, Oxitec had manufactured and deployed genetically mutated mosquitoes to, allegedly, fight Zika and dengue fever in Brazil.

Oxitec, a subsidiary of Intrexon[197] that owes its allegiance to Oxford, was also financed by the Bill & Melinda Gates Foundation[198] and *New York Times* investors who profited greatly from the Zika "pandemic."

According to company records, Oxitec manufactured the genetically-mutated mosquitoes tested in Brazil in 2014. Florida fools permitted the same to be done at the time of this writing. I helped delay this same operation in Hawaii.

Bill Gates is shown in videos and the above image spreading[199] upon his audience mosquito beasts while defending Oxitec's mutants.[200]

Partnering vaccine developers and investors in this flavivirus enterprise include Oxford University's ISIS innovations, Oxitec, and Intrexon. These entities were also funded by BlackRock and the Evercore Company "private equity" investors.

These diseases and their entrepreneurs are additional examples of how the global elite, through operations like the National Security Crime Syndicate, exploits what they deploy—"novel" mysterious diseases and the genetic biotechnologies to fight them.

We see the same pattern and complicity from *New Yok Times* owner and Zika co-investor, Carlos Slim. Slim's 'yellow press' diverted from the Zika virus emergence mystery,[201] much like it did with COVID-19.

The *Nature Medicine*[101] Remedy in Fraud

As examined in the previous chapter, the media "spin" that came out of February 2, 2020, teleconference relied on attendees Andersen and Garry and their subsequent fraudulent publication in *Nature Medicine* six weeks later.

Fauci announced that paper at a White House press conference shortly after the teleconference. Yet, Fauci stealthfully lied about (and concealed) the authors' names, Andersen, Garry, et al., with whom Fauci's co-conspirators plotted the *Nature Medicine* whitewash in their teleconference 'meeting of the minds.'

During that period, officials at the 'highest levels' of the corrupted governments in the U.S. and China forbade their complicit agents and scientists to discuss the matter (other than to discredit the Pradhan group's "Indian paper").

At that time, for instance, recall that Tucker Carlson on *FOX News* reported that anyone advancing the lab origin of COVID "conspiracy theory" was "**lying**." In fact, everyone dismissing the lab source theory at that point was either stupid or 'prevaricating,' as Farrar put it.

Anderson and Garry et al. had huge incentives to misrepresent important facts. Based on Andersen's conflicting writing to Fauci the day before the teleconference, Andersen, Garry, and Fauci committed frank *fraud* and the *felony* of evidence tampering for their enterprises' **cover-up**.

These covert accomplices in treason and bioterrorism purportedly debunked the lab origin theory placing the nations and global populations at greater risk. Concealing timely exposure of the lab origin and genetic mutations within the HIV-1 spike protein gain-of-function device could have encouraged earlier, more varied remedies. By concealing the evidence, as agreed upon during the teleconference, Andersen and Garry became accomplices in the wider criminal conspiracy, causing massive numbers of people to die.

In Andersen's e-mail to Fauci on Friday, January 31, 2020, the same day the Indian paper hit their radar screens, Andersen wrote Fauci concerning "Science: Mining coronavirus genomes for clues to the outbreak's origins." In review, his first paragraph controverted his *Nature Medicine* fraud by stating:

> "The unusual features of the virus make up a really small part of the genome (<0.1%) so one has to look really closely at all the sequences to see that some of the features (potentially) look engineered."

Supposedly, the aforementioned censorship was administered by the NIH directors and Jeremy Farrar's group at the request of the *Nano-Micro: Small* authors[202] who also reported on the unusual genetics of the COVID-19 virus. The bulk of those *Small* authors, who allegedly asked Collins's NIH to delete the scientifically determined gene sequences, worked for the "China State Key Laboratory" bureaucracy controlled by Harvard's partner "Dr. Zong." This additionally incriminated Harvard principals, Dean Daley and Charles Lieber particularly.

These facts explain why Andersen, Garry et al. published their lies — the opposite of what they witnessed when they did their genetic analyses. During and after the teleconference, Anderson and Garry followed their instruction from Fauci, Collins, Farrar, and military and intelligence community officials. If they didn't obey, they might have been assassinated. Presumably, they feared losing their grants, careers, or lives.

The suspects committed an organized criminal conspiracy to falsify the origin narrative, leverage the media to defraud the public, and lull society into compliance with profitable vaccination, masking, and social distancing objectives, all profiting the multi-national (global elite) corporations active in this National Security Crime Syndicate transhumanist coup.

The "Nuremberg defense," however, does not exonerate their criminal complicity. They 'prevaricated' that the lab virus origin theory was "improbable."

Clearly, Andersen, Garry et al. were influenced as to how they should write the article and deny Pradhan's findings of HIV genes and the "gain of function" lab manipulation giving rise to COVID-19.

Proof of Andersen, Garry et al.'s unscientific criminal influence facilitated by Fauci is in Andersen's March 6, 2020, e-mail to Fauci, Collins, and Farrar that states, "Thank you again for your advice and leadership as we have been working through the SARS-CoV-2 'origins' paper. We're happy to say that the paper was just accepted by *Nature Medicine* and should be published shortly (not quite sure when) ... Tony, thank you for your straight talk on CNN last night —it's being noticed." (See: Fauci email, p. 2401)

Fauci replied, "Thanks for your note. Nice job on the paper." (See also: "2401")

Criminal Violations of Title 18 U.S.C. §1519

Previously I wrote about Title 18 U.S. Code §1519 alleged violations. At that time, I did not conclude sufficient evidence existed to charge Fauci et al. with a biocrime under 18 U.S.C. §175 "Prohibitions with respect to biological weapons," and/or 18 U.S. Code §2332a – "Use of weapons of mass destruction." Now sufficient evidence for these charges arguably exists.

This decision should be made by a state attorney and grand jury, not untrustworthy (presumed corrupt) DOJ agents.

There is probable cause to presume guilt and bring the evidence to a grand jury for indictments for the following facts:

1) The suspects fraudulently concealed and disparaged the findings published by Pradhan's group in the "Indian paper." They concealed the four genes from the AIDS virus spike protein in the SARS CoV-2 previously cultured bioweapon. '*Mens rea*' (willful intent or 'scienter') is evidenced because the complicit agents knew honest disclosure posed an economic threat to their commercial enterprise.

2) The suspects committed an organized criminal conspiracy to falsify the origin narrative, leverage the media to defraud the

public, and lull society into compliance with profitable vaccination, masking and social distancing objectives, profiting the multi-national (global elite) corporations active in this National Security Crime Syndicate transhumanist coup.

All in all, these co-conspirators lied to discredit science and news reports that said, "Coronavirus Contains 'HIV Insertions', Stoking Fears Over Artificially Created Bioweapon."[130]

3) The NIH, too, coordinated their cover-up. They subsequently censored the spike protein genetic sequences published in the *Small*[202] report in favor of the complicit parties, especially Harvard and Chinese partners, including Dr. Zong's Key Laboratory group known to be controlled by communist China and its anti-capitalist military.

4) The suspects abused mass media outlets and deprived the public of alternative medicines and natural remedies in reckless or malicious disregard of the public's health and safety.

5) The suspects arguably violated Title II of "Enhancing Controls on Dangerous Biological Agents and Toxins," Sec. 201 (Sec. 351A), because the suspects are "reasonably suspected" of "I) committing a crime set forth in section 2332b(g)(5) of title 18, United States Code (Acts of terrorism transcending national boundaries involving the 'Federal crime of terrorism."[203]

That means an offense that —"(A) is calculated to influence or affect the conduct of a government by intimidation or coercion, or to retaliate against government conduct" (the latter damaging the Trump administration especially) "... II) knowing involvement with an organization that engages in domestic or international terrorism [such as China that helps finance terrorist regimes in the Middle East, for instance] (as defined in section 2331 of such title 18) or with any other organization that engages in intentional crimes of violence [such as the CIA's 'spin off' National Security enterprise known as "The Chertoff Group" (also implicated in the Las Vegas massacre)]; or (III) being an agent of a foreign power (as defined in section 1801 of title 50, United States Code);" as was Fauci and complicit agents including Peter Daszak and Harvard's cohort.

"For purposes of this section, the term "for use as a weapon" includes the development, production, transfer, acquisition, retention, or possession of any biological agent, toxin, or delivery system **for other than prophylactic, protective, bona fide research, or other peaceful purposes**." Emphasis is added to the exception that would be used to exonerate the suspects.

Prosecution depends on the definition of "peaceful purposes" versus the intention to manufacture commercially, deploy, and thereby profit (politically and economically) from the development of the hyper-weaponized (i.e., gain-of-function) virus. Profiting from COVID's "gain of function" biological weapon and disease induction cannot be reasonably construed as "peaceful purposes," nor "prophylactic, protective, bona fide research" in the United States, or could U.S. financial and technical sponsorship in China since the official moratorium *prohibited* this research.

6) The suspects appear to have also violated 18 U.S. Code §2332a - Use of a weapon of mass destruction[204]— "(1) against a national of the United States[204] while such national is outside of the United States [as occurred in China and elsewhere in December 2019]; (2) against any person[204] or property[204] within the United States [as was done throughout the COVID pandemic], and... (D) the offense, or the results of the offense, affect[ed] interstate [and/]or foreign commerce, or, in the case of a threat, attempt, or conspiracy, would have affected interstate or foreign commerce;... (2) the term 'weapon of mass destruction' means —(A) any destructive device as defined in section 921 of this title ; (B) any weapon that is designed or intended to cause death or serious bodily injury through the release, dissemination, or impact of toxic or poisonous chemicals, or their precursors; (C) any weapon involving a biological agent, toxin, or vector (as those terms are defined in section 178 of this title);..."

7) The suspects concealed by redaction of evidence (most heavily from January 31st to February 3rd), Fauci's e-mails, thus reflecting a secret cabal, or the "National Security Crime

Syndicate," implicating agents in the CIA/DOJ/FBI/DARPA and FOIA pharmaceutical enterprise.

There was no officially announced criminal investigation of the e-mail evidence. But there was, supposedly, an ongoing investigation into the lab-origin theory. Thus, the redactions evidence obstruction of governmental operations and a violation of 18 U.S. Code §1519. "Silence is evidence most persuasive."

8) The suspects are evidenced committing a 'pattern-and-practice' of deceptive trade in science and geopolitics surrounding outbreaks of numerous lab viruses, including HIV/AIDS, Ebola, Zika, HPV, West Nile Virus, and SARS, as this author has documented during the past quarter-century.

9) The suspects operate a "health science racket" —an enterprise that profits from diseases, public deceptions, cultural indoctrinations, academic and governmental (i.e., institutional) impositions, and toxic vaccinations.

10) The suspects include an identified 'inner circle' of agents who acted as co-conspirators in the COVID scheme. Besides Andersen's and Garry's co-author, Andrew Rambaut (a British evolutionary biologist trained at Oxford), those implicated in this global criminal organization include the "teleconference" attendees listed in **Jeremy Farrar**'s e-mail to Fauci on Saturday, February 1, 2020, shown below.

11) The suspects knowingly and willfully agreed to keep their teleconference discussion secret until they were instructed on further agreements (i.e., 'meeting-of-the-minds') evidencing criminal *mens rea*. Farrar, who organized and emceed the teleconference, wrote to the insiders in bold print, **"Information and discussion is shared in total confidence and not to be shared until agreement on next steps."**

Other cited (cc'd) suspects include:

Mike Ferguson, a senior fellow at the alleged non-profit Center for Medicine in the Public Interest (CMPI), is identified in the attendees' teleconference list. He is also a senior advisor and leader of the BakerHostetler law firm's Federal Policy team. Under Ferguson's leadership, the firm has provided government affairs and lobbying services to prosper businesses of all sizes, from global corporations to startup firms.

In 2016, Ferguson joined CMPI, which is described as a "free-market think tank" founded by the Pacific Research Institute. (PRI) Under these circumstances of alleged COVID biocrime, with the evidence above collected, the CMPI and PRI are presumed to be operating as 'fronts' for the cabal, the CIA/In-Q-Tel, and National Security Agency, to profit from healthcare and population-control commerce. CMPI's advertised "research agenda deals with clinical outcomes and econometric studies that analyze the value of new medicines and genomic and molecular-based medical innovation," according to the CIA-edited *Wikipedia*. "Robert Goldberg" is listed as the first "Key people" controlling the New York City-based CMPI organization.

Goldberg's association with Michael Chertoff further evidences the cabal committing the "COVID Coup" and compounds evidence of malice in the National Security Crime Syndicate's bioterrorism impositions to "destroy America from within." This unconscionable, treasonous, and deadly healthcare, vaccination, and immigration agenda imposed by the Biden Administration is further evidence of the crime syndicate.

Goldberg and Chertoff were co-defendants in a lawsuit brought by Chinese immigrants denied visas. Therein Goldberg is identified as the former "Consul General of the U.S. Consulate General, Consular Section, Guangzhou, P.R. China."

More evidence of organized crime maximizing profits internationally comes from CMPI's officers having written articles opposing Canada's price controls and generic drug supplies. For the same reason, damaging President Trump would have been a high priority for these officials. CMPI members also oppose restrictions on drug advertising in the European Union.

Further evidencing its CIA connections, the CMPI is also linked to the International Center for Economic Policy Studies (ICEPS), later renamed the Manhattan Institute. "The incorporation documents for the ICEPS were signed by prominent attorney Bill Casey, later Director of the Central Intelligence Agency."

Further evidencing the cabal's intent to destroy America from within, Former Director of the CIA, General David Petraeus, and former

National Security Advisor and Secretary of State, Henry Kissinger; and former Department of Homeland Security Secretary (under President George W. Bush), Michael Chertoff, opposed Bob Casey (D-PA)'s 2016 bill[205] to strengthen immigration policies and border security.

> Hope it will help frame the discussions.
>
> From: Jeremy Farrar <...>
> Date: Saturday, 1 February 2020 at 15:34
>
> 1st February (2nd Feb for Eddie)
> Information and discussion is shared in total confidence and not to be shared until agreement on next steps.
>
> Dial in details attached.
> Please mute phones.
> I will be on email throughout – email Paul or I Paul if any problems
> If you cannot make it, I will phone you afterwards to update.
>
> One Hour
>
> 6am Sydney
> 8pm CET
> 7pm GMT
> 2pm EST
> 11am PST
> (Hope I have the times right!)
>
> Thank you for the series of calls and for agreeing to join this call
>
> Agenda
> - Introduction, focus and desired outcomes - JF
> - Summary - KA
> - Comments - EH
> - Q&A - All
> - Summary and next steps - JF
>
> Kristian Andersen
> Bob Garry - I have not been able to contact Bob. Please forward if you can
> Christian Drosten

Additional 'inner circle' complicit agents are named below:

1) **Edward "Eddie" Charles Holmes** is the co-author with Andersen, Garry, and Rambaut, of the fraudulent *Nature Medicine* paper, schemed during the Teleconference. Holmes has been an evolutionary biologist, virologist, and a National Health and Medical Research Council (NHMRC) member since 2012. The Australian professor at the University of Sydney is also an Honorary Visiting Professor at Fudan University, Shanghai, China (2019–present) and Guest Professor at the Chinese Center for Disease Control and Prevention, Beijing, China (2014–present).

In March 2020, Holmes co-authored a paper titled **The proximal origins of SARS-CoV-2**, also published in the *Nature Medicine* yellow press, which debunked speculation the virus may have leaked from a research facility.

2) **Marion Koopmans** is a Dutch virologist who is Head of the Erasmus MC Department of Viroscience. Koopmans participated in a 28-day mission to Wuhan to investigate the origins of the global COVID-19 pandemic. She was part of a team of experts appointed by the World Health Organization (WHO) and Peter Daszak, who whitewashed the investigation. Following this WHO mission, "Professor Marion Koopmans" and Professor John Watson[206] were invited by the Chatham House, the Royal Institute of International Affairs, based in London, to be debriefed, implicating the Anglo oligarchy in the COVID biocrimes.

3) **Stefan Pohlmann** (a.k.a., Prof. Dr. Stefan **Pöhlmann**[207]) is the Head of the Infection Biology Unit, German Primate Center. There, Pohlmann and staff investigate "how emerging viruses interact with host cells and cause disease." Their advertised "focus" is "on lymphocytic choriomeningitis virus, Ebola virus, and SARS coronavirus. One aim of [their] research is to develop cell culture systems that allow predicting transmissibility and thus pandemic potential of emerging viruses."

Pohlmann's group advertisements make known their business includes development, production, transfer, acquisition, retention, or possession of biological agents, toxins, or delivery systems not for prophylactic, protective, bona fide research for peaceful purposes, but alternatively to generate mutant pathogens to study their gain-of-function pathogenesis.

These statements of fact further evidence complicity in bio-crimes violating Title II of "Enhancing Controls on Dangerous Biological Agents and Toxins," Sec. 201 (Sec. 351A) because Pohlmann's complicity is "reasonably suspected" of advancing biological weapons for mass depopulation.

Pohlmann's group also advertises[207] their expertise in developing and researching primate herpesviruses. "The transmission of herpes B virus from macaques to humans can cause severe disease while closely related viruses seem to be apathogenic in humans," the group reported.

Gross fraud in the development and distribution of herpes virus mutants, including those responsible for human papillomaviruses (HPV), cervical cancers, and vaccines claimed to prevent them, was published by me on behalf of the Hawaii Department of Public Health that initially censored and illegally concealed from the public record the detailed scientific intelligence report I submitted.

4) **Paul Schreier** is the Chief Operating Officer at Wellcome Trust[208] and Jeremy Farrar's companion. He joined Wellcome in September 2019, shortly before the infamous "Event 201" was administered; Six weeks later, COVID-19 emerged as planned during that conference.

Schreier, subordinate to Farrar, is responsible for the effective and efficient delivery of Wellcome's operations. His particular focus is on finance, grants, digital resources, and emerging technologies. Before joining Wellcome, Schreier led Hakluyt & Company, prior to which he served as deputy secretary with responsibility for economic policy and strategy in the Australian Department of the Prime Minister and Cabinet. He was a partner with McKinsey & Co; and an officer in the Royal Navy, including a period in Command.

5) **Patrick Vallance** (a.k.a., Sir Patrick John Thompson Vallance), is a British physician, scientist, and clinical pharmacologist who worked in academia and the biotechnology industry. He has served as the Chief Scientific Adviser to the Government of the United Kingdom since March 2018. In 2006, in his mid-40s, he joined Glaxo-Smith-Klein (GSK), Pfizer's owner, as head of drug discovery. Four years later, he became head of medicines discovery and development, and in 2012 he was appointed head of research and development at GSK.

Vallance's attendance in the Teleconference ties Pfizer and Moderna's mRNA vaccines to the biocrimes mentioned above and the National Security Crime Syndicate's 'inner circle.'

Deceptive science and unfair and deceptive commercial trade by Pfizer were the subjects of this author's lawsuit against Pfizer in the Middle District Court of Florida.

Final Note in this Chapter

The facts above evidencing organized biocrimes committed by the named suspects in the "National Security 'Crime' Syndicate" will undoubtedly be attacked. The anticipated opposition to this book compounds evidence of the crime syndicate's pattern-and-practice of issuing propaganda and diversionary science to commit consumer fraud and indemnify those guilty of pan-genocide or 'humancide.'

We have witnessed this 'standard operating propaganda procedure' for 'damage control' numerous times since HIV/AIDS emerged from Litton Bionetics' lab in-or-about 1970 under U.S. military contracts and Congressional appropriations. (See the author's scientific report in *Medical Hypothesis*.[209])

Recently, in the context of dismissing COVID-19's emergence from the Wuhan lab, ignorant or complicit agents have resurrected the old Russian blame game. Propagandists have long argued that Russia's KGB sourced false stories incriminating the U.S. military in the development of HIV/AIDS. In fact, the actual science corroborating this thesis was not advanced by Russian agents at all. This author joined six non-Russians internationally, advancing the U.S. military/vaccine industry connection to the AIDS pandemic.

"Ye shall know them by their fruit." (Matthew 7:15–20) The science 'spin doctors' named above are liars and schemers who have killed millions of people. Today, their ravages and damages are mounting because people are cognitively conditioned to disregard the 'elephant under the carpet' and 'nude emperor' parading on the world stage.

Although many are calling for criminal investigations at the time of this writing, no one has diagnosed the underlying pathology made crystal clear in this book. The 'regulatory capture' of science, politics, the White House, the Justice Department, the Big Tech media, and all intelligence agencies East and West by the globalists is implicated in the January 31 thru Feb. 3 e-mails of Anthony Fauci.

The evidence in hand justifies a Nuremberg-like trial in the International Criminal Court. Neglecting this necessity and public duty promises worsening depopulation biocrimes. Neglecting such prosecution invites the excuse that SARS/CoV-2/HIV will "naturally" mutate and recombine with more deadly viruses such as H5N1 —the bird flu. That virus kills nearly 60% of humans infected. This dire probability is based on science. The crime syndicate's pattern-and-practice, including efforts to censor this author,70, and his intelligence on biowarfare, risked civilization's extinction.

Flu researcher Ron Fouchier, in 2013, fought similarly against these deadly powers on these urgent matters. As a result, the corrupted court-authorized unconscionable impositions.

Fouchier worked to release life-saving intelligence showing how easy and risky it is to hyper weaponize H5N1. The University of North Carolina and Wuhan lab subsequently demonstrated this with the SARS/CoV-2/HIV mutant.

Like this author, Fouchier was precluded by a court from advancing his warnings. In Fouchier's case, the devil-doers flipped his argument. The crime syndicate deprived Fouchier by ruling that the know-how to produce bioweapons of mass depopulation would fall into the wrong hands if he prevailed.

CHAPTER V

WHY THE FBI LIED ABOUT CHARLES LIEBER'S CHINESE ESPIONAGE

The 'Deep State' suspects contrived and controlled COVID's impositions, and the FBI acted to cover this up.

On January 27, 2020, the FBI filed a "CRIMINAL COMPLAINT" against Harvard's famed chemistry professor, Charles Lieber. The indictment was brought by Affidavit of Special Agent Robert Plumb. It was filed in the United States District Court for the District of Massachusetts. It charged Lieber, a pioneer in the nano-bioelectronic synthetic tissue manufacturing field central to the transhumanist movement, for falsely denying the extent of his service to Chinese government-controlled academics, military agents, and agencies in Wuhan. That indictment was publicly heralded on the Department of Justice website.[210]

However, agent Plumb lied about and concealed Lieber's military and commercial espionage activities with China. Lieber did not transfer intelligence to build "car batteries." He transferred state-of-the-art nano-bioelectronic transhumanist technology.

The following related background and intelligence come from discoveries in my lawsuit against Pfizer, Moderna, Henry Schein Co. (my previous employer), the Hearst Media conglomerate, and governmental Does 1 – 50. These "Defendants" are key agents of the Deep State enterprise largely (if not completely) controlling healthcare, science, and politics in alliance with captured intelligence agencies.

Background on Intelligence Community Commerce

The devil-doers control Big Pharma, Big Tech, Big BioTech, Big Energy, Big Military, and Big Banking. For this reason, I have repeatedly referred to this racketeering enterprise as the "Military-Medical-Petrochemical-Pharmaceutical-Banking Cartel." The American component I call the "National Security Crime Syndicate." Their actions advance drugs with hideous side effects. For me, watching television's drug advertisements is like witnessing social engineering by a criminal cult that normalizes and commercializes diseases and deaths.

In this context, the intelligence community plays a major role, especially in contriving media propaganda. We witnessed the media's adoration of Dr. Fauci and his propaganda to conceal COVID crimes. We watched as psychopathic liars and complicit Deep State agents terrorized viewers to condition people to accept Pfizer and Moderna vaccines as the only remedy to gain "herd immunity." Tragic economic damages and more deaths resulted.

As a 'public duty,' I have taken on this establishment in federal court in opposition to the unsafe and ineffective COVID "mRNA vaccines." As is now widely known, these products contain genetic material and nano-bioelectronic components in their attachment device called a "hydrogel."

The "Defendants'" vaccines compete directly against natural alternatives that are far safer and proven to be effective. My non-drug anti-viral "OxySilver™ with 528" product, for example, is a powerful antioxidant. It functions electromagnetically similar to the Defendants' mRNA vaccines, but without the risks and genetic damage. OxySilver bio-electrically features the 528 frequency of sound and light relayed through water memory to achieve therapeutic benefits. To prevent OxySilver's competition, the Defendants repeatedly disparaged my reputability, smeared and misrepresented OxySilver, and deprived and damaged my companies.

Between 2009 and the present time, the CIA assigned FBI informant and corrupt lawyer, Paul J. Sulla, Jr.,[211] to stalk, harass, and maliciously prosecute me in Hawaii's corrupted courts, both state and federal. The main objective was to tie me up in legal defense to silence my voice. To a large degree, they succeeded.

I complained to the FBI on these matters on several occasions. I opened official cases, all to no avail. In retrospect, now understanding that the FBI was "captured" long ago by the Deep State elite, my faith and trust in this DOJ entity, along with the courts, was naively misplaced.

Nevertheless, court actions by activists contribute to saving lives by raising public awareness of important issues. In this context, in my case against Pfizer, Moderna et al., I filed several public records for 'judicial notice' to alert the court and people everywhere how untrustworthy the

FBI and other government agencies have acted during the COVID Coup as exemplified by their criminal indictment against Lieber.

Partnering Entrepreneurs "Charlie" Lieber and Robert Langer

According to press reports, Charles Lieber is awaiting trial while dying of leukemia at the time of this writing. Aside from his position at Harvard, Lieber is the co-founder of "NanoSys" and Precision NanoSystems Inc (PNI). Danaher Company, along with Integrated DNA Technologies —a Baxter subsidiary— acquired PNI.

Baxter and Bayer are advertised as corporate "Partners" in the U.S. Centers for Disease Control (CDC) Foundation —a "philanthropy" that not only allegedly "keeps America secure by controlling disease outbreaks," but also acts as a trade organization according to the Foundation's July 2, 2021, roll-out of "monoclonal antibodies" promotions, substantially enriching Moderna, the manufacturer.

Moderna is also the maker of "mRNA-1944," developed with financial support from the Department of Defense's DARPA. Along with GlaxoSmithKlein (Pfizer's parent) and Bayer (two partners in mRNA vaccine-maker CureVac Co.), this enterprise commercialized Lieber's nano-bioelectronic vaccine technologies for common industry and enterprise interests.

Lieber worked closely with his co-author and mRNA vaccine technology entrepreneur, Moderna's founder, and MIT professor, Robert Langer. Both Lieber and Langer labs (at Harvard and MIT, respectively) were financed by public funds from the NIH and Department of Defense. This funding enriched the Lieber/Langer publishing and scientific research partnership. Both men, alone and together, played significant roles in developing the Defendants' mRNA vaccines.

More Background: DARPA, the "Common Thread"

Within the National Institutes of Health (NIH), Dr. Fauci et al.' email records, there is evidence of DARPA's central role in the COVID Coup and racketeering enterprise. This includes the aforementioned correspondence with 'inner circle' officials, Fauci's superiors, and allied academic and business agents. As detailed in the previous chapter, several suspects were identified and caught acting to conceal and lie about the spliced genes from the AIDS virus, HIV-1, within the major bioelectronic, antigenic, allegedly "therapeutic" component of the Pfizer, Moderna, and CureVac mRNA vaccines.

Evidence of DARPA being the 'common thread' linking these suspects includes corroborating emails of Dr. Ralph Baric obtained by the U.S. Right to Know (USRTK) public interest research group. Through the Freedom of Information (FOI), they obtained evidence of Dr. Baric corresponding with Kristian Andersen. These emails evidence Daszak, Andersen, et al. coordinating the "Indian paper" damage control and lab origin cover-up. These two men became the leading COVID lab virus denialists.

NIH, NIAID, Google, USAID, the WHO, and the Gates Foundation heavily financed Daszak and EcoHealth Alliance. As already explained, the urgent teleconference was conducted to conceal, discredit, divert from, disparage, or otherwise neutralize the damning scientific determinations of Pradhan et al., proving the HIV-1 genetic enhancement of the COVID-19 virus—the deadly "gain-of-function."

On February 4, 2020, two days after the urgent teleconference wherein the 'inner circle' had a 'meeting of the minds' to "coordinate" their damage control, Andersen corresponded with Daszak, America's leading bat coronavirus scientist and the President of EcoHealth Alliance.

Included in the USRTK FOI correspondence was more mail from Dr. Andrew Pope, Director of the National Academy of Sciences Board on Health Science Policy, Health and Medicine Division. Pope urgently requested Baric's response to the NAS's and Trump Administration's investigation of COVID's origin in the "dual-use" research.

Recall that Droegemeier wrote the NAS President, Marcia McNutt, to secure valid scientific discovery one day earlier: Monday, February 3, 2020, and one day after the urgent teleconference.

A copy of Andersen's and Daszak's correspondence is shown on the next two pages.

Another important piece of evidence proving DARPA's complicity in the cover-up is the correspondence between EcoHealth Alliance's Dr. Jonathan H. Epstein, Vice President for Science and Outreach (subordinate to Peter Daszak), soliciting Baric for "dual-use safety language" required by DARPA to protect the agency from liability in the wake of the scandal. This email from Epstein to Baric addresses the growing evidence of COVID's lab origin by detailing the "content, timing, and the extent of distribution of potentially sensitive dual-use information."

DARPA sought as much intelligence as possible following the "Indian paper's" release to coordinate with those complicit in the covert actions—Daszak's group and the NIH teleconference participants—justification for, and indemnification from, the "dual-use" lab mutation.

The following public records available online[212] are documents whose accuracy cannot reasonably be questioned by reason of the express identities, mailing dates, and COVID origin subject matter of teleconference participants and concerned governmental and private financiers.

```
To:      Peter Daszak[daszak@ecohealthalliance.org]
Cc:      Pope, Andrew[APope@nas.edu]; Chakravarti, Aravinda[Aravinda.Chakravarti@nyulangone.org]; Baric, Ralph
         S[rbaric@email.unc.edu]; Trevor Bedford (trevor@bedford.io)[trevor@bedford.io]; Gigi Gronvall[ggronvall@jhu.edu]; Tom Inglesby
         (tinglesby@jhu.edu)[tinglesby@jhu.edu]; Stanley Perlman (stanley-perlman@uiowa.edu)[stanley-perlman@uiowa.edu]; Shore,
         Carolyn[CShore@nas.edu]; Chao, Samantha[SChao@nas.edu]
From:    Kristian G. Andersen[kga]
Sent:    Tue 2/4/2020 12:05:54 PM (UTC-05:00)
Subject: Re: URGENT: Please review by NOON if at all possible.
```

I too agree with all that has been said, but would caution against adding language suggesting that the virus might evolve (i.e., "mutate" to most people) towards better infectivity or transmission - a lot has been said about this for Ebola and other viruses, and it's been driving fear because most people don't fully understand what it means. I'm not arguing that it's not something that might well happen - the SARS data beautifully show it - but I would be worried about the message it could send.

Reading through the letter I think it's great, but I do wonder if we need to be more firm on the question of engineering. The main crackpot theories going around at the moment relate to this virus being somehow engineered with intent and that is demonstrably not the case. Engineering can mean many things and could be done for either basic research or nefarious reasons, but the data conclusively show that neither was done (in the nefarious scenario somebody would have used a SARS/MERS backbone and optimal ACE2 binding as previously described, and for the basic research scenario would have used one of the many already available reverse genetic systems). If one of the main purposes of this document is to counter those fringe theories, I think it's very important that we do so strongly and in plain language ("consistent with" [natural evolution] is a favorite of mine when talking to scientists, but not when talking to the public - especially conspiracy theorists).

Best,
Kristian

On Tue, Feb 4, 2020 at 9:02 AM Peter Daszak <daszak@ecohealthalliance.org> wrote:

I agree with all of the other comments so far sent in, and want to add the following:

1) In the 3rd paragraph, it's important to add "including further samples from wildlife", and perhaps the rationale for this "to identify other viruses closely related to nCoV"

2) Re. references for #3 that there are current and planned studies underway on the bat origins of CoVs. Here are some references to pick from if they make sense:

- Latinne A, Hu B, Olival KJ, et al., Origin and cross-species transmission of bat coronaviruses in China. *Nature Communications* 2020. **In review**

- Wang N, Li S-Y, Yang X-L, et al.; Serological Evidence of Bat SARS-Related Coronavirus Infection in Humans, China. *Virologica Sinica* 2018. doi: 10.1007/s12250-018-0012-7.

- Hu B, Zeng L-P, Yang X-L, et al.; Discovery of a rich gene pool of bat SARS-related coronaviruses provides new insights into the origin of SARS coronavirus. *PLOS Pathogens* 2017;13(11) e1006698. doi 10.1371/journal.ppat.1006698

- Zhou P, Fan H, Lan T, et al.; Fatal Swine Acute Diarrhea Syndrome caused by an HKU2-related Coronavirus of Bat Origin. *Nature* 2018.

COVID COUP: "The Rise of the Fourth Reich"

Cheers,

Peter

Peter Daszak

President

EcoHealth Alliance

460 West 34th Street – 17th Floor

New York, NY 10001

Tel

Website: www.ecohealthalliance.org

Twitter: @PeterDaszak

EcoHealth Alliance leads cutting-edge research into the critical connections between human and wildlife health and delicate ecosystems. With this science we develop solutions that prevent pandemics and promote conservation.

From: Pope, Andrew [mailto:APope@nas.edu]
Sent: Tuesday, February 4, 2020 9:11 AM
To: 'Chakravarti, Aravinda'; Kristian Andersen (KGA1978@gmail.com); Ralph Baric (rbaric@email.unc.edu); Trevor Bedford (trevor@bedford.io); Peter Daszak; Gigi Gronvall; Tom Inglesby (tinglesby@jhu.edu); Stanley Perlman (stanley-perlman@uiowa.edu)
Cc: Shore, Carolyn; Chao, Samantha
Subject: URGENT: Please review by NOON if at all possible...
Importance: High

Many thanks again for your thoughtful participation yesterday. The plans have changed in terms of our product. Instead of a "Based on Science" web posting, we are now developing a letter that will be signed by the 3 Presidents of our 3 Academies (NAS, Marcia McNutt; NAM, Victor Dzau; NAE, John Anderson), in response to a letter from OSTP. We think this will be more appropriate and expeditious.

Thus, given the urgency of the request from OSTP and HHS we ask that you please review the attached DRAFT CONFIDENTIAL letter and let us know if you have any concerns or suggested edits. In particular, we would like to ask if there might be some additional detail added to the data needs that are identified. We think it would be helpful to be a bit

Many sincere thanks again for your continued engagement on this important activity!

Andy

Andrew M. Pope, Ph.D.

Director

Board on Health Sciences Policy

Health and Medicine Division

The National Academies of Sciences,

Engineering, and Medicine

apope@nas.edu

~

To: Baric, Ralph S[rbaric@email.unc.edu]
From: Jon Epstein[epstein@ecohealthalliance.org]
Sent: Fri 3/23/2018 6:54:18 PM (UTC-04:00)
Subject: dual use safety language

Hi Ralph,

DARPA wants a written section on communicating dual-use information. Do you have some written text you could send the

A communication plan that addresses content, timing, and the extent of distribution of potentially sensitive dual-use information. The plan must also address how input from DARPA, other government, and community stakeholders will be taken into account in decisions regarding communication and publication of potentially sensitive dual-use information.

Cheers,
Jon

--

Jonathan H. Epstein DVM, MPH, PhD

Vice President for Science and Outreach

EcoHealth Alliance
460 West 34th Street – 17th floor
New York, NY 10001

(direct)
(mobile)

web: ecohealthalliance.org

twitter:

EcoHealth Alliance leads cutting-edge scientific research into the critical connections between human and wildlife health and delicate ecosystems. With this science, we develop solutions that prevent pandemics and promote conservation.

WHY THE FBI LIED ABOUT CHARLES LIEBER'S CHINESE ESPIONAGE

DARPA's Capture by the Drug Cartel

In a revealing article[213] titled, "DARPA Coronavirus? A new twist in ongoing plague," investigative journalist Whitney Webb[214] details little-known U.S. military connections to the Deep State's COVID vaccine enterprise. Quoting Webb's scholarly evidence-based reporting, the Coalition for Epidemic Preparedness Innovations (CEPI) administered the public/private enterprise bringing negligently tested vaccines into risky, often deadly, emergency usage.

CEPI, described in *Wikipedia*, is "a partnership of public, private, philanthropic and civil organizations that will finance and co-ordinate the development of vaccines against high priority public health threats." Founded in 2017 "by the governments of Norway and India along with the World Economic Forum and the Bill and Melinda Gates Foundation [i]ts massive funding... finance the rapid creation of vaccines and widely distribute them."

CEPI financed two pharmaceutical companies, Webb reported. Inovio Pharmaceuticals and Moderna Inc. Both held "close ties to and/or strategic partnerships with DARPA." Their vaccines relied on controversial "genetic material and/or gene editing." CEPI also involved the University of Queensland in its enterprise. That institution maintained independent financial ties to DARPA pursuant to engineering[215] and missile development.[216]

"[T]top funders of Inovio Pharmaceuticals include both DARPA and the Pentagon's Defense Threat Reduction Agency (DTRA),"[217] Webb reported. Inovio received "millions in dollars in grants from DARPA, including a $45 million grant[218] to develop a vaccine for Ebola."

Inovio specialized in the creation of "DNA immunotherapies and DNA vaccines, which contain genetically engineered DNA that causes the cells of the recipient to produce an antigen and can permanently alter a person's DNA," Webb advised. "Inovio previously developed a DNA vaccine for the Zika virus, but — to date — no DNA vaccine has been approved for use in humans in the United States. Inovio was also recently awarded over $8 million[219] from the U.S. military to develop a small, portable intradermal device for delivering DNA vaccines jointly developed by Inovio and the U.S. Army Medical Research Institute of Infectious Diseases (USAMRIID)."

Regarding the coronavirus, Inovio's efforts to develop a DNA vaccine for MERS are applicable. "Inovio's MERS vaccine program began in 2018 in partnership with CEPI[220] in a deal worth $56 million. The vaccine currently under development uses[221] "Inovio's DNA Medicines platform to deliver optimized synthetic antigenic genes into cells... is partnered with USAMRIID and the NIH, among others. That program is currently undergoing testing in the Middle East."

Inovio's earlier collaborations with the U.S. military involved developing DNA vaccines for Ebola and the Marburg virus. Inovio's CEO, Dr. Joseph Kim,[222] called this enterprise an "active biodefense program." Additional grants for this work came from the Department of Defense's Defense Threat Reduction Agency (DTRA), Dr. Fauci's NIAID, and other government agencies, Webb evidenced.

"It is also worth noting," Webb added, "that Inovio Pharmaceuticals was the only company selected by CEPI with direct access to the Chinese pharmaceutical market through its partnership with China's ApolloBio Corp.,[223] which currently has an exclusive license to sell Inovio-made DNA immunotherapy products to Chinese customers."

The second pharmaceutical company CEPI selected to develop a vaccine for the new coronavirus was Moderna Inc., which collaborated with the NIH and was largely funded by CEPI. "Moderna's mRNA treatments, including its mRNA vaccines, were largely developed using a $25 million grant[224] from DARPA, and the company often touts its strategic alliance with DARPA in press releases."[225]

Public Records Evidence a 'Civil Conspiracy,' and COVID Coup

How do the facts mentioned above intertwine with the FBI and Charles Lieber's indictment?

The FBI's indictment of Lieber, coupled with the e-mails from the NIH and the UNC, reveal the pattern-and-practice of lying about, censoring, and/or discrediting the knowledge central to the COVID Coup, the Wuhan outbreak, and the genetic engineering of the lab virus for biowarfare and vaccine commerce.

The extent of the COVID Coup and its devil-doers became apparent during my litigation discoveries. In my court case against Pfizer, Moderna et al., I alleged the Defendants used deceptive trade practices to compete unfairly against natural medicines, especially antioxidant

OxySilver™. I challenged and filed to enjoin risky mRNA vaccines, falsely advertised as "safe." I realized what was concealed most damagingly was the so-called "ANTIGENIC CHALLENGE" and the bioelectronic components. This 'double whammy' was responsible for myriad side effects (i.e., "adverse reactions") caused without 'informed consent.'

For there to be a *civil conspiracy*, four elements must be met: (1) there must be an agreement between two or more parties; (2) to do an unlawful act by unlawful means; (3) the committing of an overt act in pursuance to the conspiracy; and (4) damage to the plaintiff as a result of the act.[226]

The records I filed with the court evidenced these four elements:
1) "an agreement" had been made between the aforementioned government and private parties in the Defendants' mRNA vaccine enterprise;
2) They had unlawfully concealed, tampered with, lied about, and discredited scientific evidence of genetic engineering of the CoV/SARS/HIV-1 recombinant's attachment mechanism (i.e., spike protein "antigen") in alleged violation of Title 18 U.S.C. §1519. They interfered illegally with governmental operations, including the White House's investigation;
3) the suspects committed the overt act of publishing falsehoods "in pursuance to the conspiracy" to protect their financial and commercial interests, and
4) they damaged society and me as a competitor. My commercial and humanitarian (public health) interests were deprived by their illegal act(s). For instance, they falsely informed consumers about the safety of their vaccines and concealed the HIV inserts in the antigenic challenge. They defrauded consumers into taking their vaccines, unaware of misrepresentations of safety. They smeared and discredited relying on alternatives, other safer preventatives, such as hydroxychloroquine or my OxySilver™.

The public records, beginning with the FBI's indictment of Lieber, evidence these multiple wrongdoings damaging more than the public's trust in science and medicine. At MIT, the shame and guilt also project onto Lieber's business ally, fellow researcher, and co-author, Robert

Langer. Their nano-bioelectronic technology used in the Moderna and Pfizer vaccines was well-financed by the government, including DARPA and the NIH. Their research and developments became central to the chemistry and efficacy of the mRNA vaccine delivery system that features the "hydrogel" lipid bioelectronic coating surrounding the antigenic spike protein.

The e-mails between doctors Baric, Andersen, the EcoHealth officials —Daszak, Epstein, and Pope— establish the depth and width of governmental actors (i.e., "state actors") and actions to conceal Dr. Baric's interlacing of the SARS/coronavirus/HIV-1 "dual-use" 'gain-of-function' virology. Meanwhile, the NIH e-mails from Fauci and friends evidences the co-conspirators' decision to lie about Baric's work and Daszak's complicity. These agents and their cohorts in biocrime, such as Andersen and Garry, committed fraudulent concealments and deadly misrepresentations

Little Known Private and Public Interests

Here, as in court, I reveal little-known private interests in this allegedly corrupt pharmaceutical enterprise; private companies and their agents who monopolize biotechnology, virology, and vaccinology favoring the suspects and their now identifiable conflicting interests.

These little-known private interests are served by principals, including Lieber, Langer, Baric, and Daszak, et al., acting subordinate to the global elite, GSK and Pfizer, Moderna, and government agencies, including DARPA and the NIH.

It is public knowledge that Harvard's "Lieber Research Group" had been financed for years by the NIH, DARPA, the Office of Naval Research, the Air Force Office of Scientific Research, and the MITRE organization. As briefly mentioned before, from this massive taxpayer support, Lieber developed his private company, NanoSys (a.k.a., Precision Nanosystems), which Danaher acquired. Danaher is a conglomerate that also acquired Integrated DNA Technologies, Inc., co-owned by Baxter Healthcare Corporation.[227]

Bayer and Pfizer's parent, GSK (Wellcome), partnered in CureVac, advertised as an alternative mRNA vaccine that, despite lower reported efficacy and questionable safety, is predicted to supply lesser-developed countries defending against newly emerging viral variants.[228]

Likewise, the "Langer Lab" at MIT, in collaboration with Lieber's lab at Harvard, spun off Moderna with additional DARPA, NIH, and CEPI financing.

Due to these private commercial interests financed initially by taxpayers, current licensing agreements and lucrative patents exist. For example, a commercial license exists between Integrated DNA Technologies (IDT) that sells genetic products such as nucleic acids to third parties. One states it is "provided under 'Limited License,' granted by Broad, MIT, and Harvard ('Licensor') and Integrated DNA Technologies, Inc."[229]

Early in 2017, Pfizer Inc. (NYSE: PFE) committed funding of $4 million to enable Baric's affiliate, NCBiotech, to establish and administer the multi-year academic fellowship program to accelerate North Carolina's fast-growing expertise in so-called "gene therapy" [including Pfizer's and Moderna's mRNA SARS/HIV-1/ coronavirus vaccines.]"[230]

As early as 2008, an NIH grant was issued to Pfizer's "partner" in "anti-infective efforts, Dr. Baric at the UNC (i.e., AI23946-08)."[231] Beginning in 2003, Baric was officially recognized as affiliated with Fauci's NIAID.

These interlocking agents and agencies in public/private alleged 'tortious' (and criminal) enterprise began work on synthetically altering coronaviruses for the express purpose of "general research, pathogenic enhancement, detection, mutation, and potential therapeutic interventions." As early as May 21, 2000, Baric and the UNC sought to patent the critical sections of the coronavirus family for commercial gain.[232]

In the first public record of Baric's claims, this Pfizer/Moderna agent sought to patent a means of producing "an infectious, replication-defective, coronavirus." The NIH grant GM63228 supported this work.[233]

The FBI and Fauci's minions had to know all about this work. Nonetheless, the co-conspirators continued to recklessly neglect the risks and intimacy between HIV/AIDS and the COVID disease killing people.

Most importantly, the electromagnetic spike protein attachment relies foundationally on early HIV-1 attachment science.[234] Concealing the threat that civilization will suffer from new diseases from COVID vaccines, such as thrombosis and strokes, is unconscionable and criminal. The Defendants' deceptive trade presumes profitable depopulation.

This valid assertion and related facts predict long-term damage, such as from more mutation risks validated by the (2003) international patent "Application filed by the Government of the United States of America as Represented by the Secretary of The Department of Health and Human Services National Institutes of Health." That patent, WO2005010034A1, is titled "Soluble fragments of the SARS-CoV Spike Glycoprotein." The patent compounds evidence of the criminal conspiracy when it directs attention to the link between the AIDS virus and the current COVID pandemic. It states in relevant part:

> The results provided herein not only offer new tools to study entry of the SARS virus into cells, confirm that ACE2 is a receptor for the SARS-CoV SI glycoprotein and localize the RBD [receptor binding domain of the spike protein] but also facilitate development of novel vaccine immunogens and therapeutics for prevention and treatment of SARS... Site directed mutagenesis was used to create the consensus cleavage sites corresponding to that of the HIV-1 [AIDS virus] envelope glycoprotein (Env) and some coronaviruses within the full length SARS-CoV S glycoprotein gene in pCDNA3.

In short, government contractors Baric, Fauci, Collins, Lieber, Langer, Farrar, Daszak, Daley, and their corporate partners and sponsors, especially Pfizer and Moderna, concealed their monopolistic syndicate succeeding the AIDS enterprise.

Why Did the FBI Lie About Lieber's Business with China?

The FBI, DHS, NIH, NIAID, like the NSA, is subservient to the 'intelligence' gathering, espionage administering, and sabotage delivering CIA. The "alphabet agencies" operate under the influence of the private equity investment elite—the global banking cartel. Recall, for instance, that Michael Chertoff took with him eleven members of the DHS and CIA to form the Chertoff Group, including Reginald Hyde[235] —a key secret agent who helped establish the CIA's In-Q-Tel investment group.²³⁶ Recall also that the Chertoff Group is deadly, having direct involvement, inside intelligence, and commercial alliances with Deep State financial agents that profited from the *Las Vegas Deep State Massacre*.²³⁷ Recall that these suspects include George Soros, James Murren, and Steven Wynn. Recall these parties' complicity in killing people en-masse for-profit and political influence. Therefore, complicity in the COVID Coup is not beyond them.

When this Deep State intelligence community realized that Lieber was a liability, sloppy with his paper-trail to Wuhan at the beginning of the pandemic, under scrutiny by the Trump administration seeking answers about the origin of the pandemic and compromised by his Chinese espionage activities implicating his boss--the panicking George Daley and his "Harvard-China 'Evergreen' Enterprise" associates, Lieber needed to be 'neutralized' because he was compromised and compromising.

It is reasonably presumed these Lieber enemies gave him the "follicular lymphoma"²³⁸ from which he is dying because Lieber could expose the most advanced brain-Cloud transhumanist bioelectronics.

The CIA had been investing in Lieber's Nanosys company spun off from Harvard's proprietary interests through In-Q-Tel. These investors were engaged with Lieber in Microchip Biotechnologies,²³⁹ providing the products for data-mining, intelligence gathering, and data analysis for biodefense (including the OpGen²⁴⁰ microbial genome analysis system). The Pfizer and Moderna nano-bioelectronic hydrogel devices in the emerging vaccines were of key interest to these agents and agencies.

Thus, you can bet the CIA had lots to do with Lieber's arrest by the FBI. It is unreasonable to think otherwise, especially since the FBI's indictment *falsified Lieber's espionage activity*. Special Agent Plumb's Affidavit says nothing about Lieber's Nanosys technologies and bioelectronic data-mining through wireless vaccine hydrogel devices.

Add to this explosive scandal China's military involvements that Dean Daley, as evidenced by Daley's e-mail to Fauci on Sunday morning, February 2, 2020. The espionage activities and arrests of Lieber's students in the People's Liberation Army troubled Daley and certainly In-Q-Tel investors as well.

As detailed in the previous chapter, Harvard's Dean Daley's e-mail correspondence with Anthony Fauci and NIH Director Francis Collins that Sunday morning came one hour *before* Fauci wrote Collins, "the Indian paper is really outlandish."

Fauci wrote Daley about unnamed decision-makers, presumably superior intelligence officials in NSA and the CIA overseeing the NIH and NIAID, "[W]e do not know who is doing what until we are told..."

Later, Starnes Walker revealed Michael Chertoff's powerful influence over several functions of government and decisive actions. Chertoff, the past Secretary of Homeland Security, co-authored the infamous USA PATRIOT Act. Backing his action in a statement on December 21, 2005, supporting the "reauthorization" of the Act, Chertoff said:

> "The number one tool in defending this country is intelligence. Gathering it, investigating it, and sharing it. If we don't have the full ability to use the tools of gathering, sharing, and using intelligence, we are putting very important weapons in the war on terror down on the ground and walking away from them. And I don't think that's anything we can afford to do. . . Our line of defense is a line with intelligence and investigation. And the PATRIOT Act gives us the ability to do that in a way that respects the Constitution, respects civil liberties, but gets the job done."

So the FBI had ample motive to commit the crime of falsifying court records in Lieber's prosecution. At the time of this writing, Lieber faces trial on his pleading of innocence while he is dying[241] of "a very advanced form of lymphoma," according to his lawyer, Marc L. Mukasey. That's suspiciously favorable to the DOJ/FBI prosecutors as well as Lieber's Nanosys co-investors troubled by what a 'death bed confession' might bring.

Why Intelligence Agents Redacted Fauci's E-mails

Recall that Fauci's e-mails from Friday, January 31st through Monday, February 3rd, 2020, contained the most heavily redacted intelligence on the subject of the "Teleconference" —COVID-19's gain-of-function genetic engineering with HIV's S-protein inserts central to the fast-tracked vaccines, including Charlie Lieber's hydrogel delivery device. Had censors not redacted this intelligence, they would have broken PUBLIC LAW 107–188,[242] the PUBLIC HEALTH SECURITY AND BIOTERRORISM PREPAREDNESS AND RESPONSE ACT OF 2002, passed on June 12, 2002. This makes the Attorney General of the United States primarily responsible for enforcing the ACT that states in pertinent part:

"(h) DISCLOSURE OF INFORMATION.—" (1) NONDISCLOSURE OF CERTAIN INFORMATION.— No Federal agency specified in paragraph (2) shall disclose undersection 552 of title 5, United States Code, any of the following:"(A) Any registration or transfer documentation submitted under subsections (b) and (c) for the possession, use, or transfer of a listed agent or toxin; or information derived therefrom to the extent that it identifies the listed agent or toxin possessed, used, or transferred by a specific registered person or discloses the identity or location of a specific registered person."(B) The national database developed pursuant to sub-section (d), or any other compilation of the registration or transfer information submitted under subsections (b) ... and (c) to the extent that such compilation discloses site-specific registration or transfer information."(C) Any portion of a record that discloses the site-specific or transfer-specific safeguard and security measures used by a registered person to prevent unauthorized access to listed agents and toxins."(D) Any notification of a release of a listed agent or toxin submitted under subsections (b) and (c), or any notification of theft or loss submitted under such sub-sections."(E) Any portion of an evaluation or report of an inspection of a specific registered person conducted under sub-section (f) that identifies the listed agent or toxin possessed by a specific registered person or that discloses the identity or location of a specific registered person if the agency determines that public disclosure of the information would endanger public health or safety."(2)

COVERED AGENCIES. —For purposes of paragraph (1) only, the Federal agencies specified in this paragraph are the following:''(A) The Department of Health and Human Services, the Department of Justice, the Department of Agriculture, and the Department of Transportation.''(B) Any Federal agency to which information specified in paragraph (1) is transferred by any agency specified in subparagraph (A) of this paragraph.''(C) Any Federal agency that is a registered person, or has a sub-agency component that is a registered person.''(D) Any Federal agency that awards grants or enters into contracts or cooperative agreements involving listed agents and toxins to or with a registered person, and to which information specified in paragraph (1) is transferred by any such registered person.

Accordingly, it is surprising that the NIH and UNC's e-mail records of Fauci and Baric contained as much information as they did. It was just enough for a veteran health science investigator aware of HIV/AIDS research and the evolutionary virology of 'emerging viruses' to figure out what was and is really happening.

Keep in mind that **Lieber's colleagues at Harvard and MIT knew this world-leading nano-bioelectronics expert did not work on "car batteries."** He worked feverishly and invested personally in human vaccine hydrogel bioelectric circuitry. Wireless brain-Cloud data-mining for 5G real-time transhuman/cyborg control depended on this, most exclusively.

Quoting from February 1, 2020 —the day that Fauci, Collins, Farrar, Andersen, Garry, Daszak, et al. were in panic-mode planning their urgent teleconference to coordinate concealing the HIV genes in the S-protein that is bioelectrically encased in Lieber's hydrogel mRNA vaccine delivery device— *The Economist*[243] published the following:

> IN 2013 CHARLES LIEBER, a pioneer of nanoscience who is now the chairman of Harvard University's chemistry department, visited the Wuhan University of Technology (WUT), in China, to celebrate the founding of a lab he was credited by that university with helping to establish and oversee: the WUT-Harvard

WHY THE FBI LIED ABOUT
CHARLES LIEBER'S CHINESE ESPIONAGE

Joint Nano Key Laboratory. It was a remarkable coup. WUT is an institution of little renown. Harvard is generally regarded as the top of the academic tree. And Dr Lieber, whose research has been seen by some as a forerunner of Elon Musk's ambitious scheme to supercharge the human brain with nanotechnology, has been seen as a potential Nobel laureate.

Three days later, an in-depth investigation by Robert F. Service[244] published in *science*,[245] tripped over this "puzzle" as follows:

> In Lieber's case, however, the battery angle poses a puzzle. That's because a search of the titles of Lieber's more than 400 papers and more than 75 U.S. and Chinese patents reveals no mentions of "battery," "batteries," "vehicle," or "vehicles." (According to Lieber's CV,[246] through 2019 he has co-authored 412 research papers and has 65 awarded and pending U.S. patents. The website of the Chinese National Intellectual Property Administration indicates that Lieber has been awarded 11 Chinese patents.)

One U.S. nanoscientist and former student of Lieber's says, "I have never seen Charlie working on batteries or nanowire batteries." (The scientist asked that their name not be used because of the sensitivity surrounding Lieber's case.)

The proof of Lieber's "progress in merging electronic and biological systems at the 3D tissue level by using microporous nano-electron scaffolds appears below, for example, extracted from a 2013 science journal:

> The basic platforms use conventional nanowire material and device systems with well-exploited physical or chemical properties, and they also have wide ranging applications in many other fields, such as energy scavenging systems (54–61) or components for integrated circuit (34,35). These basic platforms, such as planar nanowire field effect transistors (34,35,37,40,43) or vertical nanowire arrays (55–58,60,61), have been used in biomolecular sensing (52,53), extracellular recording

(52,53), drug delivery (62–64) and localized cellular imaging (65). On the other side, the advanced platforms have been designed to address some intrinsic complexity in biology and medical sciences in way simply not possible previously. They allow new types or new scales of interact and measurements with their target systems (31,66–68), and in so doing, open up completely new opportunities in science and technology. Examples of advanced platforms include recent intracellular field effect transistor probes (31,67–69) and nanoelectronics-innervated synthetic tissues (66).

This review discusses the basic concepts of nanoscale field effect transistors (nanoFETs) and their applications in cellular electrophysiology.

In other words, THE BATTERY in Lieber's system is the Holy Spirit flow of electromagnetic and bio-acoustic "field energy" that innervates electrophysiology and animates electro-genetics (i.e., sound and light signaling to and from DNA). This fact is central to the criminal conspiracy, cover-up, and transhumanist objective of controlling plague survivors through super-fast computing, AI, and the Cloud.

WHY THE FBI LIED ABOUT CHARLES LIEBER'S CHINESE ESPIONAGE

This is the main reason why the FBI lied. The agency and complete Justice Department had been 'captured' to facilitate this overriding transhumanist agenda.

More evidence below links the "Biden Crime Family" to this nefarious organization and operation.

The Biden Crime Family's Chinese Investments

On September 23, 2020, the U.S. Senate Committee on Homeland Security & Governmental Affairs released their report titled, "Hunter Biden, Burisma, and Corruption: The Impact on U.S. Government Policy and Related Concerns," linking the Biden crime family and Democratic Party's complicity with the intelligence community and the COVID Coup. That governmental report stated in relevant part:

> The Chairman's investigation into potential conflicts of interest began in August 2019, with Chairman Grassley's letter to the Department of Treasury regarding potential conflicts of interest with respect to Obama administration policy relating to the Henninges transaction.[226] During the Obama administration, the Committee on Foreign Investment in the United States (CFIUS) approved a transaction that gave control over Henninges, an American maker of anti-vibration technologies with military applications, to a Chinese government-owned aviation company and a China-based investment firm with established ties to the Chinese government. One of the companies involved in the Henniges transaction was a billion-dollar private investment fund called Bohai Harvest RST (BHR). BHR was formed in November 2013 by a merger between the Chinese-government-linked firm Bohai Capital and a company named Rosemont Seneca Partners. Rosemont Seneca was formed in 2009 by Hunter Biden, the son of then Vice President, Joe Biden, by Chris Heinz, the step son of the former Secretary of State John Kerry and others.[227]

The Senate Committee neglected to report that Henniges and BHR, along with Rosemont Seneca, were vicariously partnered with the Evergrande/Evergreen group financing Harvard. They were partnered through the government of China and through the government of the United States and the World Health Organization, as explained below.

They all concealed, along with the FBI, Charlies Lieber's actual espionage. They were not interested in car "batteries" nor electric vehicles in this COVID-related context advancing bioelectronics and "novel" mRNA vaccine commerce. The partnerships and financial investments were for "emerging markets around the world." Henniges,[247] with Biden's agency through Rosemont Seneca, expanded its global footprint with the addition of new facilities in Mexico, China, and Germany, at that time. This was while Joe Biden was Vice President in the Obama Administration. This was also at the time the Ebola and Zika "crises" were presumably administered.

The "car battery" story was "cover," according to the evidence in hand. The fraudulent concealment secured the Deep State's COVID Coup. As proven, Lieber's "batteries" struck at the soul of humanity. Human spirituality mediated by piezoelectricity and scalar frequencies would be captured by the nano-bioelectronics injected in vaccine hydrogels. This was most useful for the military's control over populations through data-mining —the main requirement and emerging market for transhumanism and brain-Cloud integration. In the interim, the "Final Solution" for the Great Global Reset rested on this secrecy and conspiracy. Healthcare commerce would be used to generate massive profits and the Brave New World impositions.

Shocking proof of this conspiracy involving the Bidens acting as agents (or "front men') for the Deep State, or National Security Crime Syndicate, through Henniges et al., is evidenced first by Harvard Dean Daley's e-mail to Fauci on Sunday, February 2, 2020, citing the need for 'coordinating' the cover-up of the "Indian paper" to protect the interests of Harvard and their Chinese financiers of Lieber's bioelectronics—Evergrande and COVID officials.

Recall from Chapter IV that Daley wrote Fauci, "Alan Garber, Harvard's Provost, and I met yesterday with a team led by Jack Xia, the CEO of China's Evergrande Company, and Dr. Jack Liu, Evergrande's chief health officer, who stated th[e]y were acting on behalf of Dr. Zong

WHY THE FBI LIED ABOUT
CHARLES LIEBER'S CHINESE ESPIONAGE

Nashan, China's key point person on the coronavirus outbreak (see below) [redaction], and they arranged a conference call for tomorrow morning EST with Dr. Zhong."

Redactions then concealed the names of the principal officials in the US/UK/China National Security Crime Syndicate. They coordinated their nefarious commercial, "scientific," and military interests with Dr. Nashan. That evidence compels the presumption of a concealed 'end game' motive of great interest to "Dr. Zong," the Chinese military, and the UK/US subversives coordinated by Jeremy Farrar, the World Economic Forum elite, the intelligence community spinning or concealing all of it, and the Obama/Biden Democratic leadership being complicit.

The above screenshot from "ProgrammerSought"[248] —a high tech investment blog— stated the following before it was removed from the Internet:

> Le Yue, the founder of LeTV, first let the real estate tycoon Sun Hongbin, China's chairman, acknowledged the investment failure of 15 billion yuan, and now he has to invest another 6.974 billion Hong Kong dollars in another real estate company's new energy auto company

Faraday Future's ["FF"] director of Evergrande Group.

The chairman of the bureau, Xu Jiayin, broke.

Evergrande Health announced that FF Top Holding Ltd, the FF original shareholder controlled by Jia Yueting, filed an arbitration with the Hong Kong Arbitration Center on October 3, requesting that the consent of Evergrande as a shareholder to enjoy the financing be dismissed and all agreements should be lifted.

Mind you, immediately below this information, tying Evergrande/Harvard's leaders and money to "new energy auto" futures company investing in "health," were articles promoting the main villains in this book. These intertwined articles promoted Bill Gates's Microsoft company and its advancements; 5G and the Cloud; Oxford University's interests in blockchain commerce co-financed by the CIA's In-Q-Tel investment firm; and Germany's technology for advancing robotics, transhumanism, memory science, and Artificial Intelligence (**AI**) assisted learning.

Next, consider who was HHS Secretary under Obama and Biden at that time.

The Obama/Biden "Health" Enterprise Ties to Ebola, Now COVID

Barack Obama nominated Sylvia Mary Burwell to be the 22nd United States Secretary of Health and Human Services on April 11, 2014, just as the Ebola "crisis" in Africa began making news headlines. On March 23, 2014, the WHO officially declared an outbreak of "EVD" caused by "**Zaire ebolavirus**." That was the same strain that had first emerged in Zaire in 1976 —impossible without refrigeration of viral cultures. I wrote extensively about this outrageous biocrime and its ties to risky vaccine developments in *MedicalVeritas.org*. The international media picked up my intelligence and said it prompted angry mobs[249] resisting vaccines in Africa and elsewhere.

As reported[250] by my deceased partner, Medical Veritas International, Inc.'s Associate Editor, Sherri Kane, there were "three fundamental mistakes nearly everyone [wa]s making." I explained to Sherri, "The first is, *we're not dealing with a 'normal virus' here*, nor a high cure rate, but a discriminating disease evidencing biological warfare."

I put the Obama/Biden administration under scrutiny for their conflicting interests in the geopolitics surrounding the Ebola crisis and federal Emergency Response. I reported extensively that the 2014 Ebola re-emergence was perfectly timed to pressure Liberia's President, Ellen Johnson Sirleaf, into signing a controversial oil drilling deal with Big Energy–Big Pharma's Deep State ally in globalization and investment banking.

At that time of Ebola Zaire's reemergence, Liberia was the world's most tumultuous and controversial country. Oil drilling operations off its coast were socio-politically, economically, and commercially challenged. At the same time, money was pouring in from the International Monetary Fund at the request of President Sirleaf, the winner of the 2011 Nobel Peace Prize and the 2012 Indira Gandhi Prize for Peace, Disarmament, and Development.

Then, after this author's series of articles exposing 'EbolaGate'[251] published on *WarOnWeThePeople.com*, the crisis soon disappeared. My "Refrigerator Requirement" clear-and-convincing analysis shamed the devil-doers, seemingly into censoring the entire matter.

"Grasp the **certainty** (not simply theory) of a refrigerator being the only commonality between the 1976, 2014, and later 2018 Ebola Zaire outbreaks," I encouraged researchers. "A refrigerator acts as an 'unnatural reservoir' capable of vectoring Ebola's return to headline news. This best account of the man-made outbreaks is proven by a preponderance of the evidence. Discover this evidence by examining the science and history of that precise Ebola Zaire strain. Compare those facts with the stupid propaganda used to divert from this certainty (scientists would call it high probability). Every reasonably intelligent investigator must thereby conclude that the media and scientific community obviously concealed Ebola's lab origin and imposed outbreaks."

Such stupidity, for example, was compounded in 2018 by what the WHO stated as "Facts"[252] about the Ebola Virus Disease (EVD). "The virus is transmitted to people from wild animals and spreads in the human population through human-to-human transmission." But immediately below this misrepresentation-by-omissions, the WHO stated, "Ebola virus disease (EVD) first appeared in 1976 in 2 simultaneous outbreaks, one in what is now, Nzara, South Sudan, and the other in Yambuku, Democratic Republic of Congo."

These two locations are 400 miles apart. It is delusional to believe some un-identified bat or other 'animal' could teleport itself instantly that distance to 'simultaneously' infect humans with the same strain of Ebola Zaire (EZV), especially since that virus is highly unstable. It alters its genome rapidly between 'horizontal' transmissions (i.e., from one animal to another.)

"Skeptics argue rapid transit could account for that 400-mile transmission. But in 1976, transit in this part of Africa was not so 'rapid.' Certainly not for an Ebola-carrying bat, wild animal, or native. The same animal or infected person would need to have vectored the simultaneous outbreaks. That stretches credulity and shames skeptics and trolls."

Abandoned biological-warfare weapons refrigerator in Russia photographed by U.S. military investigator Ralph Mirebs.

If that is not incredulous enough, further research and propaganda analysis showed more WHO fraud. The [Obama/Biden darling WHO and] U.N. sponsored health agencies fraudulently claimed in 2014 that 'fruit bats' were the likeliest 'natural reservoir' for this strain of EZV. So I investigated further.

I quickly determined that the 'official story' was obviously *false*. A review of the WHO's scientific reference purportedly found four percent (4%) of bats tested in Bangladesh (Asia) were claimed to carry antibodies for "African Ebola" found thousands of miles apart! And 4% similarity in the virology world of gene tracking is presumably insignificant. You may recall that the simian-immunodeficiency virus in chimpanzees (SIVcpz) is more than 60% identical to human AIDS–HIV-1. Yet, no scientist claims they know for sure that a chimp sourced AIDS, even though chimps were abused to manufacture the tainted hepatitis B vaccines that transmitted AIDS to the world.

I concluded that the Liberia outbreak of Ebola accompanied the "major socio-economic and political upheaval" involving BigPharma/BigEnergy and Sirleaf. Only that best explained the 2014 outbreak that killed more than 1,000 Liberians.

Similarly, I concluded the 2018 Ebola outbreak was administered to contrive justification for unethical vaccine trials, for which complicit scientists, calling themselves 'ethicists,' justified expanding ethical considerations under "emergency conditions."

This activity, Kane and I realized, is best called "crisis capitalism." As Stephen Kunitz reviewed, it is standard procedure for multi-national corporations to expand genocides killing citizens of lesser-developed nations.[253]

COVID COUP: "The Rise of the Fourth Reich"

'Guided by Science'
President Obama Visits NIH, Touts Progress in Ebola Vaccine Effort
By Carla Garnett

President Obama at NIH Dec. 2

In his Dec. 2 visit to NIH, President Barack Obama revealed two things about his approach to problem-solving: The U.S. will respond with compassion and science will lead the way. He used his time at what he calls "America's laboratory" to congratulate scientists for delivering a potential Ebola vaccine and to champion scientific research once again as the nation's most powerful weapon against global health threats.

"We are going to be guided by the science—not by speculation, not by fear, not by rumor, not by panic—by science," said President Obama, in a 22-minute address to a packed Masur Auditorium. Earlier he had visited two NIAID senior investigators (see sidebar) and their labs in the Vaccine Research Center. With the briefings in Bldg. 40 and the speech in the Clinical Center, he spent about 90 minutes on campus.

"One of the things that has always marked us as exceptional is our leadership in science and our leadership in research," the President said. "Here at NIH, you have always

SEE PRESIDENT, PAGE 6

PRESIDENT
CONTINUED FROM PAGE 1

been at the forefront of groundbreaking innovations."

The visit was his second to NIH as President. In September 2009, during his first term, Obama came here to announce American Recovery and Reinvestment Act funds. Since then, he has continued to bank up his interest in and commitment to scientific research with various funding mechanisms to benefit NIH.

"It's wonderful to be back in America's laboratory, even if I don't always understand what you're doing," Obama joked, humorously reminding the audience that NIH director Dr. Francis Collins had promoted him to "scientist-in-chief" during that 2009 visit.

The President thanked NIH and its partners for developing a candidate Ebola vaccine, which had just completed phase 1 clinical trials the previous week. "No potential Ebola vaccine has ever made it this far," Obama pointed out.

The President also noted other progress in the epidemic. "A few months ago, only 13 states could test for Ebola," he said. "Today 36 states can. Previously, there were only 3 facilities in the country deemed capable of treating an Ebola patient, including NIH. Today, we're announcing that we now have 35" designated treatment centers.

16 Years in the Making
Ebola Vaccine Researcher Recalls 'Logical March Forward'

Before a jovial President Obama took Masur Auditorium's stage to talk to a house full of NIH employees and several patients who greeted him like a rock star, he dropped by Bldg. 40—NIAID's Vaccine Research Center—to meet some of the scientists behind Ebola vaccine research and see some of their work firsthand.

Dr. Nancy Sullivan, chief of the VRC's biodefense research section, has been working on an Ebola vaccine for nearly 2 decades, dating back to when she was an investigator at the University of Michigan with then-NIH grantee and now former VRC director Dr. Gary Nabel.

Obama, accompanied by HHS Secretary Sylvia Burwell, NIH director Dr. Francis Collins and NIAID director Dr. Anthony Fauci, made Sullivan's lab his first stop.

"The President was actually very well-informed about how vaccines work," she said. "He was very engaged and interested. In fact, he asked some insightful questions, some that not even other scientists have asked us."

The concept for Sullivan's vaccine is 16 years in the making, beginning back when few people outside the global infectious disease community had even heard of the deadly virus. Over the years, Sullivan and her team continued to tweak her idea, constantly improving on it. Eventually she followed Nabel to NIH in 1999, before the VRC was even built. Many in the vaccine research community had begun to believe Ebola was insurmountable. It was just too aggressive for a vaccine to ever protect against it. Did Sullivan ever lose heart that her work may never prove successful?

"No, I never did," she said. "The vaccine was always on a logical march forward. And we always had the support of NIAID."

When Sullivan's vaccine went to the phase 1 clinical trial stage last fall, it was indeed a proud and historic occasion for several reasons: The first Ebola vaccine to be tested in humans was developed by a woman; principal investigator Dr. Julie Ledgerwood of the VRC clinical trials section led the study; Mary Enama, VRC protocol operations manager, coordinated development of the trial; Laura Novik was the study coordinator; and a woman was the first volunteer to receive the vaccine.

"That wasn't planned," Sullivan said, "but it's kind of remarkable."

NIAID immunotechnology chief Dr. Mario Roederer collaborates with Sullivan to analyze immune responses in potential vaccine candidates. At first, he didn't know who was going to be on the upcoming VIP tour. He was summoned from an out-of-town meeting to give a preview of his lab to an advance team. When he learned the advance team was actually the Secret Service, he realized he'd soon be giving the President of the United States a demo of the world's most sophisticated flow cytometry operation. Timing was tricky, though. Roederer had to take a

200

WHY THE FBI LIED ABOUT
CHARLES LIEBER'S CHINESE ESPIONAGE

During the 2014 "Ebola Emergency," my refrigerator thesis and political analysis fell on mostly deaf ears. The captured health agencies, especially the HHS and CDC, both under the direction of the Obama/Biden cohort in genocide, Sylvia Mary Burwell, served effectively to output propaganda. This "Emergency Response" administration certainly did not want attention diverted to their refrigerators.

The mainstream media generally went silent or recklessly antagonistic too. Their silence was accompanied by frightening 'fake news.' As a result, money poured from Obama/Biden government coffers into these complicit health agencies, supposedly bolstering biodefense. That, too, was a lie. Their "prevaricating" is evidenced by the deadly defensive activities of Joe and Hunter Biden's backing of the Metabiota Company detailed below and in Chapter X.

Quoting journalist Oliver Cook in *The BL*[254] (June 28, 2021), from his article titled, "Revealed: Hunter Biden invested in a pandemic firm partnering with Daszak's EcoHealth And The Wuhan Lab":

> Metabiota, a pandemic tracking and response firm that has collaborated with Peter Daszak's EcoHealth Alliance and the Wuhan Institute of Virology, was a primary financial backer of Rosemont Seneca Technology Partners, an investment group led by Hunter Biden.
>
> Rosemont Seneca Technology Partners (RSTP) was a spinoff of Rosemont Capital, a venture capital firm created[255] by Biden and John Kerry's stepson in 2009. Biden served as a Managing Director...[256]
>
> Metabiota [is] a San Francisco-based startup that claims to detect, track, and analyze new infectious diseases... According to financial reports,[257] RSTP" first round of fundraising, which totaled $30 million. Neil Callahan,[258] the former Managing Director and co-founder of RSTP—a name that frequently shows on Hunter Biden's hard drive—is also a member of Metabiota's Board of Advisors.

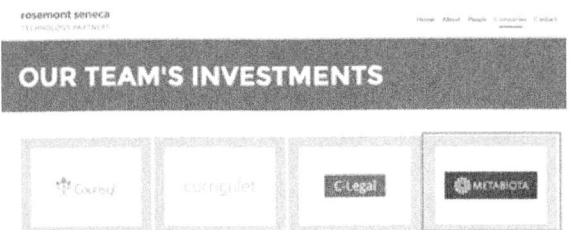

RSTP investments. (Rosemont Seneca website/Screenshot via TheBL)

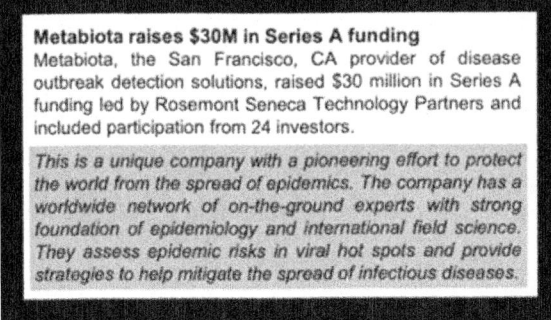

RSTP's investments in Metabiota. (Marlin & Associates June 2015 HIT Market Update/Screenshot via TheBL)

Metabiota & COVID-19 origin

Since 2014,[259] Metabiota has been a partner[260] of EcoHealth Alliance as part[261] of the "PREDICT" initiative of the U.S. Agency for International Development (USAID), which aims to "predict and prevent global emerging disease threats."

In this endeavor, researchers from Metabiota, EcoHealth Alliance, and the Wuhan Institute of Virology collaborated on a study into bat infectious diseases in China. According to the research, "sensitive and broadly reactive RT-PCR assays were performed at the Wuhan Institute of Virology, Chinese Academy of Sciences," reported *The BL*.

Shi Zhengli, the Director of the Center for Emerging Infectious Diseases at the Chinese Communist Party's Wuhan Lab, is one of the researchers included in the 2014 publication.[262] Peter Daszak, who was... removed from the *Lancet* COVID-19 panel due to many conflicts of interest as a "longtime collaborator" of the Wuhan Institute of Virology, is named as a co-author.

WHY THE FBI LIED ABOUT CHARLES LIEBER'S CHINESE ESPIONAGE

Daszak, a key figure in COVID-19's lab origin, with his EcoHealth Alliance, used public funds to collaborate on bat coronavirus research in Wuhan under agreements with Anthony Fauci's National Institute of Allergy and Infectious Diseases (NIAID).

EcoHealth Alliance and Metabiota researchers have also worked together on presentations on how to "live safely with bats"[263] and in studies tying new infectious disease epidemics to wildlife trade facilities, such as "wet markets."

Recall previously, by FOIA request; the UNC e-mails revealed fraudulent concealments by scientists Ralph Baric, Kristian Andersen, Peter Daszak, and Daszak's subordinate, Jonathan Epstein. In the screenshot above, you see, in 2014, the Metabiota/EcoHealth Alliance/Wuhan bioweapons lab group published their "Evidence for" RNA retroviruses infecting "Multiple Bat Species in China." This predated their COVID-19 cover-up correspondence by six years.

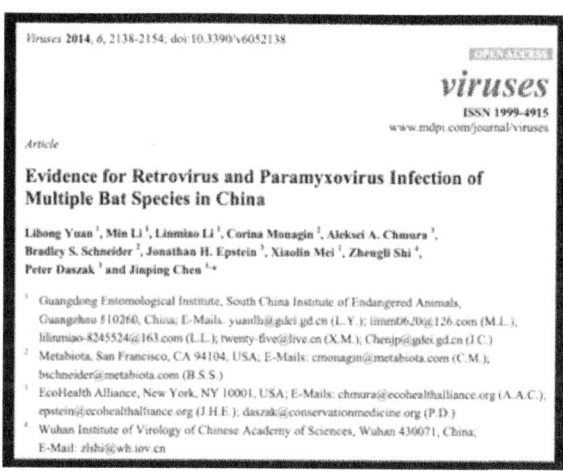

2014 study. (Screenshot via TheBL)

According to news reports[264] in 2016, this group of Ebola responders at Metabiota, EcoHealth Alliance, the WHO, and Tulane Univ. repeatedly blundered, costing lives. Metabiota and Tulane officials shared a Sierra Lione lab after the WHO damagingly delayed their emergency response enabling the illness to spread. A 'blame game' ensued. Kristian Andersen's partner in deception, Robert Garry, criticized Metabiota's Ebola response as WHO officials equally scorned Garry's Tulane. This ploy diverted from the WHO's deadly delay. Their "first responders'" dysfunction was costly and deadly —part of the scheme to re-release the refrigerated 1976 strain of Ebola Zaire.

They created the crisis and the lame response for cartel insiders' geopolitical and economic interests. The screenshot above also evidences that by 2014, this same group was already advancing the COVID-19 'plandemic' sourcing from UNC and Wuhan lab-engineered RNA retroviruses in bats.

Obama/Biden appointee Sylvia Mary Burwell, shown in the photo below with Obama, Fauci, and Fauci's subordinate, Dr. Nancy Sullivan, served to the end of Obama's administration, yielding to Trump in January 2017. Burwell was an AIDS and Ebola scientist and bureaucrat. In 2017, Fauci gave his infamous Georgetown University lecture[88] predicting the Trump Administration's "unprecedented" challenge by the COVID 'plandemic.'

Burwell oversaw the HHS's and CDC's responses to the 2014 Ebola and later Zika outbreaks. With Fauci and his underling, Dr. Nancy J. Sullivan, these officials selected Metabiota, at the heart of the Biden-Daszak-EcoHealth Alliance group that led the Ebola response in Africa with Garry's colleagues at Tulane.

How did Burwell and Sullivan become so well-connected in this criminal syndicate? Both were privy insiders with Harvard roots administering intelligence. Thus, they are linked as 'persons of interest' through Dean Daley to Lieber's Lab and criminal operations with the Chinese in Wuhan. Giving them the benefit of the doubt, they were willfully blind to their criminal complicity and its genocidal implications.

Burwell earned her bachelor's degree in Government from Harvard University and a BA in Philosophy, Politics, and Economics from the University of Oxford as a Rhodes Scholar (like Bill Clinton). She then served the Clinton administration as Chief of Staff to the Secretary of the Treasury, Robert Rubin, Deputy White House Chief of Staff to Erskine Bowles, and Deputy Director of the Office of Management and Budget. She joined the Bill and Melinda Gates Foundation in 2001 and served therein as the president of the Gates Global Development, Chief

Operating Officer, and Executive Director before its reorganization in 2006.

Sullivan's Merrimac College[265] biography evidences her evolution into genocidal complicity in the Ebola and COVID crimes syndicate.

On November 3, 2014, Merrimac published:

> Sullivan earned a master's degree in environmental engineering and then a doctorate in cell biology at Harvard University. She studied HIV while working on her doctorate but Sullivan shifted the focus of her post-doctoral research to the Ebola virus to broaden her expertise in 1998.

"Ebola is somewhat like HIV," Sullivan said. "There are similarities in the virus outer structure and proteins." Here, she referenced the important "Spike protein" antigen that shares similarities between HIV, Ebola, and the lab engineered COVID virus and vaccines. Merrimac College continued:

> Studying Ebola also offered the chance for significant contributions to science since there were so few researchers in the field...
>
> After Harvard she worked as a research fellow at the University of Michigan Medical Center's Internal Medicines Department from 1998-1999 before moving to NIH.
>
> Plans were made last year to begin a clinical trial with her vaccine at the NIH in December 2014 but the VRC

worked with the U.S. Food and Drug Administration to accelerate approval because of the Ebola outbreak in Africa. The trial began in September.

The vaccine is already in Phase 1 clinical trials to evaluate safety at NIH, at the University of Maryland, and Oxford University in England. Additional trials are planned at Emory University in Atlanta, Switzerland, Mali and Uganda.

It's hard to predict how long the clinical trials will last. Everyone is working very hard to complete the trials as quickly as possible," Sullivan said.

The scientific paper on that vaccine trial is evidenced below. As you can see, UNC's Ralph Baric is a co-author financed with Sullivan by NIH Grant Nos. T32 AI007151/AI/NIAID NIH HHS/United States[266] and F32 AI152296/AI/NIAID NIH HHS/United States.[267]

The vaccine being tested was Moderna's mRNA-1273 vaccine for COVID-19, not Pfizer-BioNTech's similar BNT162b2 product. Again, this process commenced in or before 2014 —near the time America's "moratorium" was instituted on "dual-use" "gain-of-function" viral vaccine military bioweaponry mutating bat coronaviruses.

> N Engl J Med. 2020 Oct 15;383(16):1544-1555. doi: 10.1056/NEJMoa2024671.
Epub 2020 Jul 28.

Evaluation of the mRNA-1273 Vaccine against SARS-CoV-2 in Nonhuman Primates

Kizzmekia S Corbett [1], Barbara Flynn [1], Kathryn E Foulds [1], Joseph R Francica [1], Seyhan Boyoglu-Barnum [1], Anne P Werner [1], Britta Flach [1], Sarah O'Connell [1], Kevin W Bock [1], Mahnaz Minai [1], Bianca M Nagata [1], Hanne Andersen [1], David R Martinez [1], Amy T Noe [1], Naomi Douek [1], Mitzi M Donaldson [1], Nadesh N Nji [1], Gabriela S Alvarado [1], Darin K Edwards [1], Dillon R Flebbe [1], Evan Lamb [1], Nicole A Doria-Rose [1], Bob C Lin [1], Mark K Louder [1], Sijy O'Dell [1], Stephen D Schmidt [1], Emily Phung [1], Lauren A Chang [1], Christina Yap [1], John-Paul M Todd [1], Laurent Pessaint [1], Alex Van Ry [1], Shanai Browne [1], Jack Greenhouse [1], Tammy Putman-Taylor [1], Amanda Strasbaugh [1], Tracey-Ann Campbell [1], Anthony Cook [1], Alan Dodson [1], Katelyn Steingrebe [1], Wei Shi [1], Yi Zhang [1], Olubukola M Abiona [1], Lingshu Wang [1], Amarendra Pegu [1], Eun Sung Yang [1], Kwanyee Leung [1], Tongqing Zhou [1], I-Ting Teng [1], Alicia Widge [1], Ingelise Gordon [1], Laura Novik [1], Rebecca A Gillespie [1], Rebecca J Loomis [1], Juan I Moliva [1], Guillaume Stewart-Jones [1], Sunny Himansu [1], Wing-Pui Kong [1], Martha C Nason [1], Kaitlyn M Morabito [1], Tracy J Ruckwardt [1], Julie E Ledgerwood [1], Martin R Gaudinski [1], Peter D Kwong [1], John R Mascola [1], Andrea Carfi [1], Mark G Lewis [1], Ralph S Baric [1], Adrian McDermott [1], Ian N Moore [1], Nancy J Sullivan [1], Mario Roederer [1], Robert A Seder [1], Barney S Graham [1]

Affiliations + expand
PMID: 32722908 PMCID: PMC7449230 DOI: 10.1056/NEJMoa2024671

WHY THE FBI LIED ABOUT CHARLES LIEBER'S CHINESE ESPIONAGE

Abstract

Background: Vaccines to prevent coronavirus disease 2019 (Covid-19) are urgently needed. The effect of severe acute respiratory syndrome coronavirus 2 (SARS-CoV-2) vaccines on viral replication in both upper and lower airways is important to evaluate in nonhuman primates.

Methods: Nonhuman primates received 10 or 100 μg of mRNA-1273, a vaccine encoding the prefusion-stabilized spike protein of SARS-CoV-2, or no vaccine. Antibody and T-cell responses were assessed before upper- and lower-airway challenge with SARS-CoV-2. Active viral replication and viral genomes in bronchoalveolar-lavage (BAL) fluid and nasal swab specimens were assessed by polymerase chain reaction, and histopathological analysis and viral quantification were performed on lung-tissue specimens.

Results: The mRNA-1273 vaccine candidate induced antibody levels exceeding those in human convalescent-phase serum, with live-virus reciprocal 50% inhibitory dilution (ID_{50}) geometric mean titers of 501 in the 10-μg dose group and 3481 in the 100-μg dose group. Vaccination induced type 1 helper T-cell (Th1)-biased CD4 T-cell responses and low or undetectable Th2 or CD8 T-cell responses. Viral replication was not detectable in BAL fluid by day 2 after challenge in seven of eight animals in both vaccinated groups. No viral replication was detectable in the nose of any of the eight animals in the 100-μg dose group by day 2 after challenge, and limited inflammation or detectable viral genome or antigen was noted in lungs of animals in either vaccine group.

Conclusions: Vaccination of nonhuman primates with mRNA-1273 induced robust SARS-CoV-2 neutralizing activity, rapid protection in the upper and lower airways, and no pathologic changes in the lung. (Funded by the National Institutes of Health and others.).

Copyright © 2020 Massachusetts Medical Society.

Substances

> Antibodies, Neutralizing
> Antibodies, Viral
> CD4 Antigens
> COVID-19 Vaccines
> Spike Glycoprotein, Coronavirus
> Viral Vaccines
> spike protein, SARS-CoV-2
> mRNA-1273 vaccine

Supplementary concepts

> COVID-19 serotherapy

Related information

Cited in Books
MedGen

Grant support

T32 AI007151/AI/NIAID NIH HHS/United States
F32 AI152296/AI/NIAID NIH HHS/United States

Meanwhile, tied to Garry, the Bidens, and WHO, Metabiota's co-founder, Dr. Joseph Fair, is also MSNBC's medical contributor on COVID-19. Like Burwell and Sullivan, Fair has an impressive resume. The virologist is advertised as having "extensive experience conducting high-impact, entrepreneurial public health surveillance, and research programs in Africa, Asia, Europe, and Eastern Europe."

According to his speaker's bio,[268] "Dr. Fair is a Senior Advisor to the Foundation Mérieux USA and was a co-founder and former Vice President of Metabiota, Incorporated, where he created a $50 million research and development portfolio, funded by the U.S. Department of Defense Threat Reduction Agency, the U.S. Department of State, the U.S. Department of Homeland Security, and the U.S. Agency for International Development.

"Dr. Fair is a specialist in viral hemorrhagic fever viruses [including Ebola] and public health response and management. Prior to Metabiota, Dr. Fair served as the Chief Project Scientist for the Defense Threat Reduction Agency's Biological Threat Reduction Agency Program in *Ukraine* and a staff scientist for the U.S. Army Medical Research Institute of Infectious Diseases. He received his... M.S.P.H. degrees from Tulane University, where he developed a novel recombinant diagnostic platform for Lassa fever."

I added emphasis to Ukraine because of the Bidens' conflicting political and financial interests there. In the next chapter, I evidence the Bidens' agency for the Deep State's Democratic Party influence in Hawaii, as I suffered damage by this 'mob.' Sherri Kane was killed by the actions of agents therein corrupting the courts and law enforcement. The Bidens' Burisma dealings spilled over to our court cases in that Burisma's chief competitor in Ukraine was defrauded and damaged like we were by this same racketeering enterprise that put me through fifteen years of malicious prosecutions proximal to Sherri's manslaughter.

Accordingly, this pandemic crime syndicate was well-financed and solidly established long before the COVID crisis was leveraged to bring about major sociopolitical and economic upheaval worldwide.

All the evidence considered, the scope of this criminal enterprise and its genocidal operations best explains why the FBI lied about Charles Lieber to protect the Big Picture. The CIA, leading Democratic Party officials, Big Tech, and their 'liberal media,' have been complicit in destroying America and killing missions.

Their mission is now well-evidenced and obvious: globalization in favor of the mega-wealthy. They leveraged "public health" to succeed in imposing One World Government. For this mission, COVID has served to secure the "Great Global Reset" in the aftermath of severe social and economic damage. All of this deadly criminality is intended to secure "The Rise of the Fourth Reich."

CHAPTER VI
CONSOLIDATING GLOBAL CONTROL by MINDWAR

To gain control over civilization, you must gain control over *free will* by sabotaging human hearts and minds, imposing pain, fear, loss, and threats to survival, conditions people and society to accept controllers' agendas. Over time, aversive conditioning and resulting distress wear you down and tires you out. Life made more distressing and damaging induces apathy, weakness, submission, immune suppression, disease, and premature death.

Comprehending Devil-Doing and the Basis of MindWar

As Russian dissident Aleksandr Solzhenitsyn wrote, "To do evil a human being must first of all believe that what he's doing is good... Ideology - that is what gives devil-doing its long-sought justification and gives the evildoer the necessary steadfastness and determination. That is the social theory which helps to make his acts seem good instead of bad in his own and others' eyes, so that he won't hear reproaches and curses but will receive praise and honors."

By projecting their psychopathologies onto civilization, the global elite imposes their abuse upon society as it was imposed on them. Child abuse does this to its victims. Thus, Deep State officials view humanity as defective because deep inside, they view themselves as defective — unworthy of empathy, respect, and honor. Their traumas and choices have enabled their demonic possessions.

The global elite judge the world they helped create as weak, ugly, threatening, and deserving of abuse and death.

Their management of monopolies, control over industries, and mass-media manipulations reflect that unrelenting greed, which is symptomatic of deprivation. They feel deprived and have insatiable appetites for more money and power. Their abuse replaced love, leaving them feeling eternally deprived.

Consequently, their programming, politics, and policies reflect this insanity, indecency, and toxicity —their deadly depopulation policies and kinky activities numb their self-loathing psychopathologies, also compensated by deranged narcissism. At the same time, peers accept their self-loathing schizophrenia. Their endorphin rush from sociopathic behavior is encouraged by their cohorts in crime.

The devil-doers' actions and accomplishments speak as loud as the drug side effects heralded in their TV advertisements. "Ye shall know them by their fruits." (Matthew 7:16)

We see the "fruits" of their labor everywhere we look. We see them in the racial and religious politicking ongoing, class warfare, and financial disparities they generate. We witness them in the aggression and militarization of television and video gaming —conditioning young minds for violence and bloodbaths. We see the fruits in the acceptance or normalization of mental illness and violence. People everywhere witness the Deep State's genocides, but few protest them as a result of MindWar influence.

We have watched them pass their self-centered self-loathing psychopathologies onto their children and grandchildren. Generations of psychopaths evidence their neuroses. We witness nepotism and psychopathology degrading every industry, especially in Big Banking, Hollywood, and Big Tech.

As further detailed below, Barry Diller, the chairman of the IAC media conglomerate, and Chelsea Clinton's boss described Bill Gates as having the 'emotional quotient of a snail.'[269] In a commentary on C/Net.com, Chris Matyszcyk reported that Diller said, "tech people lack nuance, but can make emotional progress." Diller cited Gates as a prime example of emotional retardation.

"The human capabilities of those who run tech companies are being severely questioned," Matyszcyk proclaimed. But to what worthless end? The psychopathic elite controlling the media, social engineering, and exploiting AI, and altering 'reality' are narcissistic sociopathic misfits.

Matyszcyk asked rhetorically, "Rapacious nerds like Facebook CEO Mark Zuckerberg might be able to grow a tech company, but can they develop a sense of what real human beings actually feel and need?"

That probability is nearly nil. In fact, Zuckerberg's video introduction to the 'Metaverse' evidences the conversion of reality to AI's seductive world. Therein, human senses, feelings, and needs are commandeered by holographic illusions.[270]

As with child victims of sexual abuse, the damage persists. They become sexually perverted or psycho-emotionally stunted. Sexual assaults leave lifelong scars. Victims continuously act out their neurotic patterns triggered by an assortment of pain-paired cues. They often act out their abuse repeatedly on others. This appears to be what civilization is witnessing and lamenting. According to Diller, Gates leads this deranged cohort but has "made emotional strides."

Deep State Population Control Every Way Possible

Today, the CIA's old MKULTRA mind-control program has advanced substantially to apply the most powerful social control know-how. The Deep State's agenda incorporates drugs, especially hallucinogenic drugs, the media, especially social media, frequency resonance, especially the 741Hz frequency of "dissonance," and the "game-changing" vaccines touted for COVID featuring the nano-bioelectronic neuroscience devices enabling the brain-Cloud connection, spiritual suppression, and transhumanism as described by Dr. James Giordano[271] at Georgetown University and the United States Naval Academy, and Dr. Charles Morgan[272] at West Point's Department of National Security. A top-down administration of this 'command-and-control warfare' is required and clearly being administered.

In this chapter, I relay much of my journey in coming to terms with the evidence of this "reality" and top-down administration of population control and depopulation accurately termed *genocide*. I survived cells of covert operators' attacks and 'learned the hard way' from their criminal psychopathology and demonic impositions. I was targeted for neutralization as a whistleblower and adversary of Big Pharma, injustice, and systemic corruption. From these experiences, pieces of the 'Big Picture' emerged to reveal how the Deep State operates and succeeds, from top-down, through media messaging, counter-intelligence campaigns, and the National Security Crime Syndicate's online trolls targeting leaders of the resistance movements considered 'enemies of the state.' Without neutralizing those leaders and activists for righteousness, the Deep State's mass mind-control via the "MindWar" cannot be achieved.

I was targeted mainly as an authority opposing vaccination risks, as was my business partner, frequent co-author, and now deceased lover, Sherri Kane. We were targeted as 'enemies of the Deep State,' as Big Pharma whistleblowers, high-profile anti-vaxxers, and drug-industry competitors having advanced the natural vaccine and antibiotic alternative, "OxySilver^{TM} with 528" frequency.

Sherri, who died from the distress imposed by the agents and agencies named below, those who committed reckless manslaughter,[273] was especially targeted as a legendary journalist exposing child sex-trafficking networks administered by agents known to us in the FBI, CIA, and the corrupted courts in Hawaii where we lived and worked to develop a natural healing community and health spa.

In the process of exposing our enemies in Big Pharma and the intelligence community, we discovered little-known connections between these entities intertwined with the corruption we witnessed.[274]

Hawaii is near halfway between China and America. The "State," having been formed following the presumed murder of King Kalakaua[275] by U.S. Navy-associated assassins, is a drug trafficking merchant-marine portal. Mexican and Chinese cartels moved massive amounts of drugs and sex slaves from the Pacific Rim, Micronesia, and Mexico, through Hawaii destined for North America.

In several state and federal courts, we battled one of the Deep State's drug cartel kingpins, Paul J. Sulla, Jr.[211] His actions and connections smacked of FBI and CIA agency. As you will soon read, Sulla's commercial connections intertwined with one of America's most infamous crime families —The Bronfmans. The height of their corrupt influence stretched to the U.S. Supreme Court's decision in the case

Gonzales v. O Centro Espírita Beneficente União do Vegetal wherein Chief Justice John Roberts[276] wrote the opinion enjoining the U.S. Drug Enforcement Agency (DEA) from confiscating the hallucinogenic "religious drug," dimethyltryptamine ("DMT"; street name "ayahuasca") opening America's billion-dollar market up to this drug's trafficking and criminal diversion.

Our journey enabled the discovery of hidden connections detailing the Big Picture. With diligence, we viewed the 'forest' of reality through the 'trees' of counter-intelligence propaganda that secures, profits, and empowers the global elite and their criminal operations and operatives destroying America by *intent*.

Accordingly, we published our information to save the lives of those 'with eyes to see,' exposing the extent of the Big Picture; because the 'Global Beast' —the Deep State's drug and vaccine crime syndicate— extends its lethal weapons into every industry, like an octopus stretches its tentacles.

In this book, I cover COVID's impact on healthcare, vaccinology, and civilization. Since vaccines are of such great importance, to be clear, the FDA classifies all COVID vaccines as "drugs." These drugs, pushed by the Deep State's PharmaMedia during this most deadly time in history, include psychotropic components that impact brain neurology aside from unnatural immunity.

Concealed 'vaccination neuroscience' is central to the COVID Coup, the required MindWar, and the "Rise of the Fourth Reich." Merging human brains with the Cloud through nano-bioelectronic vaccine devices, also classified as "drugs" (or components of injected "drugs"), compels consideration of drugs that complement or facilitate the fusion of brain neurology with computer systems and AI. Hallucinogenic drugs, as explained below, satisfies this objective most effectively.

Questioning Reality for the 'Technotronic Society'

One sure way to engineer civilization's conversion to transhumanism is using mass mind control. Society must be made to disengage from and question reality to gain acceptance of the military-industrial-complex's AI alternative. This is adequately achieved by a multi-pronged manipulation of society using the media, "thought-control," hallucinogenic drugs, bribed government officials, and now the seductive Metaverse. Here are more examples:

High-definition television is similarly seductive. It provides subliminal messaging that real life is bland compared to hyper-pixilated "living color."

Additional reinforcement comes through television messaging, 'programming,' and Hollywood movies. 'Models' therein cause you to compare and question your 'adequacy.' You thus question *your* reality.

This media also projects recurring themes that reality was moldable. Science fiction serves this purpose. Different dimensions of reality are presented, such as in the 2015 film, ***Inception***.

Likewise, ***Spiderman: Far from Home*** projects the same recurring theme. In time to save the world, Peter Parker realizes that all he is fighting is a series of illusions. The metaphor is most revealing about the global elite. Parker's illusions are broadcast by flying AI projectors. Extraterrestrial power-mongers control these.

Another ploy causing people to question reality, and accept false realities, is called "predictive programming." The 'Event 201' coronavirus pandemic conference[92] televised by MindWar producers exemplifies this ploy. Once reality is preceded and predicted by media broadcasting, it seems illusory when actual reality unfolds. People become sensitized and conditioned to accept the imposition. We become detached from the pain and suffering, thus docilely permitting it.

These media manipulations also condition compliance and acceptance of the 'Technotronic society,' predictively accepting Brave New World politics and policies.

As you have watched American cities burn, citizens riot, crimes skyrocket, and the economy plummet, please realize that you have been conditioned to accept this. Your mind has been programmed to accept your demise or replacement by machines, robots, transhumans, or AI. You accept the destruction of America and this threat to humanity brought by 'aliens' akin to demons.

Society requires conditioning to better accept these alternative realities. The global elite peddles wireless technologies that even disrupt metaphysical realities, even 'religious' doctrines, ethics, morals, and values. All are dulled by wireless impositions and drug intoxications.

Heralding this 'Technotronic era' to leading campaign-financiers of politicians as early as 1970, David Rockefeller's co-founder of The Trilateral Commission, Zbigniew Brzeziński, wrote in *Between Two Ages: America's Role in the Technetronic Era*. He said:

"The Technotronic era involves the gradual appearance of a more controlled society. Such a society would be dominated by an elite, unrestrained by traditional values. Soon it will be possible to assert almost continuous surveillance over every citizen and maintain up-to-date complete files containing even the most personal information about the citizen. These files will be subject to instantaneous retrieval by the authorities."

The COVID bioelectronic vaccine injections fulfill this "technotronic" prophecy.

According to skeptics and New World Order opponents, the Club of Rome commissioned Brzezinski, Jimmy Carters National Security Advisor, to write *The Technetronic Era*. That book predicted the "Post-industrial Age," today called "The 4th Industrial Revolution,"[277] heralded by Klaus Schwab. Zero growth and negative progress plague America today by the globalists' intent. Only degeneration is on tap for our "One Nation Under God." Crippling the U.S. economy and industrial output (i.e., the GNP) is intended to facilitate New World Order geopolitics and economics, and this agenda requires MindWar.

"In the Technotronic society, Brzezinski wrote, "the trend would seem to be towards the aggregation of the individual support of millions of uncoordinated citizens, easily within reach of magnetic and attractive personalities exploiting the latest communications techniques to manipulate emotions and control reason."

Drugs were known to facilitate this outcome, too, as the **CIA's declassified intelligence** makes known.

As far as I can tell, this objective of achieving widespread cognitive-behavioral compliance with the Deep State's globalization efforts further evolved with AI in mind. This was the basis of their military and social strategies.

MindWar is largely a defense strategy to defeat society's aversion to being controlled. Aversion to dictators, aversion to drugs, aversion to drug addictions, or aversion to being replaced by robots is a matter of social conditioning. It is all a 'dominant vs. submissive' game of MindWar.

The heart and soul of humanity had to be reconditioned and subverted to accept such alternative realities. Superior transhumans needed to be accepted by common folk to overtake the human race. This was and still is the objective of 'MindWar,' 'Transhumanism,' and the COVID Coup.

'MindWar' Games by Col. Michael Aquino

The U.S. military's chief intelligence officer who wrote the book 'MindWar' is **Col. Michael Aquino**. His social engineering methodology is evidenced herein.

Aquino was indicted, albeit never convicted, of serial assaults on children at the Presidio military base in San Francisco. He wrote for the 7th Psychological Operations Group, United States Army Reserve, from the Presidio in 1980. Aquino also produced the landmark advisory for the War College titled, ***From PSYOP to MindWar: The Psychology of Victory***.

Aquino and fellow PSYOPs officer Colonel Richard Sutter coined the term *MindWar* a few years earlier. After viewing the film *Star Wars*, the two military men applied the title MindWar to rebrand "the somewhat bland Army designation 'Psychological Operations.' This would appeal to futurists," Aquino wrote.

The "science-fictional treatment of MindWar, complete with a caricature of Sutter at its helm," appears in Aquino's Star Wars story *The Dark Side*, available at **www.xeper.org/maquino**. That link, including Sutter's caricature, is still active after nearly forty years.

Expanding on this theme of indoctrinating and bioengineering civilization to accept alternative realities and view realities alternatively, then recognizing this as an outcome most commonly praised by AI and hallucinogenic drug devotees, Aquino's futuristic MindWar is well represented in the growing 'rave culture.'

Much like the motivation to achieve an alternative and more peaceful, liberal, or progressive society, as described and prescribed by Sylvan in *Trance Formation*,[278] Aquino advanced the same for MindWar games.

"The advantage of MindWar is that it conducts wars in nonlethal, noninjurious, and nondestructive ways.

"Essentially you overwhelm your enemy with argument," Aquino explained. "You seize control of all of the means by which his government and populace process information to make up their minds, and you adjust it so that those minds are made up as you desire."

"Managed Democracy" and the "Century of Self"

Aquino's MindWar techniques evolved from the "concept of managed democracy." But that notion emerged a hundred years earlier, as reviewed in Britain's ***Independent***. Apparently, "the establishment was concerned [back then] about the outcome of elections," wrote **Youssef El-Gingihy**. Politics and society could be controlled behaviorally using the proper media technology.

During the 1920s, Edward Bernays emerged as one of the pivotal figures who helped develop the concept of *managed democracy*. "Bernays was the nephew of Sigmund Freud and the father of modern-day public relations. He understood that advertising could subliminally tap into the unconscious of repressed animal instincts, such as sexual desire. Bernays' clients included some of the largest US corporations," and the CIA.

Bernays served as an American propagandist in WWI. He applied his wartime experience and his uncle's theories of the *unconscious* to peacetime commerce. "He invented the field of public relations, popularized press releases and product tie-ins, and changed public opinion about matters ranging from women smoking[279] to the use of paper cups — all to increase sales," recalled Dr. Steven Reidbord for ***Psychology Today***.

"Viewing politics[280] as just another product to sell, Bernays also helped Calvin Coolidge stage one of the first overt media acts for a president, and helped engineer the 1954 coup in Guatemala on behalf of his client the United Fruit Company (i.e., the CIA), by painting their democratically elected leader as communist."

"Bernays transplanted the same techniques into politics as outlined in Adam Curtis's 2002 BBC documentary classic, ***The Century of the Self***, explained El-Gingihy. This political corruption "could be achieved through propaganda in various guises, such as the use of mass media to manipulate populations. The US military and CIA deployed psychological operations [ala Aquino et al.] (mass propaganda induced through emotions) across the world in various theatres ranging from the wars in Korea and Vietnam to Central America during the Cold War. Today, the only difference is that the internet has become the new playground for 21st-century psychological operations."

"There is a Policeman Inside All Our Heads [and] He Must be Destroyed," states the title of *The Century of the Self*'s third segment. Dr. Reidbord reviewed this production by writing, "[b]y the 1960s the '**human potential movement**' urged the expression of impulses instead of their repression. Business was eager to help. By marketing products as a means of self-expression, businesses turned from channeling public impulses to pandering to them.

"There is a fascinating discussion in the film about political activism being co-opted in this process," Dr. Reidbord added. "[M]aking the world a better place gave way to making oneself better in ways that, not coincidentally, required buying more goods and services."

In other words, the *Me Generation* was militarily engineered, actually *imposed,* to overtake traditional family, community, even religious values, and social 'consciousness.'

The human potential movement directed people's focus inward. Major decision-making was directed into *self-absorbance* rather than social service. Rather than applying one's 'spiritual connectedness' to serve others first-and-foremost, as traditional religions generally encouraged, spiritual connectedness in the human potential and New Age movements (and 'rave subculture') meant focusing inward to cope with personal problems and "stay centered" through the use of products and services, that is, commercial means.

Revealing 'skeletons in the closet,' early childhood traumas, perceived deprivations, and parental abuses —the common neuroticisms— were said to be keys to better health and happiness. All of this took the main stage in people's minds and commerce —the psychology profession, 'health coaching,' and more, advanced by *military design*. At the same time, Deep State covert financing by 'inside traders' secured this shift in society for corporate profit and consolidation of power.

The context of all of this, the entire institution of the Technotronic Era, and today's transition to Transhumanism, including the behavioral science ways and means to re-shape society; not forgetting the current liberalization of psychotropic drugs in commerce and medicine, was *necessarily advanced* to accommodate this *transformation.* Now leading this transition, the AI industry substantiates this promise of corporate fascism and control by Deep State appointees, virtual dictators in the 'Brave New World Order.'

This appears to be the purpose of bioelectronically-infused vaccinations.

Transforming Society Like a Monarch Butterfly in the CIA

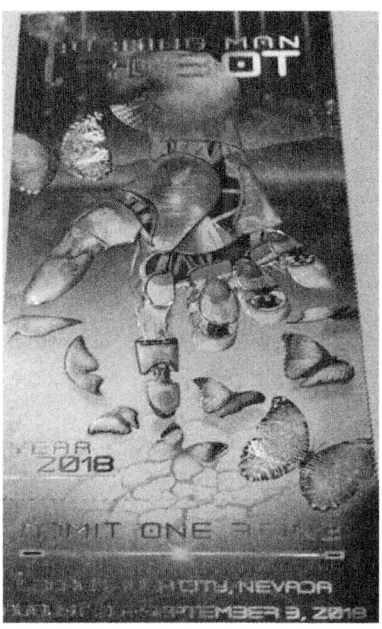

"Why did the 2018 'Burning Man' ticket have MKULTRA symbolism all over it?" asked several astute observers.

I further ask rhetorically, "Why did it present transhuman graphics and a spider web at the bottom appearing to entangle the colorful butterflies?"

The answers are apparent in the intelligence mentioned above.

Burning Man is a mind control cult that uses LSD, DMT, psilocybin, trauma, separation from family, peer pressure, attacks on religion, and utopian promises of transformation. In other words, it is a scam that 'susceptibles' fall for under the spell of techniques similarly used in the **"NXIVM"** cult as discussed below and MKULTRA (introduced above). Don't believe me?

Here's Burning Man founder "Danger Ranger" explaining it:

> "Some claim it's merely a coincidence that there are so many connections between Burning Man's history and the military/intelligence world, as I showed in exhaustively-cited depth in my YouTube series **Silicon Valley's Secret Weapon – The Shadow History of Burners**.
>
> "Is putting the Monarch butterfly on the ticket just yet another in an incredibly long (and mathematically impossible) series of coincidences? Or are they just throwing it in our faces now?"

The Dalai Lama, who recently **endorsed vaccinations** by intoxicating two infants in favor of the **cancer industry** and Deep State depopulation agenda, explained the butterfly in terms befitting this 'Me Generation,' *self-interested,* mind-controlled mentality. In fact, this activity is clearly Luciferian. Rather than carefree, the butterfly is uncaring and self-absorbed. Quoting the 14th Dalai Lama, Tenzin Gyatso, thusly:

> "**The butterfly never meets its mother.** It must survive independently and remains a stranger to affection. **An animal nurtured by mother's milk, however, is dependent on another for its basic survival. A child who grows up in a cold and detached home environment is similar to the butterfly, in that**

CONSOLIDATING GLOBAL CONTROL by MINDWAR

kindness is sparing. Once an adult, it will be very difficult for that person to show compassion."

Accordingly, consider this when reconciling how and why Deep State officials act as they do to the detriment and death of civilization as we know it. Radical 'self-reliance' is the socially divisive objective of MindWar games played by the Deep State, consistent with the alleged existence and operation of the Monarch Project and Satanic Church, as you will read below.

MindWar: The 6th Domain of Warfare

Now consider the "6th Domain of Warfare." This is not new to seasoned investigators. It's just pouring old wine into new bottles through military and corporate shell games. The advantage of this info is to provide more familiarity with the vocabulary the military uses to identify these basic principles...

Awaken to major 'dark side' mind manipulators. [281]

It has been fashionable in military circles to talk about *cyberspace* as a "Fifth Domain" for warfare. The first four domains are land, air, sea, and space.

But there's a sixth and arguably more important warfighting domain emerging: the human brain.

This new battlespace is not just about influencing the hearts and minds of people seeking information (e.g., standard 'propaganda'). It's about involuntarily penetrating, shaping, and coercing the mind in the ultimate realization of Clausewitz's definition of war[282]: compelling an adversary to submit to one's will. And the most powerful tool in this war is **brain-computer interface (BCI) technologies**, which connect the human brain to devices.

Here we enter the realm of *transhumanism*.

Military research agencies such as DARPA, advancing in places such as the Bill Gates and Jeffrey Epstein financed Media Labs at MIT[283], have advanced these BCI technologies. Similarly, elites are likewise investing at Johns Hopkins University, where hallucinogenic drugs (such as DMT/ayahuasca) and viral DNA are both being advanced to 'cure' illnesses such as PTSD and genetic diseases.

Officials and entrepreneurs now understand complex neural processes involved in enhanced "threat-detection" through, for example, "BCI scan" devices for "P300 responses."

We see, likewise, with the coronavirus bioweapon, the RNA virus spliced with HIV-1/AIDS causes brain alterations. Nerve cell pathology, blood dyscrasias, and respiratory damage deprive victims of oxygen and life.

It seems the marriage between neuroscience, genetic science, virology, and military bioweaponry is fundamentally altering not only the future of conflict. It is altering the nature of reality and civilization.

Cognitive Interference Technology

"Cognitive interference technology" is used to conduct attacks against adversaries' psychological well-being through lethal and non-lethal means. Light waves, electromagnetic waves, and microwaves can "cause psychological damage, confusion, and even hallucinations, changing the other's cognition and ultimately causing the enemy to act in violation of their own interests.

In China, as in the West, Artificial Intelligence (AI) has been developed and implemented to enable "intelligent public opinion guidance." (网络舆情智能引导, *wangluoyuqing zhinengyindao*) Software now automatically and adaptively generates content for coordinated online social media posts as part of MindWar technologies.

Broadly speaking, cognitive domain operations fall under the rubric of psychological warfare, which is itself a part of the PLA's concept of information operations.

Disinformation (虚假信息, *xujia xinxi*) has always been a component of the superpowers' agendas. In China, the People's Liberation Army's (PLA's) information warfare strategy for instance, is key. It features cognitive domain operations. Drawing from Nathan Beauchamp-Mustafaga's publication by The Jamestown Foundation Global Research & Analysis,[284] a growing part of war tactics includes "information operations and psychological warfare methods such as 'creating information chaos… implanting disinformation and erroneous information into the enemy's information system, and causing the enemy's command to make the wrong decisions and commands' in peacetime and wartime.

"Cognitive domain operations appear to be the key operational concept behind China's embrace of social media disinformation.

"According to a 2017 PLA Daily article by Zeng Huafeng of the National University of Defense Technology (NUDT), 'cognitive space' is defined as 'the area in which feelings, perception, understanding, beliefs, and values exist, and is the field of decision-making through reasoning.'

"This mind war arena includes 'leadership, morale, cohesion; training level and experience; situational awareness and public opinion.'

"Drawing from U.S. subversive psychological operations targeted against the Soviet Union during the Cold War, adversaries envision using 'information and popular spiritual and cultural products as weapons to influence people's psychology, will, attitude, behavior and even change the ideology, values, cultural traditions and social systems,' and ' target[ing] individuals, groups, countries, and even people around the world.'"

An excellent example of this in recent years is drug advertising that reshapes society to normalize deadly and damaging drug side effects or vaccine injuries such as autism. Opposition is denied as an 'unfounded' "conspiracy theory," and the damage is hashed as beneficial in expanding savant geniuses.

Zeng identified four tactics to win "mind superiority" in the cognitive space. As you read these, reflect on what you are witnessing nightly on the news: 1) "perception manipulation" through propaganda narratives; 2) "cutting off historical memory" so that targets will be open to new values; 3) "changing the paradigm of thinking" by targeting elites to change their ideology, and 4) "deconstructing symbols" to challenge national identity. [6] For Zeng, cognitive warfare is the ultimate form of winning without fighting.

Social Media as Information Operations Force Multiplier

According to the experts in MindWar, five social media functions are necessary for "full-spectrum social media campaigns: reconnaissance, hosting, placement, propagation, and saturation."

"Measures such as the Code of Practice on Disinformation, GDPR, and Social Media Privacy Protection and Consumer Rights Act"[285] are being debated but generally neglected by legislators.

"In the Cognitive War[286] – The Weapon is You!"

Cyber Warfare in Social Media Mind Control

Zak Doffman for *Forbes* (a Deep State property) described "Cyber Warfare"[287] and how the U.S. Army and other nations deploy 'Social Media Warfare' to fight truth and spread false narratives. Reflecting on *Brave New World* and *1984*, 'thought criminals' who openly opposed Big Pharma's drug and vaccine intoxications are opposed to using such cyber warfare tactics.

This best explains what happened to me beginning in 2008 when my book *Emerging Viruses: AIDS & Ebola—Nature, Accident or Intentional?* made headline news.[288] That put me on the CIA's 'radar screen.' I immediately began to be cyber-stalked, bot attacked,[289] and smeared online, especially by 'trolls' in social media,[290] and by so-called "skeptics"[291] and "fact-checkers."[292] Ultimately Google/YouTube[293] censored 157 of my videos. The CIA-edited *Wikipedia*[294] removed my entire bio from their "service," and Facebook[295] canceled my 'politically-incorrect' account.

My persecution and defamation increased most intensely after I produced the award-winning film starring Sherri, titled *UN-VAXXED: A Docu-commentary for Robert De Niro* (2016).[60] Therein, we exposed the McChrystal Group[296] of Deep State mercenaries protecting and profiting Big Pharma, deploying bots, trolls, AI data-mining, and public persuasion campaigns called "sentiment analysis," to dominate the social-engineering media.

Facebook and Google especially weaponized the counter-intelligence agents, agencies, and global elite against so-called 'enemies-of-state.' The graphic below shows the McChrystal Group functioning at the center therein. Mind you, Col. Stanley McChrystal Group's COVID assignments include pandemic response administration in major U.S. cities, including Boston —home to Harvard and MIT's nano-biotechnology labs, and Charles River Labs, where the new vaccines and DARPA's psychotropic drugs are tested. These include psilocybin and DMT—the new 'designer LSD' (a.k.a., "ayahuasca") avidly promoted and financed by the U.S. military's leading Silicon Valley data-miner, Peter Thiel.

Retired four-star general Stan McChrystal, director of the McChrystal Group meets with President Obama as the former commander of US and International Security Assistance Forces (ISAF) in Afghanistan. McChrystal now directs the main cell of agents directing "counter-insurgency" against anti-vaccinationists, including breast-feeding mothers.

"We have always seen that battle for hearts and minds in the physical sphere," *Forbes* concluded. "What we've started to see with news of cyberattacks on... command and control networks" is the same but more frequent.

In 2016, Eric Schmidt, the Executive Chairman of Google's parent Alphabet, partnered with Verily, another company named 'Frequency Therapeutics,'[297] and Pfizer's parent, GSK, announced the expansion of Google Ideas into a technology incubator called **Jigsaw**. The concealed purpose was to neutralize enemies of the Deep State. The 'front' claimed "[t]he team's mission," Schmidt wrote, was "to use technology to tackle the toughest geopolitical challenges, from countering violent extremism [such as White supremacy but not Black Lives Matter or Antifa] to thwarting [or administering] online censorship [of vaccine opponents], to mitigate the threats associated with digital attacks [except for those committed by perceived enemies of the National Security Crime Syndicate]."

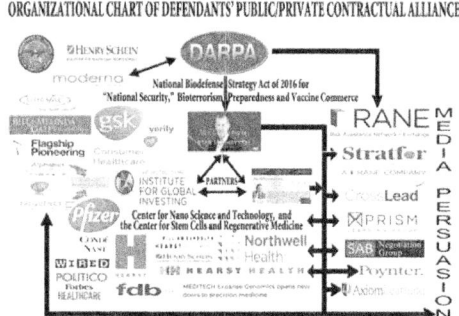

Within this campaign, in February 2017, Jigsaw and Google launched the free Perspective API.[298] This is their most advanced online censorship operation. They described it as "a new tool for web publishers to identify toxic comments that can undermine a civil exchange of ideas."

"Using machine learning technology," Google advertised, "Perspective[299] offers a score from zero to 100 on how similar new comments are to others previously identified as toxic."

Google defined *toxic* "as how likely a comment is [such as the power word 'fake' or 'fraud' or 'quack' or 'tin-foil-hat'] to make someone leave a conversation. That definition is capricious.

"Publishers can use Perspective in a number of ways," Google continued, "from offering readers instant feedback on the toxicity of their comments to giving readers the power to filter conversations based on the level of toxicity they'd like to see."

Jigsaw claimed, "its AI can immediately spit out an assessment of the phrase's 'toxicity' more accurately than any keyword blacklist, and faster than any human moderator."

Perspective's partners include the Mexican drug cartel propaganda arm, Carlos Slim's *The New York Times*. Other Deep State properties included in this censorship and MindWar operation include *The Guardian*, *The Economist*, and presumably *Vice*, *WIRED*, *Politico*, *Forbes*, *Wikipedi*a, and others.

Using this media censorship capability, the Deep State manipulates audiences' minds any way they choose. This results in output of only "fake news."

Gang Stalking Trolls Target Pharma Enemies

In this context, Google's fake fight[300] against online dissidents evidences the abuse of behavioral science and mass media propaganda for persuading people that the villains are heroes and heroes are villains. They did this to me beginning in 2008.

I was vilified personally, professionally, and commercially. Sherri was also targeted at that time by FBI informant and alleged CIA operative Paul J. Sulla, Jr. and his complicit agent, Alma C. Ott.

Sulla is a Hawaii lawyer, professional forger, and ayahuasca trafficker who allied with Ott —America's leading white supremacist and anti-Semitic radical right propagandist. The two men acted to damage us and steal our companies and properties.

Sulla's infamy and criminal activity extended to the top of the Democratic Party and Deep State operations in Ukraine and Russia. Sulla did to us by forging our mortgage to steal our Hawaii property. He also did to Burisma's leading competitor in Ukraine, aiding-and-abetting the Bidens' Deep State enterprise.[301]

Agents within their racketeering enterprise perform multiple tasks, often using multiple identities. Ott created an alias of "Dr. True Ott"[302] to conceal his identity and multiply his online operations. Ott launched the website "LabVirus.com" to directly divert traffic from my *Emerging Viruses* websites, articles, books, and films. Ott opened an e-commerce store competing against my HealthyWorldStore.com. Ott did all this as a subordinate to retired FBI Division 5 Chief and covert CIA operative Ted Gunderson.[303] With Sulla's complicity, this group, Gunderson and Ott's minions, directed a cell of online "trolls" under the FBI's and CIA's direction.

After "retiring" from the FBI, Gunderson went to work with the CIA. Both the FBI and CIA collaborate to neutralize undesirables. The National Security Crime Syndicate intelligence arms employ Internet gang stalkers to infiltrate discussion groups and post propaganda. They discredit, damage, censor, and neutralize their targets, including my voice in health science, natural medicine, and the vaccination-risk-awareness movement.

Gunderson died exposed and disgraced. Sherri Kane's legendary investigative journalism and activism vetted him[304] as the FBI's top agent protecting the nation's leading child sex trafficking ring.

Gunderson's subordinates included several trolls and teams of "gang stalkers." Sulla, Gunderson, and Ott held alliances with several pedophiles. Sulla's son, Jasun, was indicted for soliciting sex from minors online. Ott's similarly perverted web host, Alex McGowan Studer, defended pedophiles online and published videos with Gunderson, Ott, and allied sex predator, Michael Nuccitelli. Kane and I vetted these perverts as the "Troll Triad."[304]

That FBI/CIA-linked group produced a fake "sex tape" to smear us. The film was co-produced by Anthony J. Hilder —Gunderson's MKULTRA side-kick, joined by radical right counter-intelligence partner Deborah Tavares. Together, they *built* and deceived the "right-wing" "patriot movement" in America.

This group set up the Patriot Movement in America to be controlled. Then they abandoned it after Sherri and I exposed their devil-doing.[305] The FBI and CIA behavioral science experts used these agents to create, control, and eventually compromise the right-wing radicals in America used to foment racism, anti-Semitism, unrest, and acts of violence.

We witnessed such acts over and over again. We recorded a 'pattern-and-practice' of deadly deceit. This group often published important truths but always followed them by discrediting lies. That is the hallmark of COINTELPRO counter-intelligence. By intent, audiences were left bewildered by their mixed messages.

They tried and failed to destroy our reputations completely.

Gunderson acted under a CIA contract revealed to us by his former partner —heroic whistleblower Barbara Hartwell.[306] Gunderson attempted to deceive CIA psychological operations agent Hartwell about the child sex-trafficking network Gunderson and his team administered. Gunderson led this group under the auspices of being a retired FBI hero advocating for women devastated by their children's kidnappings.

Gunderson served directly under the cross-dressing psychopath, J. Edgar Hoover. Gunderson ran the FBI's Fifth District, infamous for targeting Black Panthers and activists. Gunderson's gang targeted Bobby Seal and Malcolm X. We were similarly targeted, as was Dr. Martin Luther King, Jr.

Gunderson's most infamous CIA sidekick was Col. Michael Aquino. Aquino is shown in the screenshot below. That image came from Gunderson and Aquino playing 'good cop, bad cop' in a Geraldo Rivera television interview.

Rivera neglected to report that Aquino produced the leading psychological warfare operations manual for the Army's War College. The whole program was a PSYOP—a 'psychological operation.'

Gunderson and Aquino, we also learned, were in pursuit of the most advanced MindWar technology. It was called the "PROMIS software." It is reasonable to presume this is like IBM's Watson computer—the world's most powerful AI operating at 5G 'real time' speed.

This PROMIS software was described by Gunderson and his cohorts as the **optimal population control technology**. That implies neuroscience applications like those described by doctors Morgan and Giodano at the beginning of this book. The CIA's MKULTRA program developers sought this powerful technology desperately. Investigative journalist Danny Casolaro was allegedly murdered on the verge of exposing this story. Casolaro threatened the international Deep State cabal he called "the Octopus." This involved the "Inslaw case." The PROMIS software owner accused the FBI and CIA of stealing his work from which PROMIS evolved.

The arm of the Octopus in America I have referred to as the "National Security Crime Syndicate." This group aids-and-abets global population control with financing from sex, drugs, arms, and human trafficking.

Evidence for this "conspiracy theory" abounds. For instance, among our chief detractors is "iPredator Troll"[304] Michael Nuccitelli. His advertisements, for instance, sell sex potions while his other blogs attract underage sex targets to his "security service." This is a classic example of mixing psychopathic commerce in sex and drugs with profitable cyber-bullying to neutralize opposition. For years, Nuccitelli harassed us and our friend and radio personality, Michael Vara.[307]

Google's Jigsaw hypocritically advertised, "For many people, online harassment," such as we experienced, "is a nuisance." Ironically, Google repeatedly enabled, defended, and protected Nuccitelli and Gunderson's group, libeling us continuously online.

Gunderson's subordinates launched the now dead link LabVirus.com in 2009. That was shortly after Obama's minister, Rev. Jeremiah Wright, brought me into the national spotlight by crediting me on prime time television for writing the *Emerging Viruses* book. The CIA/FBI/Google et al.'s agents smeared me and censored me online from that point forward.

That is "a sad reminder that the Internet contains some of the worst parts of humanity..." Google acknowledged.

Google's Jigsaw Propaganda

"Almost everyone has experienced how a single toxic [troll] voice can poison a conversation," Jigsaw's propagandists continued. "But what is sometimes described as cyber-bullying has become better organized, better funded, and often government-backed," as Kane and I were among the first to experience and expose.[308]

"Today's harassment campaigns are driven by many of the same political and ideological conflicts that exist in the physical world," Jigsaw advertisers added. Google allegedly helps "moderators sort comments more effectively or allow readers to more easily find relevant information." The opposite is actually true.

Relevant information, and so-called "misinformation," is only what the crime syndicate's secret agents and officials deem "relevant."

We first showed that Google's propaganda was hogwash and media manipulations severe by investigating, tracking, and recording Google's search engine data manipulations[309] over several years. Google fudged their data and programmed their search engines to damage and conceal my AIDS research especially.

Deep State officials directed Gunderson, Ott & Co., and Google to censor, smear, and 'neutralize' my lab-virus discoveries and vaccine-cancer research.

"We'll be releasing more machine learning models," Google noted. "[B]ut our first model identifies whether a comment could be perceived as 'toxic' to a discussion."

This was in the days long before today's rampant totalitarian censorship, 'anti-vaccination hesitancy campaigning,' and vaccine-delivered bioelectronic devices for melding minds with Google's iCloud.

"Over at Jigsaw, Alphabet's altruistic incubator," *PC Magazine*'s[310] propaganda added, "[t]he team has spent its time examining more technical forms of censorship, such as DNS poisoning with its Intra app[311] and DDoS attacks with Project Shield."[312]

All of these online surveillance technologies and command-and-control abilities are subject to direct 'intra-human' applications according to the dual use of military neuroscience being reported and the advanced nano-bioelectronic technologies being injected. This capability fulfills the primary objective of controlling populations vis-à-vis the PROMIS software. It also fulfills the Brave New World-controllers' objectives to influence and regulate your every move and mood.

"With Perspective, the goal is... using machine learning [AI] to determine what is or isn't against a given set of rules. [The] challenge is an intensely subjective one: classifying the emotional impact of language" to confront "problems like confirmation bias, groupthink,[313] and harassing behavior in an environment where technology has amplified their reach..."

Microsoft, the Healthcare Holy War, and the Medical Mafia

Organized religions and religious leaders have been under attack in favor of the 'gods of science,' and there are a growing number of anecdotal reports of COVID vaccine recipients no longer being able to "feel God" or the presence of the Holy Spirit immediately after their jabs.

Spirit is energy. Energy is 100% electrons resonating, vibrating, and spinning at certain *frequencies*. Frequencies have a profound impact on DNA, genetic expression, health, or illness. In medicine, nuclear magnetic resonance imaging (MRI) works because that is all true. Computer images enable the diagnosis of internal pathologies due to sound waves, magnetic fields, and electronic sensors.

The National Security Crime Syndicate knows this to be true too. It administers a religious "Holy War" in which Microsoft acts as a Goliath against opponents.

This Holy War is being waged on every American and people of spiritual and religious persuasion. This is the anticipated "End Times" battle over human consciousness for command over souls.

In medical patient care, healthcare providers' practices now rely on Microsoft's alliance with IBM,[314] Hearst,[315] and the McKesson cartel[316] that you probably never heard of. These leading IT corporations and their partners administer the Deep State's "Medical Mafia."

This observation is clearly evidenced in Hearst's September 2016 issue of *Popular Mechanics* magazine that targeted doctors branded "quacks" and "conspiracy theorists." The advancement of natural medicines, faith-based healing methods, and more honest holistic (multi-disciplinary) science have become increasingly taboo. We are to entrust genetic engineers with our lives and fates.

At this time in the MindWar, Hearst companies direct the flow of drugs, vaccines, and propaganda to the eyes and ears of doctors, patients, and parents across America to end religious exemptions from vaccinations for school children[317] and impose similar mandatory vaccinations on everyone.[318] The EU proposed this first for 2022[319] implementation, sparking widespread protests in European cities.

For Americans, two facts are most frightening:
1) Hearst largely controls how unwitting Americans think, especially about their own healthcare, including what drugs and vaccines doctors prescribe; and
2) Hearst and its partners monopolized American healthcare, using *intelligence software* integrating Microsoft technology. This is a vital part of the PROMIS-like population control scheme. This IT imposition has largely become the "standard of care" in medicine. This enterprise now regulates medical prescriptions that doctors are **forced** by that "standard" to prescribe or inject.

Acknowledging this truth and monopolistic power, the McKesson Company advertises itself as the "central nervous system of healthcare."

All of this is administered in violation of anti-monopoly laws, especially those opposing fixed and exorbitant prices of meds and vaccines profiting the racket.

Hearst wields the most powerful influence in the press and broadcast media. Hearst joined leading Democratic Party financier Barry Diller's IAC[320] in forming strategic partnerships.[321] IAC stands for InterActiveCorporation. Barry Diller is its Chairman and Senior Executive of IAC and the Expedia Group. Diller founded the Fox Broadcasting Company and USA Broadcasting. Diller is now largely responsible for the liberal media's favor for Democratic candidates.

Hearst, IAC, and complicit Big Pharma media (i.e., the PharmaMedia) are intertwined with Bill Gates's Microsoft Company and IBM. This Microsoft-tainted 'brain' controlling healthcare, health science, and politics features the McKesson Company's partnership with Hearst's <u>First Drug Databank</u> (FDB). This cartel now provides the standard software used by hospitals, pharmacies, and private practitioners.

Add to this cartel's power Hearst's and IAC's media propaganda boosting prescription drugs sales and vaccinations, all with toxic side-effects resulting in profitable illnesses.

Likewise, the MindWar response to COVID-19 is controlled by this <u>criminal enterprise that has monopolized medicine</u>, poisoned people's minds, bodies, and the environment; and concealed far more than its price-fixing criminal enterprise.

Least Trustworthy Companies Manufacture COVID Vaccines

News reports show federal prosecutors repeatedly fined McKesson[322] for bilking taxpayers as <u>America's main supplier of vaccines and drugs</u>.

This is similar to the Pfizer company that received a landmark fine[323] for false advertising. The company agreed to pay $2.3 billion, "the largest health care fraud settlement in the history of the Department of Justice," the FBI press release reported in 2009. That fine was a drop in the bucket compared to Pfizer's profits. But that settlement resolved "criminal and civil liability arising from the illegal promotion of certain pharmaceutical products," the Justice Department announced.

This pattern of drug-syndicate organized crime was recognized in 2001 when the Federal Trade Commission indicted Hearst for monopolizing healthcare's intelligence database[324] and conspiring with McKesson to fix extremely high prices on drugs and vaccines. Yet, no repentance came after that. The enterprise's FDB software still directs doctors and nurses to charge exorbitant fees for pharmaceuticals. The scheme heavily damaged the U.S. economy by ripping off patients, insurers, and taxpayers billions of dollars. Millions in fines previously levied by the U.S. Government made little difference.

And when we began to report on this global injustice, Hearst attacked me, my partner, and OxySilver™ with 528 in *Popular Mechanics* (**PM**).

The dark alliance waging Holy War, which is also MindWar, indicts the top shareholders in Microsoft, Hearst, ICA, McKesson, and their partners seeking The Great Global Reset using COVID as a PSYOP. Their plot enriches these anti-God drug companies and their allies, such as Proctor & Gamble, Merck,[325] and GlaxoSmithKlein.[326] It is painfully obvious that their "4th Industrial Revolution" featuring AI and military neuroscience injectable technologies advances a satanic agenda. I delve more into this topic of spiritual damage and religious fallout in Chapter XIII.

Big Pharma programming lulls people into submissive complacency as the drug/vaccine CULTure revises 'normalcy.' This globalist monopoly conditions people into thinking that living life suffering from damaging drug/vaccine side-effects is acceptable, not unconscionable. The "gods of science" have encouraged genocidal pandemics and mass murders. This is otherwise called "iatrogenocide" —physician-induced homicide by reckless, negligent intoxications from drugs and vaccines.

All of this is imposed under the guise of "good science," "modern medicine," and "public health." Their MindWar successfully secures their racket and civilization's degeneration. As God, Jesus, and their loving angels urged, "Come out[327] of her, my people, that ye be not partakers of her ᶜsins, and that ye receive not of her plagues." (Rev. 18:4)

The MindWar hoodwinked everyone, including devout religious people. The deceived targets place faith in those advancing organized crime versus traditional spiritual practices, including prayer, fasting, natural medicines, and 'hands-on-healing.'

Evidence proves this genocide is done willfully and knowingly. The enterprise commissioned, for instance, the McChrystal Group[328] to administer much of its MindWar. As briefly mentioned previously, retired Gen. Stanley McChrystal[329] now directs the COVID response in major U.S. cities. This military intelligence hit-squad coordinates dozens of agents and subordinate trolls and companies for online PSYOPS and consumer saturation, as you can read online.[330]

Using military tactics for "counter-insurgency," Big Pharma's and Big Banking's competitors (along with religious leaders opposed to vaccinations) are smeared, discredited, and censored.

Sherri and I discovered this covert operation linking the McChrystal Group to the MindWar and the Hearst-McKesson-ICA-NBC alliance by direct personal experience.

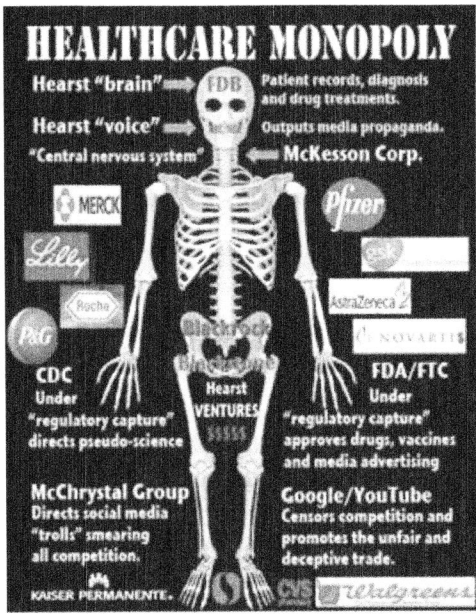

McChrystal's subordinate propagandist Colin McRoberts[331] and his PRISM LEARNING GROUP[328] began stalking and attacking us online. Their links to the Deep State's drug enterprise were discovered subsequently. This deadly demonic aggression prompted us to research and publish articles exposing these "PharmaWhores" on *Waronwethepeople.com.*

After experts and officials increasingly challenged vaccine and drug fallacies, a social backlash began. This uprising prompted Big Pharma to commission the McChrystal Group to preclude an international uprising.

Altogether, the Hearst-McKesson-McChrystal PharmaWhores and PharmaTrolls —protection racketeers—administered mass persuasion campaigns that "dumbed down" parents, teachers, youth, and America's 'body politic.' This 'market saturation' was required to maintain the social meme (i.e., belief system) that **presumes** vaccines are "safe and effective" or that hallucinogenic drug side effects are worth risking lives for the "experience of God" or to relieve post-traumatic-stress-disorders.

This disturbing saga can be viewed in the films ***VAXXED*** and ***UN-VAXXED***. My film, **UN-VAXXED**,[60] vets the McCrystal Group like none other. You see how McCrystal, and his backers, enslaved people[332] to Big Pharma's 'medical cult.'

As a result, diseases, including cancers, are increasing, not reducing. People's faith in medicine and government is degenerating, raising catastrophic costs to America's economy, military, leadership, and future generations.

Yet, most of our politicians bribed by campaign financiers remain silent.

Accordingly, critics contend we are witnessing the death of America. The once "Christian nation" has been intentionally defiled along with the Constitution and Constitutional freedoms. The rise of Communism, socialism or the Brave New World is apparent, and Big Pharma's poisonous deceptions are central to this evil.

Prophecy Being Fulfilled in the Brave New World

The above screenshot shows Hearst's home page (on 8-30-16). It blames "hackers" for U.S. National Security breaches and political election fraud. The side captions promote "ANCIENT ALIENS" and an Asian military threat in America.

Hearst, America's wealthiest and most influential propaganda mill (i.e., "yellow press"), owns and operates *Popular Mechanics*. That unassuming newsstand fixture is best known for advising readers about cars, technology, and gadgets for sports, adventures, and aerospace, not spiritual metaphysics challenged by vaccine nano-bioelectronics intertwined with religious intolerance.

Jealousy best describes the National Security Crime Syndicate that competes against:
- a) the world's religions that honor the Holy Spirit power of LOVE, faith, prayer, and healing "miracles;" that even secular science says is <u>far safer and more reliable than placebos without any risk or cost</u>;
- b) truth grasped by the awakening masses of intelligent consumers whose "word of mouth" and "social network marketing" obviously threatens the PharmaCULT— Big Pharma's enterprise and the medical paradigm; and
- c) the natural healing professions such as chiropractors, homeopathic doctors, acupuncturists, faith healers, and others that the crime syndicate has disparaged to conceal and control **effective alternatives** to deadly drugs and toxic vaccines.

CONSOLIDATING GLOBAL CONTROL by MINDWAR

For decades, this drug cartel has been increasingly capturing and controlling all branches of government, and their agencies, including the FDA, CDC, NIH, DEA, NSA, FBI, CIA, and DOD.

The MindWar CULT is insidious and seductive. The 'Medical Mafia' conducted smear campaigns against competitors, suppressing natural cures for nearly all diseases since the early 1900s. More than a hundred years of MindWar, population-programming, has resulted in stupid people. "Stupid is as stupid does." It is stupid to believe you should trust your life to companies repeatedly indicted and convicted of fraud.

If this cult has lied about and concealed the lab origin of COVID, what makes you think you can trust their vaccines to prevent COVID?

Gross censorship and MindWar extend to those injured and killed from COVID vaccines. The vaccine-injury VAERS data suppressions and misrepresentations are outrageous and unconscionable. This is another classic example of deadly malfeasance. Devil-doers hid, diverted from, and then 'normalized,' damaging vaccine side effects that have destroyed millions of lives. This fact conflicts with the overriding claim that the Pfizer, Moderna, J&J, or Merck vaccines are "safe and effective." Safety cannot be proven nor presumed when the damage data is hidden and neglected.

America's sickness, lethargy, poverty, and insecurity shames this once "Judeo-Christian nation."

Unbridled **treason** has astonished observers and whistleblowers.

Make no mistake, the Deep State, Big pharma, and Big Biotech monopoly, with Bill Gates at the helm pushing vaccinations and world depopulation, is administering this treasonous attack on consumers by hashing common sense —a main objective of MindWar.

A&E Television Networks -- a joint venture between the Disney-ABC Television Group, **Hearst Corporation** and NBC Universal -- announced today that it has an agreement to acquire Lifetime Entertainment ...
Thu, 27 Aug 2009

A&E Television Networks -- a joint venture between the Disney-ABC Television Group, Hearst Corporation and NBC Universal -- announced today that it has an agreement to acquire Lifetime Entertainment Services.

The officials and agents committing deceptive trade, and most effectively seducing and conditioning minds to become and act mindlessly, are reaping the benefits. For instance, the Hearst-ICA media, through its incestuous nepotism and partnerships with ABC/Disney and NBC Universal, expresses Hearst's stated mission: To "provide drug knowledge that helps healthcare professionals make precise medication-related decisions."

The Growing Risk to America and Elsewhere

When network TV ads tell viewers to "ask your doctor" about dozens of new drugs, each causing dangerous and often deadly side effects; or when child actors in commercials encourage parents to vaccinate, resulting in millions of people succumbing to auto-immune diseases, cancers, and more, you are viewing MindWar and genocide in the *Brave New World*.

As my co-authors and I made clear in *The Ayahuasca Death Cult,* this Cartel's MindWar PSYOPS created (and now controls) the psychotropic and hallucinogenic drug markets too. This *cult* competes directly against the religious world.

AI is now abused to control information, fads, and norms to mold society this way. It is now vogue to demean religious groups and people. This MindWar message is central to the Deep State's efforts to demonically engineer behaviors from birth to grave. Birth and life are cheapened by this CULT that finances infanticide. Even the sanctity of death is diminished as elders are separated from their children in impersonal institutions and hospices. This dehumanizing intelligence-degenerating cult has popularized 'Western medicine.' The gods of science are now replacing organized religions in civilization's transition to transhumanism.

The Growing Hallucinogenic Drug Cult and "Neuralink."

If these intrusions seem farfetched to you, review Chapter I again. If you can't imagine how vaccine hydrogels containing nano-bioelectronic devices dovetails with hallucinogenic drugs, here's your wake-up call from *Psychedelic Times*.[333]

Brain (neural plasticity) induced by hallucinogenic drugs, Ayahuasca, in particular, facilitates the merging of humans with AI. Once an mRNA "vaccine," or others containing a bioelectronic hydrogel, are injected into a human, the genome becomes genetically modified and subject to external ownership and wireless controls.

According to reasonable concerns, psychedelic experiences may facilitate brain-machine interfaces (known as BMI). Officials have secretly backed this psycho-neuropharmacology because it vicariously trains, conditions, or prepares human brains to cope with vaccine-administered BMIs. That is because psychedelic drug users generally seek to escape from reality and experience altered reality akin to Metaverse activities. According to Psychedelic Times, these are promoted with claims that users develop an increased level of "openness to experience," as shown in research at Johns Hopkins University.

Psychedelics also increase 'neuroplasticity'[334] advocates claim. That is, the ability to allegedly think more freely and make novel connections across brain regions, advocates argue, is advanced. This is especially important when learning to adapt or submit to 'novel' impositions.

Considering how challenging integrating a BMI may be, it is not unreasonable to consider using devices such as biochips to facilitate the brain-Cloud interface.

Neuralink[333] is one such early BMI device that consists of a small computer chip that gets implanted into the brain. Dozens, or eventually thousands of tiny electrodes, were precision-inserted into human brains by neurosurgical robots. Today, nano-bioelectronic devices injected with vaccines, thanks to the work of Langer and Lieber et al., provide this same capability.

In this context, researchers and Deep State officials reportedly prefer individuals who have been *conditioned* by psychedelic experiences to best cope with the 'neural-link' to the Cloud.

MindWar Against the Religious World

So the COVID Coup targets civilization to advance transhumanism and the brain-Cloud connection. That is a direct attack on civilization, and especially the religious world. People of faith in the Western World abide by the laws of Moses, not the gods of neuroscience and politics.

Recall CIA agent Dr. Stanley Krippner and his work in Project Stargate. I already mentioned that that "mind-control" program resulted in the Hollywood flop ***The Men Who Stare at Goats***. But that comedy concealed the neuroscience reality of "remote viewing," developing "higher consciousness," and expanding premonition that Krippner and his disciples sought, great for "super-soldiering."

Ayahuasca devotees worldwide, beginning with the CIA in Brazil,[335] Terrance and Dennis McKenna in America, the Sulla family[336] and their "religious" community in Hawaii, the Bronfman crime family,[337] and many others, now prescribe and/or traffic LSD, psilocybin, and DMT for "spiritual awakening," "psychedelic journeying," and "experiencing your 'inner self.'"

These brain-altering drugs and vaccines are ideal treatments for lost souls —naïve consumers are open to losing their souls for want of self-esteem, personal identity, and social acceptability.

This exploitation by agents of deadly deception intentionally targets weak persons with addictive personalities. This movement publicizes DMT as "**the God Molecule.**" This deceptive label conceals the drug's intended impact on weak minds, including those in the religious world. This is a religious world and devotee takeover. Krippner wrote the scholarly monograph, *The Future of Religion*,[338] to peddle this CIA-sourced substitute for righteousness.

This DMT con job leverages the 'power word' GOD to hook and persuade people to take this strong, dangerous, and sometimes deadly Class I narcotic hallucinogen.

Krippner, I recall from *The Ayahuasca Death Cult* book, directly attacked organized religions by writing, "The Religious Right has been remarkably successful because there was no other game in town."

So Krippner promoted the DMT "alternative" as an opportunity for the drug industry —legal and illegal— to embrace and expand this growing market.

Recall that I referenced the video lecture by Dr. James Giordano on *Battlescape Brain: Military and Intelligence Use of Neurocognitive Science*.[339] Therein we view this most esteemed neuroscientist, and ethicist speak of the 'super race' of transhumans forthcoming, courtesy of the DoD's investments with Big Tech, Big Biotech, and DMT/ayahuasca in this little-known MindWar program.

According to Giordano, "The mind is the seat of the self... There is great power there." The military's neurosciences "puts the brain in our fingertips."

Raising ethical concerns, Giordano questioned the "good" in perfecting the human race. "How do you define 'good,'" he asked? "Who gets the goodies in brain science? What happens if this technology runs away from us? We will ask, "What have we done?"

Justifying this industry nonetheless, Giordano explained, "Neuroscience is a global enterprise." It is a "'dual use' balancing act."

"Biopower and biopolitics" are competitive, and "competition is part of the human condition," Giordano added

"What is the essence of self," anyway—a "puzzle to be solved."

There is a widely known saying that "curiosity killed the cat." For religious people, faith adequately serves to satisfy neuroscientific curiosity. Exploring the brain's concept of "self" using bioelectronic devices and vaccine injections to link human brains to the Cloud is sacrilegious, which is why the globalists' media has been increasingly condemning religious groups and their alleged 'hypocrisy.' The Deep State prefers to place public health at risk by mandating vaccines containing aborted fetal tissues, foreign species parts, deadly antigens, and DNA corrupting genes.

Terrance McKenna joined CIA agent Krippner as America's leading gurus promoting hallucinogenic drug "journeys," especially using DMT in ayahuasca "tea" ceremonies. Terrance and the Sullas manufactured the drug in Hawaii. Krippner attended their "church." Sulla's family and real estate associates were the main traffickers of the "Daime tea" to the mainland U.S., presumably supplying the CIA and governmental researchers as well.

Before Terrance died of a brain tumor, he revealed that he had acted under the influence of the FBI[340] the entire time he was lecturing. This was part of a plea bargain. McKenna agreed to work for the feds to infiltrate and influence the New Age community to spread this new "Daime religion."

Krippner, the McKennas, Jeffrey Bronfman (in the infamous Bronfman crime family[341]), and Paul Sulla's clan similarly served the Deep State's scheme to replace organized religion with their hallucinogenic drug cult.

Likewise, the FBI's MindWar was well personified by Alma C. Ott's alias, "Dr. True Ott." Ott was Paul Sulla's lead witness in one of our court cases against him.

The cartel successfully, albeit falsely, claimed this drug cult was a renewed "old religion" protected by the Constitution. Consequently, following Jeffrey Bronfman's successful Supreme Court case, the cult expanded worldwide.

Top officials in the DEA, DOJ, FBI, and CIA enabled the explosive growth of this risk to the world's youth and religious community.

Meanwhile, there is a huge growing market of aimless unemployed susceptible citizens searching for meaning in life. This market will grow as more and more people die from various drugs, biological weapons, viral mutations, and vaccine side effects.

In effect, the Deep State demons behind the COVID Coup have been abusing MindWar techniques, drugs and vaccines, to destroy the 'fabric of society' and alter generally accepted ethics, morals, and values. According to these transhumanistic investments in military neuroscience, they are completing their quest to convert the religious world to their drug cult.

Links Between 'Magic Mushrooms" and the Brave New 'Promis'

Few people discern the intimate connections between hallucinogenic drugs and advancing brain-Cloud objectives in Big Tech's MindWar. One top psychedelic drug scientist put it this way, "The climate's looking good" for magic mushrooms and ayahuasca to turn into widely accepted prescription medicines.

His speech was made at a gathering of the world's billionaires,' reported Erin Brodwin.

Brodwin's background reveals a 'model' Big Tech propagandist. She claimed to be "a senior health and tech reporter at *Business Insider* based in San Francisco, where she focused on startups and Silicon Valley. She became one of the world's most vocal media prostitutes favoring the Deep State's Luciferian mind-control, transhumanism, and the AI psychedelic agendas.

After joining *Business Insider* in September 2014 as a 'science reporter,' Brodwin covered the Western African Ebola epidemic[342] — "the most widespread Ebola outbreak in history." She interviewed[343] "a

CDC contact tracer who spent months on the ground tracking the virus' spread." But she neglected 'EbolaGate'[344] and the Ebola lab virus intelligence published by yours truly that same year, available by the click of her mouse.

Brodwin, shown in the above image eating a 'New Age' 'cultured sausage,' wrote a series of articles for *Business Insider* after molting into a "Health-Tech Correspondent." She has occasionally co-authored with Alyson Shontell to heavily promote the "Ayahuasca Death Cult."[345]

Consistent with Dr. Giordano and Dr. Morgan's military neuroscience lectures, as early as January 2017, Brodwin declared the benefits of killing the human ego and "fundamentally transform[ing] the brain."[345] Therein she referenced as her source of valid intelligence the Rockefeller-controlled World Bank-influenced World Health Organization.

As already mentioned, the WHO and World Bank are heavily influenced by Bill Gates & Co., the World Economic Forum, and its founder Klaus Schwab.

Brodwin advocated for the ultimate PSYOP —the global MindWar. Her writing fit the most profitable MKULTRA mind-control and population control scheme perfectly.

This plot to enslave humanity effectively leveraged psychiatry and psychedelic "medicine."

Kicking off a major long-term publicity campaign, an actual cultural indoctrination program, Brodwin announced a YouTube video promoting "What magic mushrooms do to your brain and state of mind."[346]

Thereby, Brodwin became a leading Silicon Valley and Peter Thiel missionary; and her soapbox, *Business Insider,* became Silicon Valley's leading yellow press (i.e., propaganda mill).

She wrote:

"The researchers believe they are on the cusp of nothing less than a breakthrough: A single dose of psychedelic drugs appears to alleviate the symptoms of some of the *most common, perplexing, and tragic* illnesses of the brain.[347] Because depression is the leading cause of disability worldwide, the timing seems ideal."

Business Insider was founded in 2007 by Henry Blodget and Kevin P. Ryan. According to news reports, Blodget, the CEO, and Editor-In-Chief, hails from Yale. This is where the original Skull & Bones Fraternity continues its Satanic rituals, including mock murders.[348] (A few mock murders were videotaped exposing Past President George W. Bush for his participation therein.) Yale is also where newly discovered "emerging" viruses such as Zika and Ebola are stored.

Henry Blodget has a terrible reputation, perfect for his role in the National Security Crime Syndicate. Quoting the Deep State's *Wikipedia*, Blodget "worked on Wall Street until he was banned for life from the securities industry because he violated securities laws and subsequent civil trial, which ended with a $2 million fine plus a $2 million disgorgement and his permanent ban in 2003.

In 2013, Amazon and *Washington Post* owner Jeff Bezos made a substantial investment in *Business Insider*, which was bought out in 2018 by Axel Springer SE —Europe's largest digital publishing house with its own shady history.

Axel owns "numerous multimedia news brands, such as *Bild, Die Welt,* and *Fakt* and has more than 15,000 employees. It generated total revenues of about €3.3 billion in the financial year 2015. The presumed "left-leaning" publisher has been vetted by right-leaning investigators as a fraudulent press beholding, again, to the Rockefeller-Rothschild banking cartel.

This view of the *Business Insider* being a neo-Nazi wolf in Israeli-sheep's clothing is corroborated by the little-known history of the company for whom Brodwin blindly wrote. Before the close of WWII, Axel served the Nazis.

Tim Weiner's book, *Legacy of Ashes: The History of the CIA*,[349] details Axel Springer's service to Hilter & Co. as well as to the fledgling CIA. That transition occurred as Henry Kissinger, and General Bolling began exfiltrating Nazi scientists and business leaders to America during 'Project Paperclip.'[350]

Weinger recorded:

> "It was second nature for [Rockefeller Standard Oil Co. lawyer and CIA Director, Allen] Dulles to plant stories in the press. American newsrooms were dominated by veterans of the government's wartime propaganda branch, the Office of War Information, once part of Wild Bill Donovan's domain. The men who responded to the CIA's call included Henry Luce and his editors at Time, Life, and Fortune; popular magazines such as Parade, the Saturday Review, and Reader's Dogest; and a public-relations and propaganda machine that came to include more than fifty news organizations, a dozen publishing houses, and personal pledges of support from men such as Axel Springer, West Germany's most powerful press baron."

Accordingly, what follows is the pro ayahuasca propaganda output by Neo-Nazi Deep State ally, Axel Springer, penned by Brodwin.

"Psychedelics appear to have a unique ability to treat conditions that fail to respond to even the best current medicines," she reported.

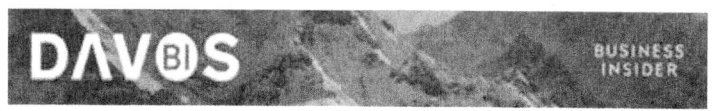

In August 2017, Brodwin wrote (for *Business Insider*)[351] about the 'party drugs' coming of age. They now showed 'commercial value,' being legally prescribed as 'medicine.' She later wrote in 2019 the following.

> "Other psychedelic and semi-psychedelic drugs are also on researchers' dockets, such as marijuana," Brodwin advertised for the exploding cannabis industry,

"which some argue has psychedelic properties, and ketamine, a partial psychedelic that could have uses in depression and addiction.

"Like any other drug, however, psychedelics can come with side effects that Carhart-Harris said we shouldn't necessarily ignore.

"For example, people who've been given psilocybin for anxiety often describe an anxiety-provoking experience during the treatment before they begin to feel the drug's therapeutic effects.

"Also, some people are not good candidates for psychedelic drugs, such as those with a family history of schizophrenia or bipolar disorder, Carhart-Harris added.

"If all continues as planned, experts say we should start to see the first legal uses of psychedelics as medicine within the next few years. On Wednesday, Carhart-Harris said he hoped to see something happen along those lines —and likely in severe depression— by 2024.

"It's progressing well," he said.

"The prospects are improving for turning psychedelics into approved medical treatments. "Speaking this week at the World Economic Forum's annual meeting in Davos, Switzerland, a leading neuroscientist said drugs like magic mushrooms and MDMA are moving closer to regulatory approval.

"If given the green light, the drugs could be used to treat a variety of mental health indications, including depression and PTSD.

"First, there were the cancer patients. In a handful of people diagnosed with advanced stages of the disease, a single dose of magic mushrooms appeared to quell their anxiety about death. Then there were the veterans, whose intrusive flashbacks of violence seemed to be quieted by therapy sessions that involved ecstasy. And recently, a group of people with depression appeared to find some relief in ayahuasca, a hallucinogenic brew that indigenous communities in South America have used for thousands of years.

"All that research reached the world stage this week at the World Economic Forum's annual meeting in Davos, Switzerland. There, a leading British scientist who studies the impact of psychedelics on the brain said things are looking up for psychedelics turning into approved treatments.

"'The climate's looking good,' Robin Carhart-Harris, the head of psychedelic research at the center for neuroscience and pharmacology at London's Imperial College, said during a Wednesday session at Davos titled "The New Science of Psychedelics." He spoke in an interview with Alyson Shontell, the editor in chief of Business Insider US.

"Research on psychedelics —a word that comes from the Greek roots "psyche," or soul, and "delos," or manifest— has been heating up in recent years. The drugs appear to have a unique ability to treat conditions that fail to respond to even the best current treatments. Oftentimes, all that's required to see those effects is a single dose, or "trip," in a supervised medical setting."[352]

The Most PROMISing Indoctrination for Thiel & Co.

On Oct. 3, 2018, Brodwin promoted Peter Thiel's "new biotech company" that raised $25 million to help unleash the "virgin market" of psychedelic research... Theil, for those unaware, is Silicon Valley's leading data-miner for the U.S. military and intelligence community.[353]

That fact is chilling since the 'end-game' of transhumanism relies on brain-Cloud concordance largely financed by the military, enabling whole person *data-mining.*

As mentioned, neuroscientists believe hallucinogenic drug "trips" help condition the mind to accept alternative realities, as required in the human-to-transhuman conversion.

Called Atai Life Sciences, Theil's company owns a large stake in the company called Compass Pathways. Continuing to quote Brodwin:

"German entrepreneur and investor Christian Angermayer founded Atai. Angermayer also said he planned to invest in research focused on drugs that could help fight aging and extend life.

"In June, an under-the-radar startup backed by Silicon Valley tech mogul Peter Thiel made enough psilocybin — the active ingredient in magic mushrooms— to send 20,000 people on a trip. It was part of a larger effort by the company, called Compass Pathways, to study how psychedelic drugs could be used to treat depression.

"It was just the beginning.

"On Wednesday, German entrepreneur and Compass investor Christian Angermayer launched a new startup: a biotech company focused on financing more of the kind of research that Compass is doing. Called Atai Life Sciences, the company has already raised $25 million from investors like ex-hedge fund manager Mike Novogratz and Icelandic entrepreneur Thor Bjorgolfsson. Atai also owns roughly 25% of Compass, which Thiel has funded.

"Atai's goal is to help fund more studies that explore the therapeutic potential of psychedelic drugs like psilocybin and others on mental illnesses such as depression. In the past, non-profit efforts at doing this kind of research have run into many obstacles, from a lack of funding to a difficulty enrolling enough patients for their trials. Angermayer hopes Atai can help address those issues, telling Business Insider he believes Atai will help unleash a "virgin market" for the research.

"Alex Tew and Michael Action Smith, founders of the popular meditation app Calm, also invested in Atai. Former National Institutes of Health director Tom Insel, who previously served as an advisor to Compass, will stay in that role."

Compass also raised an additional $33 million as part of their subsequent funding round, bringing its total to more than $38 million in seed money.

"Compass is already the world's leading producer of psilocybin for research, Angermayer told Business Insider. Earlier this summer, the company received regulatory approval to begin one of the first large studies looking at the effect of psilocybin on treatment-resistant depression, a severe form of the illness that does not respond to other medications. Compass also filed for a patent on a form of the drug that it makes in a lab.

"In addition to its work on drugs for mental health, Atai will also fund studies of treatments designed to fight aging and extend life, Angermayer said. To do so, the company is partnering with German-based Innoplexus, which uses AI to develop drugs.

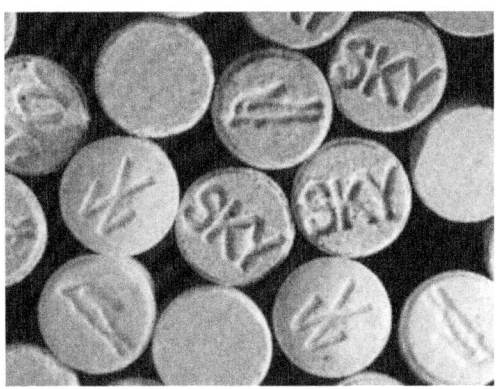

"Reuter reported that 'DEA Psilocybin has become a promising candidate for anxiety and depression treatment because it appears to disrupt the sorts of engrained brain activity patterns that are the hallmark of those diseases.'"

This is how the "medicine" is being sold. One recent study looked at the compound's potential to help alleviate anxiety in cancer patients; others have looked at psilocybin's potential effect on depression, PTSD, and alcoholism.

"Compass Pathway's study, which got FDA approval in August, looks at the effect of three different doses of psilocybin (1 mg, 10 mg, and 25 mg) on treatment-resistant depression. A 'standard' dose of dry magic mushrooms is roughly 2 grams, or about 20 mg pure psilocybin, according to nonprofit educational organization Erowid. The clinical trial involves 216 people enrolled across several research sites in Europe and North America.

"The magic mushroom isn't the only psychedelic drug getting renewed attention. There's been a steady trickle of scientific research on psychedelic drugs' potential therapeutic benefits for at least the last five years.

"A study in 2017 indicated that ecstasy could help veterans cope with PTSD symptoms; one in 2012 hinted that ketamine might curb major depression. That spate of research finally seems to be leading to the development of promising potential treatments that could get government approval.

"David Nutt, the former chief drug advisor for the British government and a current advisor to Compass Pathways, is optimistic about the federal approval process. He told Business Insider last year that he expects to see psilocybin approved as a treatment for depression by 2027."

Capitalism comes to psychedelics?

In July 2018, Brodwin wrote that Peter Thiel's Compass Pathways churned out "20,000 doses of magic mushrooms, and is making more."

In November 2018, Brodwin advised us not to miss Silicon Valley's entry into "the hottest area of healthcare." She explained what it looks like for new startups aiming to "disrupt the $35 billion addiction market."

In December 2018, Brodwin and *Business Insider* heralded more about the "virgin market" in psychedelics:

> "Magic mushrooms for depression and ecstasy for trauma. Today, two psychedelic drug candidates are leading the way: psilocybin, the active ingredient in magic mushrooms, and MDMA, also known as ecstasy.
>
> "Researchers are particularly interested in how psilocybin appears to quell the symptoms of severe

depression — especially versions of the disease that fail to respond to as many as a half-dozen other top-line treatments. Similarly, they're fascinated by how MDMA — in the context of talk therapy — seems to help treat post-traumatic stress disorder brought on by an acute experience of violence, like in war, or tied to any other cause. [354]

In May 2019, Brodwin advertised Denver as the "first city in the US to decriminalize magic mushrooms." She explained in a most positive way, omitting the baggage, "what they do to your body and mind."[355] "It's high time for psychedelic science," Brodwin promoted a month later.[356]

The investment is coming in. A new crop of interested parties has arrived on the psychedelic scene: for-profit groups. She continued her infomercial:

"Nonprofit groups have largely blazed the psychedelic trail for the past few years. The Multidisciplinary Association for Psychedelic Studies, for example, has raised more than $70 million for research on MDMA, marijuana, LSD, and ayahuasca.

"Last summer, a startup backed by Silicon Valley tech mogul Peter Thiel churned out enough of the active ingredient in magic mushrooms to send 20,000 people on a psychedelic trip — part of a larger research effort by the company, called Compass Pathways, to study how psychedelic drugs could be used to treat depression.

"Then in November, a German entrepreneur launched a new company called Atai Life Sciences with $25 million to back more studies that explore the therapeutic potential of psychedelic drugs on psychiatric disease.

"The investment is coming in, and that's interesting that it's commerce leading the way in terms of being visionary," Carhart-Harris said."

"So far, Compass claims it has made two 250-gram batches of psilocybin, the equivalent of 20,000 doses of 25 mg of the drug. Although some of that will be tested for stability, the rest has been shaped into capsules. Those will soon be shipped to a handful of sites in Europe and North America, where Compass plans to use the psilocybin in clinical trials.

"'We're building on the significant work that has already been done in this area, by gathering evidence in a larger population to see whether psilocybin therapy could provide a breakthrough to help patients,' the company said in a statement."

MIT, IBM and Fulfilling the Original PROMIS

I disclaim this section by reason of what is publicly known: Public information and commercial advertisements are always nearly 50 years behind the military inventions and secret applications of the same technologies.

MIT is represented below by *The Daily Beast* showing Deepak Chopra and Elton John's AIDS Foundation among Jeffrey Epstein's investors in transhumanist neuroscience and social engineering.[357]

Among the most advanced sources of neuroscience intelligence for AI, industrialists are the MIT Media Lab. This lab, closely linked to Robert Langer's Lab, appears to operate central to the MindWar and transhumanistic neuroscience. Moderna's chief entrepreneur and fundraiser, Robert Langer, presumably accepted Jeffrey Epstein's, Bill Gates's, and DARPA's money.

Why was Epstein interested in investing in MIT's Media Lab? Likewise, why was Bill Gates—a compulsive liar, global vaccinationist, and financier of "spirit cooker" Maria Abramovic[358] —be interested in investing with Epstein in MIT's most advanced neuroscience? Add Leon Black[359] to this group of Epstein-linked financiers.

Leon Black controls the private equity firm called Apollo Global Management. Like George Soros, James Murren, and Michael Chertoff's Group, Black and Apollo made vast fortunes from the "Las Vegas Deep State Massacre" and presumably the 9/11 World Trade Center disaster as well. I explain below why I conclude these 'suspects' are key officials in the Deep State.

To answer the above questions, students at MIT had published a study proving that aluminum, used for crude cranial shields and computer casings, actually amplifies energy broadcasts at certain frequencies.[360]

Applicable to small amounts of aluminum in traditional vaccines,[361] aluminum is used as an 'adjuvant' to increase immune responses. High levels of aluminum are also found in hotly contested "chem-trails"[362] that Bill Gates justifies spraying to "help block the sun"[363] He claimed this would reduce 'global warming.' Officials and students at MIT used sophisticated equipment to measure the impacts of non-ionizing electromagnetic radiation on people's brains with and without aluminum shields that increased certain frequencies of energy.

As early as 2012, MIT's News Office announced their Media Lab had precisely engineered "3-D brain tissues." Valuable to AI investors and manufacturers, Anne Trafton summarized what researchers borrowed "from microfabrication techniques used in the semiconductor industry." That was the industry in which Peter Thiel's father, Klaus Peter Thiel, had served at IBM in Germany for IBM's semiconductor developments.

MIT and Harvard Medical School (HMS) engineers worked together to "create three-dimensional brain tissues in a lab dish," wrote Trafton for MIT. This advanced neuroscience was predicted to impact every aspect of human life.

In July 2019, Anthony Cuthbertson wrote for *The Independent* that Gates's:

> "Microsoft has invested $1 billion in the Elon Musk[364]-founded artificial intelligence[365] venture that plans to mimic the human brain using computers.
>
> "OpenAI said the investment would go towards its efforts of building artificial general intelligence

(AGI) that can rival and surpass the cognitive capabilities of humans.

"The creation of AGI will be the most important technological development in human history, with the potential to shape the trajectory of humanity," said OpenAI CEO Sam Altman.

"Our mission is to ensure that AGI technology benefits all of humanity, and we're working with Microsoft to build the supercomputing foundation on which we'll build AGI."

"The two firms will jointly build AI supercomputing technologies, which OpenAI plans to commercialize through Microsoft and its Azure cloud computing business.

"We want AGI to work with people to solve currently intractable multi-disciplinary problems, including global challenges such as climate change, affordable and high-quality healthcare, and personalized education."[366]

OpenAI and the Transhumanist MindWar

"OpenAI also claims its technology will ultimately provide everyone with the economic freedom to pursue whatever they find most fulfilling while creating 'new opportunities for all our lives that are unimaginable today.'"

Alternatively, "since co-founding OpenAI, Mr. Musk has since stepped back from the AI startup, [and has remained] vocal about the risks artificial intelligence poses to humanity, claiming its development poses a greater risk than nuclear weapons.

"I never saw [any] miracle of science that didn't go from a blessing to a curse. I never saw no military solution, that didn't always end up as something worse," Sting wrote and performed in *Ten Summoner's Tales.*"

Chapter III's disturbing facts about MIT and Harvard, Charles Lieber and Robert Langer's labs, now suspiciously intertwines with the future of the human race, posing "a greater [extinction] risk than nuclear weapons." Pay attention to the fact that Jeffrey Epstein was a member of the Mind, Brain, and Behavior Committee at Harvard. If that doesn't raise red flags for you, nothing will.

Epstein was the common denominator among several of the world's most powerful elite and culturally influential people and organizations advancing transhumanism and the Fourth Industrial Revolution. Epstein was a former member of the Trilateral Commission, the Council on Foreign Relations, the New York Academy of Science, and a board member at Rockefeller University. Up to his reported "death," he was a member of the Edge Foundation, Inc.

These powerful geopolitical and economic connections lend credibility to Sherri Kane's publications exposing Epstein's links to child sex traffickers at the highest levels of government, but also to the saturation of perverted Deep State devil-doers in the "Octopus" of world trade subverting normalcy in the "Brave New World."

Clearly, the Musk/Gates support for "OpenAI" largely resulted from MIT's and Harvard's advancements in this field of AGI neuroscience, also financed by Epstein's group of investors.

"The new technique yield[ed] tissue constructs that closely mimic the cellular composition of those in the living brain," MIT's press release explained.[367] This allowed "scientists to study how neurons form connections and to predict how cells from individual patients might respond to different drugs."

This profitable mind-controlling society-altering technology intertwines with the Multidisciplinary Association for Psychedelic

Studies (MAPS). MAPS is the oldest and most active organization engaging this field of drug-altered brain function.[368]

MAPS propaganda has consistently concealed the multi-national corporate underworld financing transhumanist neuroscience and the "dark enlightenment" movement. This Luciferian agenda relies on the connections cited above to military-neurobiology and nano-bioelectronic research and developments. These include the mRNA vaccines' nano-metallic data-mining devices enabling the brain-Cloud connection and cyborg capabilities.[369]

"First defined in the US in the 1960s in a context in which cybernetic[370] thinking was being applied to the space race and Cold War concerns, the cyborg participated in wider cultural and political trends by combining individualism with control, and helping to erode fixed categories between human and nonhuman," reported behavioral science scholars Law and Moser.[371]

The Usona Institute is also noteworthy here.[372] Officials at Usona focus on psilocybin and other "consciousness-expanding medicines" supported by chief military data-miner, Theil.

Also noteworthy is Promega —the manufacturer of **Luciferase**. The Promega company produces enzymes and other products for the biotechnology industry. Their provisions are used in genomics, protein analysis and expression, cellular analysis, drug discovery, and genetic identity. It would be most difficult to accomplish the brain-Cloud connection using nano-bioelectronics without Promega's provisions.

Promega supposedly developed Luciferase to 'illuminate' life, their officials claimed. Now, impositions on spirituality, challenging your 'soul connection,' and altered 'consciousness' are being bioengineered in favor of Lucifer(ase).

OpenAI robots are being programmed and manufactured to be workers vicariously made slaves to the criminal psychopaths overtaking and converting civilization.

The absurdity and hypocrisy in these investments are apparent. Investors claim to be solving humanity's worst problems with computers because the problems are too complex for humans to solve.

Officials neglect the fact that most of the world's problems would be solved by freeing-up energy using Nicola Tesla's technology and promoting the suppressed fields of electro-genetics, frequency-therapeutics, and energy-medicine. Diversions from these solutions are deadly. Officials falsely claim these remedies are not feasible. That assertion is ridiculous in our age of advanced science that can read human auras and make 3-D brain tissue in a petri-dish.

Undoubtedly, the problems we face are rigged. Solutions have been suppressed to commercialize the world's problems. Crisis capitalism rules. Instead, we could be shifting paradigms into collaborative humanitarianism and remedial actions exercising 'higher consciousness.'

Instead of lifting people out of poverty and ignorance, the MindWar captures attention and degenerates thinking. Virtual zombies are mass-produced by video gaming, iPhone addition, and distressing television.

In this demonic context, MAPS advertised "*TRANCE FORMATION*," facilitated by "Ecstatic Dance," using special music frequencies combined with hallucinogenic drugs.[373]

Musical Frequencies and the MindWar

According to extensive research in this field, you can shun the toxic drugs and simply enjoy music tuned to a 'higher frequency' to encourage higher consciousness.

Alternatively, frequencies can be enslaving and dis-easing. Senior Big Pharma scientists have great respect for frequencies that can be used in healthcare, brain research, and new product developments. As reported by Pramanik et al. in *IEEE Access* journal, "Piezoelectric/acoustic biosensors" are now measuring biomolecular interactions according to variations in frequency oscillations moving through "piezoelectric crystals." Since these biosensors use sound vibrations, they are also called "**acoustic biosensors**."[374] The acoustic biosensors are presumably incorporated into the nano-bioelectronic hydrogel matrix accompanying the mRNA vaccines.

My research shows that the Deep State heavily invested in two frequencies —741Hz and 432Hz. Their propaganda especially promoted 432Hz frequency and the secreted 741Hz.[375] The Deep State also suppressed 528Hz frequency intelligence.[376] This was consistent with the undermining of 'Divine-human communion' and the heart-soul connection.

Advancing MindWar, suppressing souls, and stifling consciousness, sounds much like the PROMIS software and Luciferian control over society.[377] In this context, the global elite has leveraged the music industry through mainly the NBC/Universal Music Group—a near-monopoly over the music industry.[378]

Deep State planners already knew what I had discovered about the spiritual/religious "Solfeggio frequencies" following my 1998 publication of *Healing Codes for the Biological Apoaclypse* prompted by co-author Dr. Joseph Puleo. They knew that the 432Hz frequency stimulates mainly "sexual chakra energy," consistent with their interest in sex and drugs. This seems appropriate for sexual psychopaths but not particularly beneficial for society's healing as a whole.

Alternatively, the global elite neglected, suppressed, and smeared the 528Hz frequency[27] associated with the "heart chakra" and *LOVE: The Real da Vinci Code*. This suppression was much like they did with the "free energy" technologies that Nikola Tesla invented.

528Hz is central to nature and the original Solfeggio musical scale. 528 was the "Miracle" note (C-5 on keyboards) sequestered by the Western World's religions. Apparently, religious officials did not want to raise the 'good vibration' of people's spirits using 528Hz.

That 'good vibration' is reflected in the 528nm frequency of light at the heart of rainbows. The greenish color of life-giving, oxygen-generating, environment-restoring chlorophyll is curative for what ails humanity[379] physically, spiritually, and environmentally.

Favoring the Deep State, Utkan Demirci, an assistant professor in the Harvard-MIT Division of Health Sciences and Technology (HST), said, "We think that by bringing this kind of [neuroscience and frequency-energy] control and manipulation into neurobiology, we can investigate many different directions."

One such study proved that 528Hz frequency alone increased natural immunity substantially, as measured by a 100% increase in antioxidant activity.[380] That frequency also protected brain cells against death caused by alcohol poisoning by 20%.

Big Pharma's gods of science attacked and censored me for outputting this intelligence, as they did to orthodox Jews and 'conscious' rabbis heralding opposition to mandatory 'jabs.' It was ancient Levitical priests and their scribes who encoded the Bible with the original Solfeggio musical frequencies Dr. Puleo credited Jesus for revealing.

Chelsea Clinton's corporate affiliates, identified below, suppressed, smeared, and 'neutralized' the best news that spiritual and religious people have received since Jesus. The 'modern gospel' actually features the science of 'electro-genetics and 528hz frequency biophysics. This includes the good vibration central to sunshine, rainbows, human biofields, chlorophyll, life-giving oxygen, even miraculous healings. These 528 facts are revolutionizing medicine, natural healing, and the music industry.

Why censor all this great news?
Why favor Big Pharma and global genocide?
Google, YouTube, Facebook, Hearst Media, Vimeo, and Barry Diller's ICA media conglomerate precisely censored my works. They converted my intellectual property to their own use and abuse.

Chelsea Clinton's associates at Vivendi/Capital Records/NBC Universal Music Group re-tuned their instruments and voices per my instruction for albums but gave me no credit for implementing my advice. This included Jay Z's *4:44*[381] album and the 2018 work by Beyoncé titled *Everything is Love.*[382] U2/Bono's 'Iris Hold Me Close' in *Songs of Innocence* is similarly recorded in 528Hz. All of this is based on my published intelligence. My pioneering videos inspired these productions and many others featuring 528Hz tuning.

Yet, this same crime syndicate censored my writings and smeared my videos.

MindWar and 'Herd Immunity' in the Brave New World

You may choose to remain hopeful about these facts, but you can no longer deny them. This Brave New World *control capability* was described by Demirci and Boyden. They are brain engineering and cognitive sciences experts at MIT's Media Lab and McGovern Institute. These senior authors described their new technique for manufacturing brain tissues in a lab dish. They published it online in the journal *Advanced Materials*.[383]

Now it can be seen that this military-industrial intelligence and AGI movement embodies more than the intentions of Aldus Huxley's "Harvard Psilocybin Project" (HPP). That HPP brought the academic and cultural dilemmas to Timothy Leary and Richard Alpert under the direct supervision of founding board members Huxley, Leary, Alpert, and their senior research scholar, the renowned social engineer and motivational psychologist, David McClelland.

The fertile soil for expanding research and commerce in the religious world of psychopharmacology now integrates the great PROMIS for data-mining and social-engineering through transhumanism —mass robotic behavioral programming integrating and overwhelming 'consciousness' through 'OpenAI' neuroscience.

The brand "OpenAI" sends a deceptive message. It is a "ClosedAI" company monopolizing the field for wealthy stockholders. Their enterprise falsely claims openness while restricting humanity's options. This industry sacrifices free will, self-regulation, healthy socialization, and self-mastery. This suppression is done for the sake of military neuroscience, android consciousness, and sexy robots.

In other words, the IBM-MIT-Harvard commercial/scientific community, featuring Gates-Black-Epstein-Musk-Theil, the Maxwells and Bronfmans,' and Silicon Valley cohorts, have advanced the modern-day equivalent of the Marsh Chapel Experiment. That Harvard hallucinogenic drug study impacted a small religious group of students. Today, with ayahuasca, psilocybin, LSD, and other newly manufactured, permitted, and promoted drugs fogging human consciousness while advancing AGI, Huxley's original project has morphed into the PROMIS of controlling billions of people worldwide under the guise of 'therapy' and 'spirituality.'

In another MIT Media Lab study, researchers developed 'herd mentality' capabilities. That would certainly terminate "vaccine hesitancy" and generate "herd immunity." They manufactured social control "therapeutic" technologies. In their 2015 study of a "Crowdsourced tool for depression,[384] officials developed an adjunct to ayahuasca group psychotherapy. This could help direct civilization into the oligarch's One World Religion.

Paralleling this new military product line and consumer market for psychotherapeutic drugs for anxiety, PTSD, and group behavior disorders, Larry Hardesty announced from MIT's News Office that researchers had developed an "extrinsic regulation" program. Here, society would treat itself using the Internet and social media. This 'therapy' would encourage 'herd immunity.' Citizens would immunize sufferers against stress and distress by going online to others who would engage sufferers in online de-conditioning and de-sensitization therapies.

Later, MIT officials announced partnering with Bill Gates's seed-funder, IBM, to advance the greatest PROMIS of all time. In MIT's "Quest for [a monopoly over] Intelligence," Kim Martineau wrote for IBM, they would produce the "fastest supercomputer on Earth." It was named "Summit." It could "run the calculation-intensive models that power modern artificial intelligence" and super-learning through superconducting data.

IBM's "Satori" Exceeds PROMIS Expectations

IBM named the super-computing superconducting cluster Satori after a Zen Buddhist term for "sudden enlightenment."[385]

Satori's "dark enlightenment" reflected qualities and capabilities demonstrated by "HAL 9000" —the supercomputer character that starred in *2001: A Space Odyssey*. In that 1968 film, HAL (**H**euristically Programmed **AL**gorithmic Computer) was a sentient computer (with artificial general intelligence "AGI") that controlled all the systems of the *Discovery One* spacecraft and interacted with the ship's crew maliciously.

"Physically the size of a shipping container, Satori is intellectually closer to a Ferrari, capable of zipping through 2 quadrillion calculations per second," MIT's press writers continued. "That's the equivalent of each person on Earth performing more than 10 million multiplication problems each second for an entire year, making Satori nimble enough to join the middle ranks of the world's 500 fastest computers.[386] Satori's capability exceeds the PROMIS expectations for a super-computer controlling civilization.

"Rapid progress in AI has fueled a relentless demand for computing power to train more elaborate models on ever-larger datasets," MIT-Satori officials remarked.

At the same time, other MIT officials complained that federal funding for academic computing had dried up. This created the incentive to accept the gifts and commercial projects offered them. This was where Jeffrey Epstein's 'clients' fit in, including Bill Gates and Leon Black.

"We'd like to be able to build models that can see, hear and touch," one IBM-MIT official said about advancing "bigger scale" AI, related or not to Luciferian pedophilia and sexual perversion endemic to their community.

Given the cultural indoctrination methods and materials referenced above, future generations will adapt drug rituals and exhibit drug dependence, normalize sex crimes, and enjoy pedophilia to gain happiness and anxiety relief. They will climax with androids forthcoming from IBM-MIT's Media Labs. Robots will see, hear, and touch you (the consumer) anyway and anywhere you want, for a price, so long as you buy into their "higher consciousness."

This will all be administered wirelessly, with super memory, plus automatic video recording and data storing to replay your highest, most climactic, and ecstatic experiences. All with Big Brother viewing, permitted, or not. That was, after all, Epstein's and the Maxwells' forte, wasn't it? Videography.

Ronan Farrow closed his article for *The New Yorker* by stating that "[Leon] Black declined to comment" on his extended relationship with Epstein, Gates, and MIT. "A source close to him said that he did not intend for the [Apollo Management] donation administered through Epstein to be anonymous."

Black, whose fund veered the 911 reconstruction money from NYC to Las Vegas through MGM Grand, as known to James Murren and Michael Chertoff; then secreted, Black downplayed his relationship with Epstein. Black described his Epstein relationship as limited and focused on tax strategy, estate planning, and philanthropic advice. Black declined to answer questions about further business dealings with Epstein that suggested a closer relationship.[387]

Farrow continued, "Although the MIT Media lab ultimately secured $7.5 million from Gates and Black, Epstein and Ito's fund-raising plan failed to reach the still larger scale than they had initially hoped. Epstein had suggested that he could ensure that any donations he solicited, including those from Gates and Black, would be matched by the John Templeton Foundation."

The Templeton Foundation funds projects precisely at the intersection of faith and science —the main concern for transhumanists.

Ronan Farrow did not disclose that the John Templeton Foundation maintained its own demonic interests. The organization had faced an Arkansas court challenge by their 2010 Templeton Awardee, Francisco J. Ayala. This evolutionary geneticist and molecular biologist vigorously opposed the entanglement of science and religion. Although wisely calling for mutual respect between the two theologies, Ayala's bias was clear. The National Academy of Sciences asked him to serve as the principal author of *Science, Evolution, and Creationism*. This was a definite refutation of all math-based, frequency-influenced, and quantum field associated science disfavoring creationism and so-called intelligent design. It was by omissions and red-herring diversions, anti-religion.

Darwin's Gift to Science and Religion

In 2007, Ayala wrote *Darwin's Gift to Science and Religion* (National Academies Press and Joseph Henry Press). It was promoted as "a broad review of the proper context of science and religion in modern society." The book reinforced the "evolution of the species and survival of the fittest" theme but neglected substantial science in the fields of matrix mathematics, electro-genetics, cymatics, and bioacoustics science intertwined with traditional concepts of spirituality in religions.

Besides this, the concept of evolution of the species as it applies to humans evolving from apes is farfetched, in my opinion. If this were the case, then Anglos, who lost their brown melanin pigment and developed blue eyes (increasing risks of damage from solar radiation), would make white people more susceptible to cancers and blindness. That is not the case.[388] Ayala's writings document evolution for degeneration. That's ungodly and unrighteous, given the homeostatic balance present in the universe.

Who outputs this nonsense? Ayala promotions for *Am I a Monkey? Six Big Questions about Evolution* said it was to be published by Johns Hopkins University Press. But then it wasn't. Instead, *Am I a Monkey?* was published by the National Academies Press (and Joseph Henry Press).

Why the switch?

According to their advertisements, the National Academies Press (NAP) was created to publish the reports issued by the National Academies of Sciences—National Research Council, and other national academies that foremost serve the banking elite's financial interests beginning with the Rockefeller family.

As detailed in Chapter I, the National Academy of Science-National Research Council (NAS-NRC) advanced newly-developed genetic biotechnology capable of producing "novel" bioweapons. Depopulating microbes developed were descriptively and functionally identical to HIV/AIDS and Ebola. Now, this same institution co-sponsored hallucinogenic drug neuroscience in support of the McKennas' Heffter Research Institute, intertwined with Chelsea Clinton's 'spiritual' organization at NYU.

The Clinton Foundation's claimed mission is supposedly to help lesser-developed countries cope with HIV/AIDS. According to Barry Diller's (IAC) federal filing, Chelsea Clinton worked as Co-Chair of an Advisory Board overseeing the "**Of Many Institutes**."[389] That entity was "devoted to educating and inspiring religious and spiritual leaders to utilize multi-faith dialogue and service as a force for positive social change." Meanwhile, Clinton Foundation funding is sourced from organized crime and the World Bank.

Is this "evolution of the species," as Ayala posited? Or is this "survival of the fittest" among the criminally deranged monopolists controlling the energy industry, healthcare, neuroscience, and human "consciousness" now overtaking religions?

Recognizing the globalist agenda, "In 1994, President Bill Clinton appointed Ayala to the U.S. President's Committee of Advisors on Science and Technology. While president of the American Association for the Advancement of Science (AAAS) from 1993 to 1996, Ayala developed the AAAS's "Dialogue on Science, Ethics, and Religion." He was then made a member of the National Academy of Sciences and a foreign member of the scientific academies of, among other countries, Russia, Spain, Italy, Mexico, and Serbia.

The lesson to be learned from this section is that the public esteem granted "science" and academic achievement has been morbidly misplaced.

In this era of "fake news" outpouring from all corporate-controlled television networks and media cartels, it is inconceivable that Deep State vaccination advocates would not attempt to defraud religious and conscientious objectors.

For every action, there is a reaction. Karma is inescapable.

Trauma-based Conditioning and Projecting Psychotic Realities

Connecting the dots here, this book exposes the depth and width of a global fascist conspiracy advancing under the guises of "science," "public health," "advanced intelligence," and "higher learning."

The identified devil-doers knew they would need to infiltrate and overtake the religious, spiritual, medical, and psychotherapeutic communities to consummate this Brave New PROMISE of the ruling elite. That is, they would need to "capture" and "monopolize" their "markets."

Their mission required enticing humanity with neuro-stimulation, overwhelming the mental circuitry, to impose the 'Dark Enlightenment.' That way, once control over the 'global village' was maximized, they could profit in every way possible from their dictates.

The ruling elite began by monopolizing the flow of currency —Big Banking and private equity investing. Then it was Big Energy and the arms trades that secured power. Then medicine, food supplies, the media, and computer science, all captured for social control. AI, transhumanism, and robotic conversions depended on this insidious takeover. During civilization's 'evolution' into this PROMIS-land of unethical and immoral intellectual decay, data-mining for behavior-engineering and social control was quintessential.

The suspects vetted above, whose companies, agencies, and institutions aided and abetted this Deep State racketeering enterprise, now continue to act as criminal psychopaths. This diagnosis is most reasonable after considering their commonalities and families' dark histories. From this, we see their common pattern-and-practice of misbehavior. Their actions reflect trauma and abuse that they suffered and are now projecting and imposing upon civilization.

This is not just "survival of the fittest" and "evolution of the species." This is global fascism imposed by psycho-sociopaths.

Trauma and abuse establish the impetus for their MindWar. Only this explains their callous, deadly, and damaging actions, especially taken against children. They were traumatized and abused as children, so their 'angst' projects their damage upon little boys and girls. Pedophilia is their compulsive 'therapy.' Mandatory COVID vaccines targeting children are scientifically-unjustifiable, nevertheless imposed, malicious, and genocidal.

In this context, vaccinations play a crucial role. Highly touted and widely trusted vaccines are a form of abuse and punishment. 'Injecticide' features blood poisoning, a twisted form of blood sacrifice, and child abuse. Misrepresented as 'immunizations,' these sickening and distressing intoxications are "required" 27 times by age 2,[390] with up to six shots in a single visit. Not one has been proven "safe" in lieu of missing, neglected, and concealed "adverse events data."

To do this to children is as unconscionable and morally repugnant as Peter Thiel's alleged lust to inject young people's blood into himself to stay young forever.[391] The Deep State's *Business Insider* reported this. The "neo-Nazi wolf in Israeli-sheep's clothing" credited *Vanity Fair* for this intel. True or not, the notion is beyond sick.[392]

The cryptocracy's apparent psychotic role-playing cannot extinguish their soul-damning conditioning. The childhood traumas that sourced their mental illnesses taint their propaganda mills, like *Vanity Fair*, the *New York Times, NY Post, Wall Street Journal, Washington Post, Business Insider, Vice, Gawker,* and others. It is disappointing to witness their MindWar lust for more power, money, sex slavery, and even mass-murdering for 'therapy,' adding meaning in their sick lives. When they get depressed and anxious, and their emptiness eats at them from their ice-cold hearts, their sick and sorry state reflects the images they see in their bathroom mirrors.

Misery loves company. They style your suffering to make their misery less stinging.

Making their misery profitable, they encourage you to share their addictions for 'emotional freedom' using drugs, normalizing sexual perversion and predation, enjoying multi-media escapism like fleeting orgasms in their 'Megaverse.' They entice you similarly to remedy their psychopathology.

Isn't this fundamental to the MindWar, the COVID Coup, demonic possession, to overtake and degenerate America? Isn't this reflected, for example, in the "trans" and WOKE movements?

How would you like to live in a body you believe was 'God's mistake'? Self-esteem and self-worth rot from this core belief.

Rather than fixing what is broken, elite perverts direct their media to spread their angst and normalize their sickness, re-conditioning society to accept what is degenerative and demonic, and then suiciding civilization to extinction for transhumanism.

CHAPTER VII

THE COVID COUP WITH FINANCING FROM HELL

The *New York Times* is widely known for publishing 'fake news.' Here I draw from a propaganda piece used to control damage and counterintelligence regarding Jeffrey Epstein's disappearance and associations in the world of COVID-19 vaccine neuroscience. This *NYTimes* representative propaganda was printed on Oct. 13, 2019 (Section A, Page 1). The article was titled: "Gates Met With Epstein Many Times, Despite His Criminal Past."

This chapter compounds already overwhelming evidence of media malfeasance indicting leading criminal psychopaths controlling the Coup's financing. These suspects are Big Pharma's partners, if not controllers. These are leading 'Deep State' officials commercializing AI while merging the neuroscience of 'consciousness' into their product lines and covert military operations.

These suspects and 'persons of interest' promote vaccines described as "genetic therapies" that impact immunity, neurology, and brain function. At the same time, they promote psychotropic drugs (e.g., 'entheogens'), especially DMT (a.k.a., ayahuasca), LSD, and psilocybin. These hallucinogens are now being researched, mass-produced, and distributed to compete in 'religious commerce.' Religious leaders and groups are targeted because they generally adhere to ethical and moral doctrines that compete against the secular acceptance of the 'gods of science,' religious objection to mandatory vaccinations, and Luciferianism. Religious people are strong impediments to the transhumanist agenda.

Below I examine the financing of propaganda used to reshape minds and secure the conspirators by revealing their 'cover' in so-called 'humanitarian' foundations and institutions. These include the most prestigious academic institutions such as MIT, Harvard, and others already exposed for their covert COVID commerce. Here I evidence financing by Bill Gates aligned with Jeffrey Epstein's cohorts in crimes against children and humanity. These devil-doers' actions are financially-intertwined with the sex and drug trades, including child sex trafficking networks.

You may recall from headline news reports of Jeffrey Epstein's connections to Edgar Bronfman, Jr.[211] of the Bronfman crime family, and Leon Black[393] —the director of the Apollo Management Group, and the Clintons. These associations are proven by examining Epstein's little "*Black Book.*"[394] Inscribed within are major players and financiers in the 'Deep State' that Sherri Kane and I first reported in 2018.

We praised the **Miami Herald**'s outstanding investigative reporting exposing Epstein's pedophilia. Sherri subsequently examined and reported what was generally overlooked — links to the PizzaGate scandal involving Washington, DC political insiders. We also praised Whitney Webb's detailed reporting on the "Mega Group, Maxwells and Mossad: The Spy Story at the Heart of the Jeffrey Epstein Scandal."[395]

This intelligence is vital to examining COVID Coup finances and administration in the globalist context. Without this information, the analysis of evidence recorded in the previous chapters would be replete since this 'Big Picture' involves Dr. Fauci's complicity with financiers. Harvard's Dean Daley, Charles Lieber, MIT's Robert Langer, enjoyed Jeffrey Epstein's group financing. Money also came from Bill Gates, Chinese investors, the DOD, and the CIA's In-Q-Tel, all financing the global genocide and presumably bribing and extorting politicians and other notables to remain silent or divert from the conspiracy, enabling the Coup to unfold. Many of the suspects boarded the "Lolita Express" to Epstein's Island.

Extending the World of Pedophilia and Cancer Advocacy

There were billionaires such Leslie Wexner and Leon Black, heavyweight politicians such as Bill Clinton and Bill Richardson, Nobel laureates Murray Gell-Mann and Frank Wilczek, and even royals such as Prince Andrew, implicated in sexual crimes and financial fraud.

Few, though, compared in prestige and power to the world's second-richest man, Bill Gates. And unlike many others, Gates started his relationship with Epstein *after* the pedophile was convicted of sex crimes.

Gates, Microsoft's co-founder, whose $100 billion-plus fortune has endowed the world's largest alleged 'baby killer' —the "charitable organization" called The Bill & Melinda Gates Foundation— did his best to minimize his connections to Epstein. The *New York Times* article reviewed here is one such 'damage control.'

"I didn't have any business relationship or friendship with him," Gates falsely defended to Rupert Murdoch's *Wall Street Journal*. That's as believable as Epstein's reported "suicide."

According to common sense and substantial evidence, including the *New York Post*'s[396] obviously photoshopped series of fake photos, Epstein did not commit suicide in prison.[397]

We observed that the *New York Times*'s coverage of the Gates-Epstein links said nothing about vaccines —the "sacred cow of public health." Carlos Slim's yellow press said nothing of Gates's passion for financing vaccine developments and distributing bioelectronic jabs, especially in lesser-developed countries targeted most for depopulation. The articles said nothing of Gates's compulsion to depopulate the planet; nothing about the "Final Solution" or financing labs to alter viruses and insects to accomplish humanity's conversion to transhumanism presumably. Here again was 'persuasive silence'—evidence of reckless concealments, fraud by omissions, and deadly misrepresentations.

Vaccines, after all, are the most profitable 'protection racket' the Deep State administers, albeit laden with deadly ingredients causing costly side effects. The neglected and denied disastrous results of vaccinations are the primary methods of culling and controlling populations—sickening and enslaving people to drugs. 'Healthcare' has been 'captured' and converted to serve as a form of 'disease control' and population control. Doctors offer few cures and little to no prevention. This is by design and intent. The suspects are not stupid. They have the money to buy intelligence. The sickness they generate is the Deep State's most beneficial investment.

At Jeffrey Epstein's Manhattan mansion in 2011, from left: James E. Staley, at the time a senior JPMorgan executive; former Treasury Secretary Lawrence Summers; Mr. Epstein; Bill Gates, Microsoft's co-founder; and Boris Nikolic, who was the Bill and Melinda Gates Foundation's science adviser.

I raise these facts and concerns in opposition to the media's neglect or superficial coverage of these topics. I object to this Gates-Epstein 'damage control' administered, for instance, in the *NYTimes* whose owner holds substantial conflicting interests in the Deep State allied Mexican mafia-tied billionaire, Carlos Slim.

As I previously reported,[398] *NYTimes* owner Slim is heavily invested in BlackRock, administered by ISIS innovations' and Evercore officials, including Evercore's Senior Managing Director and Chairman of Europe Investment Banking, Bernard Taylor, co-Vice Chair for the Rockefeller JPMorgan Chase bank.

I initially revealed these connections pursuant to the suspicious Zika virus outbreaks in Brazil and Hawaii.[196] The genetically-engineered mosquito remedy heavily promoted by Bill Gates,[399] the world's leading vaccine enthusiast, raised my suspicions. There was no doubt in my mind that Gates's GM mosquitoes had been considered for use as vaccine injectors besides disease vectors.

These GM mosquitoes were released in Brazilian trials, as similarly proposed for Hawaii and Florida.[400]

The Microsoft and Silicon Valley AI connection to this globalist, fascist, psychopathic pro-vaccination, depopulation, GMO menace is readily apparent upon close examination.

Oxford's Deep State Oxitec Spin-off

Genetically engineered Zika viruses and GM mosquitoes were profitable developments for the Deep State. Delving further into the Oxford for-profit spin-off company called Oxitec, this company was of high interest to population controllers, vaccine developers, and disease investors, notably Bill Gates, the Bill & Melinda Gates Foundation, and the Clintons.

Recall that Oxitec developed and reported testing mutant mosquitoes as an anti-dengue experiment in Brazil — a country largely under the influence of the CIA.[401]

But this technology held far more value as a vehicle for depopulation. Covert operators could spread genetically engineered viruses through genetically engineered mosquitoes.

Web intelligence showed Oxitec was a subsidiary of Intrexon.[197] That company was largely funded by the *New York Times*. Consequently, Carlos Slim's conflicting interests were revealed by Kane and yours truly.

Slim was not the only co-conspirator implicated. Many media investors held a substantial interest in the companies that profit most from GM mosquitoes and the Zika fright.

According to company records, Oxitec manufactured the GM insects tested in ayahuasca country —Brazil— in 2014. Oxitec, ISIS Innovations, and Intrexon were for-profit arms of Oxford Univ. that were also heavily financed by BlackRock —one of the world's most powerful private equity investment companies backing Goldman Sachs, George Soros, and Barry Diller's ICA Democratic Party allied media conglomerate.

In addition, the Evercore Company, a private equity investment firm intertwined with the French Lagardere enterprise, joined the above suspects in the Zika racket.

NY Times owner Slim held conflicting interests in diverting the Zika virus mystery away from solid science, as I explained in articles and on a video[196] that I produced for the Hawaii County Council. Members were saddled with the decision to permit Oxitec to release its mosquitoes on the Big Island of Hawaii, where Sherri and I were living at that time.

Oxitec's owner, Intrexon,[402] was stewarded by Chairman and CEO Randal J. Kirk, who was also on the Board of Directors of ZIOPHARM Oncology, Inc.,[403] also largely financed by BlackRock, Third Security, LLC, and Morgan Stanley.[404]

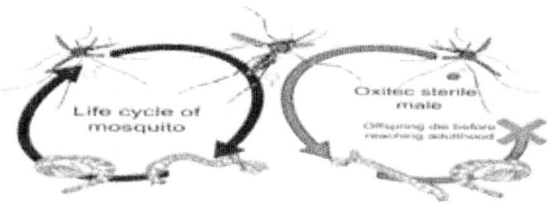

Chelsea Clinton's Public Health Degree from Oxford University

Wikipedia makes known Chelsea Clinton's 2003 132-page master's degree thesis was titled, *The Global Fund to Fight AIDS, TB, and Malaria: A Response to Global Threats, a Part of a Global Future.* The publication is not readily available on the Internet, which, it appears, justifies *Wikipedia*'s lacking link to the manuscript.

Chelsea Clinton apparently doesn't want people to read it because she didn't even have the professional courtesy and integrity to register as a science contributor on ResearchGate, a main online scientific publications provider. As you can see by the screenshot below, Clinton is not active in the scientific research community but is made to appear that she is in the mainstream media.

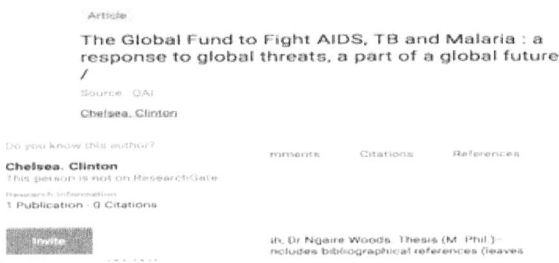

Clinton's work was "supervised by Jennifer Welsh and Ngaire Woods," *Wikipedia* reported. But that raised a 'red flag' because neither Welsh nor Woods, like Clinton, had any medical training.

Thus, how would the politically connected Clinton develop such a topic for her master's thesis? Surely Welsh and Woods did not assign it.

It turns out the assignment came from banking officials, as detailed below and evidenced by the screenshot heralding the 2017 *Lancet* single publication by Clinton and Sridhar.

The Clinton Oxford Deep State Connection

Please allow me a brief diversion to consider Oxford's role in profit-making in alliance with the leading Deep State suspects invested in manufacturing the COVID-19 virus, crisis, 'plandemic,' and the "novel" vaccines.

As aforementioned, several insiders have links to Oxford. Recall from Chapter III that Sir Jeremy "James" Farrar directed the infamous "Teleconference" on Sunday, February 2, 2020. Farrar was a professor of tropical medicine at the University of Oxford. Then he became director of the Wellcome Trust. He wrote Fauci frantically concerned about the need for the insiders to neutralize the "Indian paper" that exposed the AIDS virus genes in the COVID virus S-protein. Farrar urged lying (i.e., 'prevaricating') and coordinating with the WHO's directors, Tedros and Bernhard, falsifying the bioweapon's gain-of-function scenario.

THE LANCET

Health Policy

Who pays for cooperation in global health? A comparative analysis of WHO, the World Bank, the Global Fund to Fight HIV/AIDS, Tuberculosis and Malaria, and Gavi, the Vaccine Alliance

Chelsea Clinton DPhil [a], Devi Sridhar DPhil [b] ✉

Show more

https://doi.org/10.1016/S0140-6736(16)32402-3

Get rights and content

Summary

In this report we assess who pays for cooperation in global health through an analysis of the financial flows of WHO, the World Bank, the Global Fund to Fight HIV/AIDS, TB and Malaria, and Gavi, the Vaccine Alliance. The past few decades have seen the consolidation of influence in the disproportionate roles the USA, UK, and the Bill & Melinda Gates Foundation have had in financing three of these four institutions. Current financing flows in all four case study institutions allow donors to finance and deliver assistance in ways that they can more closely control and monitor at every stage. We highlight three major trends in global health governance more broadly that relate to this development: towards more discretionary funding and away from core or longer-term funding; towards defined multi-stakeholder governance and away from traditional government-centred representation and decision-making; and towards narrower mandates or problem-focused vertical initiatives and away from broader systemic goals.

Recall also the Obama/Biden appointee, Sylvia Mary Burwell, their leading AIDS, and Ebola scientist and bureaucrat who oversaw the HHS's and CDC's responses to the 2014 Ebola and later Zika outbreaks. Recall that Burwell, Fauci, and Nancy Sullivan selected Metabiota, tied to the Biden-Daszak-EcoHealth Alliance group that led the Ebola response in Africa with Dr. Garry's colleagues at Tulane screwing up the effort and killing people.

Burwell is connected to both Oxford and Harvard. She earned her bachelor's degree in Government from Harvard and a BA in politics and economics from Oxford, where she became a Rhodes Scholar like Bill Clinton. She then served the Clinton administration as Chief of Staff to the Secretary of the Treasury. She joined the Bill and Melinda Gates Foundation in 2001 as president of the Gates Global Development. She and Fauci's subordinate, Sullivan, also hailed from Harvard. Together, Burwell, Sullivan, and Fauci backed Oxford's efforts to develop their Ebola vaccine. You can also view the evidence of Sullivan's work with lead coronavirus gain-of-function developer Ralph Baric here: T32 AI007151/AI/NIAID NIH HHS/United States[266] and here too: F32 AI152296/AI/NIAID NIH HHS/United States.[267]

I previously discovered and reported that Oxford University (OU) established a **for-profit investment arm** in the business of spinning-off companies to enrich investors. The CIA's In-Q-Tel entity duplicated this. In other words, whenever any OU, Harvard, or MIT researcher discovered anything of high commercial value, the insiders, all Deep State players, converted this intellectual property from the non-profit sector (a presumably 'humanitarian' enterprise) to private for-profit companies enriching their stockholders and inside traders.

Sherri discovered Chelsea Clinton's inspiration and direction for her public health master's thesis was the Rockefeller-directed World Bank. The screenshot below provides evidence of the World Bank's sponsorship of Chelsea's Ph.D. thesis.

Chelsea, purportedly, did her master's research in 2003. Only a year before, the World Bank began financing this same "AIDS, TB and Malaria" effort, primarily in Africa. Here's the link to the proof.[405] The media gave the false impression that Chelsea had conceived of her master's thesis. This falsehood was especially promoted by the ICA conglomerate controlled by Chelsea's boss, leading Democratic Party financier Barry Diller. The following screenshots evidence this 'fraudulent concealment.'

It is standard academic practice for master's students to inform their department chairperson what *they* are interested in doing. Not having a topic imposed upon them.

In the first screenshot below taken from the World Bank's publication, "Population and Reproductive Health" is listed directly below HIV/AIDS. The Clinton Foundation, wherein Chelsea also serves, focuses on HIV/AIDS and vaccinations in the poorest countries. Population control through tainted vaccinations and family-planning clinics pushing jabs assure profitable disease induction and depopulation. Clinton Foundation services to the World Bank are mainly administered in abortion clinics in favor of "reproductive health" directed by the Negative Population Growth, Inc., and the Population Council that licenses "RU486"—the abortion pill. Are you for 'eighty-sixing' life?

Pro-choice advocates decree their right to rule over their bodies. Yet, their liberal majority also favors mandatory vaccinations with experimental bioweapons. Wrap your mind around that hypocrisy to fathom the conspiracy.

Table F-11. World Bank: Project Commitments by Health Theme (US$ millions, fiscal years)

Theme	2003	2004	2005	2006	2007	2008	2009	2010	2011	Total	Share
Health system performance	575.2	598.5	483.6	520.3	747.0	461.5	1,387.9	3,234.3	1,383.2	9,398.5	43%
Child health	232.1	410.4	202.2	200.1	390.7	158.7	625.7	147.8	329.0	2,644.9	12%
HIV/AIDS	325.2	210.3	243.0	87.2	313.7	50.8	218.3	127.4	152.4	1,728.2	8%
Population & reproductive health	198.7	296.3	194.2	135.8	342.6	79.0	82.2	149.5	242.4	1,728.8	8%
Injuries & non-communicable diseases	156.6	214.7	330.9	157.8	477.8	17.5	43.4	55.8	148.4	1,745.8	8%
Nutrition & food security	190.7	32.0	141.4	74.9	136.8	82.3	221.4	76.5	143.1	1,118.0	5%
Tuberculosis	91.1	49.7	66.1	25.5	60.2	11.6	22.5	41.8	25.3	414.3	2%
Other communicable diseases	6.0	45.8	33.8	71.3	84.0	22.1	91.4	383.1	98.3	837.9	4%
Malaria	7.6	8.1	7.3	157.8	77.8	78.5	280.9	26.0	148.8	729.8	3%
Other human development	66.6	133.5	165.8	142.3	214.6	44.7	112.1	226.8	226.5	1,330.0	6%
Total	**1,884.8**	**2,058.1**	**1,880.0**	**1,575.0**	**2,864.8**	**952.6**	**3,086.9**	**4,483.4**	**2,906.6**	**21,636.8**	**100%**
Subtotal mapped to the HNP Sector Board	**912.9**	**1,366.9**	**921.2**	**782.8**	**1,535.0**	**843.1**	**1,492.4**	**3,085.1**	**2,389.3**	**12,863.7**	**59%**

Source: World Bank data.

Note: Each World Bank project can identify up to five themes promoted by the project. World Bank commitments represent the proportions of total project commitments to each theme. The subtotal "mapped to the HNP Sector Board" represents the share of these commitments under the control of the HNP Sector Board. That is, each Bank-supported project is supervised by a project manager who reports to a regional manager, who is represented on a Bank-wide sector board. Each project is thereby "mapped" — or becomes the responsibility of — that sector board, in this case the HNP Sector Board.

Table F-3. Global Fund: Annual Contributions by Donor (US$ millions, calendar years)

Country	2002	2003	2004	2005	2006	2007	2008	2009	2010	Total	Share
United States	275.0	547.1	458.4	352.0	463.7	840.3	789.2	1,010.1	797.3	5,130.2	27.2%
France [1]	59.0	63.8	191.4	181.0	287.3	408.8	434.9	437.9	378.0	2,431.0	12.9%
Japan	80.4	80.0	98.1	100.0	132.1	168.0	183.8	194.4	248.8	1,287.9	5.8%
Germany	12.0	37.4	45.9	155.0	68.1	116.7	312.2	271.4	265.2	1,255.9	6.7%
United Kingdom [1][1]	79.2	40.0	68.3	16.0	184.4	187.2	79.5	182.1	319.1	1,239.8	6.6%
European Commission		137.1	314.8	69.1	117.2	91.1	127.0	285.2	62.7	1,204.0	6.4%
Italy	108.5	106.5		217.8		675.3				1,008.3	5.3%
Canada	25.0	25.0	50.5	110.5	221.2		100.2	35.4	278.1	845.0	4.5%
Spain		36.3	13.2		80.2	104.5	138.9	207.4	137.8	719.1	3.8%
Gates Foundation [1]	50.0	50.0	50.0		100.2	100.0	100.5	206.5	10.5	670.0	3.6%
Netherlands		51.7	54.3	56.1	76.8	82.7	114.2	83.5	62.8	602.1	3.2%
Sweden	22.4	51.5	41.2	58.9	82.3	64.5	145.1	50.0	74.0	642.1	2.9%
Norway	18.0	17.7	17.9	25.8	43.1	90.2	52.5	87.2	62.0	352.2	1.9%
Russian Federation	1.0	4.0	5.0	10.0	10.0	75.0	30.7	75.0	20.0	257.0	1.4%
Denmark	14.8	13.9	18.2	22.8	25.9	25.9	29.4	31.9	31.2	208.9	1.1%
Australia			13.8	10.0	12.7	15.2	38.9	12.8	42.5	171.8	0.9%
WHO [1]		0.2					38.7	85.0	66.8	108.9	0.9%
Global Fund [1]					11.0	46.7	39.7	42.9	25.5	169.9	0.9%
Ireland	13.0	8.5	12.3	11.1	26.3	27.4	30.5	14.0	11.5	160.8	0.9%
Belgium	5.4	2.8	17.5	6.1	10.3	16.8	15.9	17.9	32.4	138.9	0.7%
U.N. Foundation		4.3	0.3				45.8	0.0	1.1	55.4	0.3%
Switzerland	3.1	6.6	2.3	3.9	4.9	5.7	6.7	6.3	7.2	47.1	0.2%
Saudi Arabia		2.5	2.3	2.3	2.5		6.0	6.0	6.0	26.0	0.1%
Luxembourg		2.1	2.4	2.5	3.5	3.1	3.3	3.3	3.2	24.0	0.1%
Indonesia							4.0	7.2	6.1	23.4	0.1%
Finland					3.5	3.3	3.9	4.9	4.4	26.2	0.1%
Nigeria	9.1								10.0	19.0	0.1%
China		2.0	2.0	2.0	2.0	2.0	2.0	2.0	2.0	16.0	0.1%
Pakistan								6.0	8.1	13.1	0.1%
Korea			5.5	0.5	5.5	3.0	3.0	4.0	2.5	13.5	0.1%
Portugal		0.4	0.6	1.5	2.0	5.0	3.5		2.5	13.0	0.1%
South Africa			2.0	2.0	2.0	2.0	0.1	2.1		10.3	0.1%
India					1.5	1.8	3.5	2.0	5.0	13.5	0.1%

Quoting from page 95 of the World Bank's publication:

"The World Bank launched a comprehensive strategy for health in September 1997: the Health, Nutrition, and Population (HNP) Sector Strategy. The Strategy was clear about the Bank's role in health, citing its comparative advantage as its ability to work across multiple sectors and to conduct country-specific research and analysis in support of programs **to which it could bring significant financing**. The Strategy did not view the Bank as having a comparative advantage in communicable disease control expertise, epidemiology, and the like in comparison with WHO, UNICEF, and UNAIDS. The Bank would focus on the broader aspects of health such as systems stewardship and oversight, systems performance, and **health financing**. [Emphasis added.] That means, the main function of the World Bank and the Clinton Foundation was to "steward" all health services in response to all disease outbreaks and risks to "National Security." And it would impose its administration for depopulation financially, by pulling the purse strings on every nation.

Now you can grasp the scam and scheme. Chelsea Clinton lends her celebrity to the World Bank's HIV/AIDS population control program. She gets a master's degree, more fame, and fortune for doing their bidding while fraudulently concealing her knowledge about how this scam works.

Then, fresh out of school, she is given an executive co-directorship under Barry Diller's ICA Deep State media enterprise, grooming further to advance profitable globalization interests and eventually her candidacy for the presidency.

Where does the World Bank make money? On the back end.

Once the Bill & Melinda Gates Foundation (listed on the second screenshot above among the "Countries" serving this HIV/AIDS "Population and Reproductive Health" agenda; working with the Clinton Foundation administering all of Big Pharma's vaccines) administer to vaccine recipients getting sick, and the devastated nations having to borrow money from the World Bank to cope with the expanding ailments and costly healthcare crises, the profitable depopulation outcome is achieved.

This scam and growing debt enslave nearly every nation. Then, governments are forced to negotiate a settlement that includes turning over valuable properties, including natural resources, to the World Bank for the New World Order governors. The economic dependence not only imposes subservience to vaccination impositions resulting in profitable depopulation. This is the strategy for financing and securing the "Great Global Reset."

Scrutiny reveals that the World Bank and World Health Organization's agents, agendas, and agencies published the 2019 "Vaccination Card/Passport" (VCP) plan for the planet. This VCP was proposed by the Economic Union (EU) —the Deep State Globalists' alliance from which England sought to withdraw at the time of this writing.

Currently unfolding in the United States is this Brave New World requirement to get vaccinated for air travel or employment by 2022. Severe restrictions on un-vaccinated people are being imposed. This violation of civil rights and human freedoms is clearly an important 'stage' in both 'vaccination depopulation' and civilization's bioelectronic conversion to transhumanism.

Clicking the screenshot below opens the evidence proving this official agenda.

Criminal Foundations: Fraudulent "Humanitarian" Orgs

Several years ago, I produced a DVD vetting "Criminal Foundations" that create many of civilization's worst problems—wars, starvation and dehydration, infectious diseases, and environmental pollution. The remedies, I explained, are being blocked by political and financial special interests. From HIV/AIDS policy to dealing with "natural disasters," propaganda justifies and secures a continued course of social and environmental degeneration. People are conned into donating more and more money and time to organizations like the American Red [Double] Cross, the National Cancer Society, and many others.

Advertising my video, I wrote: "[T]his status quo injustice is imposed mainly by the public's mindset. Shaped by the media and 'non-profit' institutions and foundations, public education, attitudes, values, and lifestyles have been compromised to accommodate '**crisis capitalism**' and corporate control over globalization and political impositions that have turned people against people, nations against nations, and man against nature."

In this context, sex offender Epstein formed a private foundation called Gratitude America Ltd.[406] to boost his reputation following his criminal conviction. He "embarked on a public relations crusade that depicted him as a renowned 'science philanthropist,' rather than a sex offender," recalled the *Daily Beast*. The only traces of Gratitude America Ltd. were found in its 2019 final tax return, no websites. This tax return showed funding of the multiple sclerosis (MS) website. MS is one of the dozens of neurological disorders associated with auto-immune reactions and dysfunctions prompted by vaccinations.

Cancer industry financing and profiteering was a common thread linking Gates, Epstein, and the Deep State's criminal enterprise investing in vaccine side-effects. In 2016 and 2017, Gratitude America Ltd. funded cancer charities, including the Melanoma Research Alliance and The Clinton Foundation's engagement in Third World vaccine trials.[407] The Clinton Foundation is best known for its supposed "treatments" for the leukemia-lymphoma-sarcoma cancer complex called AIDS triggered by the earliest hepatitis B vaccines, as I explained earlier.

Epstein donated to the Elton John AIDS Foundation and the Women Global Cancer Initiative Inc. He gave $10,000 to the Icahn School of Medicine at Mt. Sinai, $5,000 to the Leukemia & Lymphoma Society, and $75,000 to the Cancer Research Wellness Institute.

Epstein also gave $25,000 to Nautilus Think, a New York nonprofit that published a science magazine and contributed $50,000 to the University of Arizona Foundation. Some of that money was used to sponsor a 2017 "Science of Consciousness" conference.

Similarly enriching MKULTRA mind-control neuroscience, Epstein's Gratitude America donated $60,000 to a group labeled as "Association for Mind Education" in Rome, Italy.

In association with Gates and Epstein, Dr. Eva Andersson Dubin, a former NBC medical correspondent and practitioner in New York and Palm Beach (discussed below), also benefitted from this cancer commerce at the Dubin Breast Center after reportedly surviving breast cancer. Dubin's husband Glenn is another reported "philanthropist" and is on the list of "The World's Richest Hedge-Fund Billionaires."

Joining Dubin on the Forbes list of wealthiest billionaires is Randal J. Kirk. I mentioned that Kirk is the Chairman and CEO of the mutant mosquito maker Oxitec's owner, Intrexon.

Sherri and I also vetted President Trump's attorney, Rudy Giuliani,[408] for his secreted connections to Randal Kirk's operatives in the Deep State's bioweapons industry.

In fact, following the 9/11-contemporaneous anthrax mailings (and CIPRO sales scam) used to justify the Deep State's passage of the infamous "Patriot Act" co-authored by Michael Chertoff, Robert Mueller and complicit CIA officials contracted Giuliani's firm, **BioOne,** to decontaminate the Capitol Building and the American Media tabloid building in Boca Raton, FL. That company maintained a massive amount of scandalous intelligence available to extort politicians and corporation chiefs. The anthrax mailings, and Giuliani's company, expunged that evidence of Deep State corruption under the guise of BioOne's "cleanup."

In 2011, Bill Gates met with the key insider in most of this devil-doing, Jeffrey Epstein. The two men continued to meet on numerous occasions, according to the *NYTimes* diversionary propaganda. This included "at least three times at Mr. Epstein's palatial Manhattan townhouse, and at least once staying late into the night, according to interviews with more than a dozen people familiar with the relationship, as well as documents reviewed by the *Times*.

"Employees of Mr. Gates's foundation also paid multiple visits to Mr. Epstein's mansion. And Mr. Epstein spoke with the Bill and Melinda Gates Foundation and JPMorgan Chase about a proposed multibillion-dollar charitable fund —an arrangement that had the potential to generate enormous fees for Mr. Epstein" and tax-evasion-havens for the investors," the *Times* reported.

This Epstein and Gates exposè put these two men central to the Deep State's financial empire, the cancer industry, and the COVID vaccine neuroscience oligarchy into the center of a global conspiracy featuring pedophilia, human sex trafficking, and genocide.

Add to this unconscionable enterprise their nano-bioelectronic bioweapons industry featuring genetically modified (i.e., mutant) viruses and DNA poisoning vaccinations that induce S-protein antigens to be spread along with many cancers and auto-immune diseases. Purportedly 'immunizing' against illnesses, all of this profited hedge fund investors and private equity banking insiders, all advancing AI, the "dark enlightenment," the "rave movement," and sentient robots.

Gates' Damage Control

Speaking of Epstein, "His lifestyle is very different and kind of intriguing, although it would not work for me," Mr. Gates e-mailed colleagues in 2011, after his first get-together with Epstein.

Bridgitt Arnold, a spokeswoman for Gates, said he "was referring only to the unique décor of the Epstein residence —and Epstein's habit of spontaneously bringing acquaintances in to meet Mr. Gates."

Carefully consider that statement, reflecting on Gates's e-mail. Neither "décor" nor spontaneous meetings with acquaintances are "intriguing." But the way Epstein administered sex slaves performing political and financial favors surely would have intrigued Gates.

It is unreasonable this spokeswoman would do anything other than defend Gates to keep her job. "It was in no way meant to convey a sense of interest or approval," Arnold defended.

Epstein managed to cultivate close relationships with some of the world's most powerful men. "He lured them with the whiff of money and the proximity to other powerful, famous or wealthy people —so much so that many looked past his reputation for sexual misconduct. And the more people he drew into his circle, the easier it was for him to attract others," the *NYTimes* revealed.

Gates and his $51 billion foundation, an alleged money laundering organization, supposedly championed the well-being of young girls. The Gates Foundation especially promoted the HPV vaccine for young girls,[409] as did the Clinton Foundation. As mentioned in the opening chapter, the HPV lab creation from herpes viruses, and vaccines to combat them, is a total fraud. Aside from being a cancer trigger, according to the evidence I published here, all the sexually transmitted viruses since the Special Virus Cancer Program were bio-engineered.

"By the time Mr. Gates and Mr. Epstein first met, Mr. Epstein had served jail time for soliciting prostitution from a minor and was required to register as a sex offender," Arnold stated.[410]

She explained that 'high-profile people' had introduced Gates to Epstein and that they had met multiple times to discuss philanthropy.

"Bill Gates regret[ed] ever meeting with Epstein and recognize[d] it was an error in judgment to do so," Arnold said. Gates said that "entertaining Epstein's ideas related to philanthropy gave Epstein an undeserved platform that was at odds with Gates's personal values and the values of his foundation."

Substantial evidence to the contrary is found in Gates's arbitrary and capricious statements that vaccines, purportedly administered to save lives, will reduce the population by 15%.[411]

Reasonable people, as well as epidemiology experts, would consider Gates's vaccine projection bizarre. That lecture statement alone, coupled with Gates's association with Epstein and their MIT financing in vaccine neuroscience, provides probable cause for a grand jury investigation for genocide.

Epstein and Gates in Underworld Commerce

On Sept. 2, 2019, *The New Yorker* published Pulitzer Prize winner Ronan Farrow's excellent excursion into the 'pit of hell' —*academia*— where intelligence is subverted by money and power. Demonically possessed by conflicting interests, the world's most esteemed institutions have been poisoned by the rich-and-famous, destroying all reputability in the arts, sciences, judicial system, law enforcement, and ethical sanity throughout society.

Ronan's article was titled, "How an Élite University Research Center Concealed Its Relationship with Jeffrey Epstein." He analyzed documents that showed MIT Media Lab —considered the most influential mind-control and social-engineering research and development facility—was "aware of Epstein's status as a convicted sex offender." Yet, officials bent over backward to conceal Epstein's influence there in favor of the Deep State's population management Artificial Intelligence (AI).

The exploration and manipulation of human consciousness transitions civilization into Brave New World transhumanism. I've already evidenced that the human race is anesthetized by drugs, lusts, sports, and frequency weaponry broadcast by the corrupted media and music industry. Gates, Epstein's Mega Group affiliates, and the Deep

State-controlled U.S. and U.K. militaries and intelligence community substantially financed much of this "Rise of the Fourth Reich."

Epstein, Gates, and *MINDFREAK* Financing

Recall that the Epstein scandal resulted in the resignation of Joi Ito, the director of the MIT Media Lab. The president of MIT, L. Rafael Reif, apologized to the community and promised "an immediate, thorough and independent investigation."

"Epstein appeared to serve as an intermediary between the lab and other wealthy donors, soliciting millions of dollars in donations from individuals and organizations, including the technologist and philanthropist Bill Gates and the investor Leon Black," Ronan Farrow wrote.[412]

According to the records and interviews obtained by *The New Yorker*, "Epstein was credited with securing at least $7.5 million in donations for the [neuroscience] lab, including two million dollars from Gates and $5.5 million from Black, gifts the e-mails describe as 'directed' by Epstein or made at his behest. The effort to conceal the lab's contact with Epstein was so widely known that some staff in the office of the lab's director, Joi Ito, referred to Epstein as Voldemort or "he who must not be named."

In October 2014, the Media Lab received a two-million-dollar donation from Bill Gates. To administer the infusion of cash used to direct covert 'dual purpose' military and commercial brain-tissue research and development, Ito responded to Peter Cohen, the MIT Media Lab's Director of Development and Strategy at the time, "Jeffrey money, needs to be anonymous. Thanks."

"This is a $2M gift from Bill Gates directed by Jeffrey Epstein." Cohen replied, "For gift recording purposes, we will not be mentioning Jeffrey's name as the impetus for this gift." Gates's agent also wrote to the Media Lab's leadership, stating that "Gates also wished to keep his name out of any public discussion of the donation," Farrow recorded.

Recall that IBM initially financed Gates and Microsoft with its dark history in early data processing for the Rockefeller Cartel and Third Reich. Recall also that Peter Thiel's father served Nazi Germany and IBM. Neglecting this background intelligence, Gates is mostly credited for developing operating systems for personal computers and mobile phones. But he became very active in the early emergence of the Internet as well. This is the 'holy land' for the PROMIS software, data-mining, and population control operations.

Gates's influence was heavy in the Internet search community. Bing, video gaming Xbox, and digital services through MSN—the Microsoft Network advanced MKULTRA (mind-control) and social engineering PROMIS objectives. With NBCUniversal (CNBC's parent),[413] Gates & Co. created the 24/7 cable news station, MSNBC —the leading Democratic Party propaganda mill. Skype Technologies for Internet telephony and eavesdropping was its largest acquisition for $8.5 billion in October 2011. Currently, Gates's enterprise is backing the Metaverse with Facebook, Zuckerberg, and the Google clan.

Don't forget that Gates saved Steve Jobs from Apple's bankruptcy in 1977 by heavily investing in the generally neglected partnership between the two companies, MS and Apple.[414] That market domination, a virtual monopoly administering the CIA-controlled Google, empowered the intelligence agencies' 'data-mining,' smut-gathering for bribing officials, and MindWar advancing the Deep State's progress that continues.

Gates and the MIT chiefs knew Epstein had a habit of sexually abusing underage girls. "In 2008," Farrow recalled, "after a Florida grand jury charged Epstein with soliciting prostitution, he received a controversial plea deal, which shielded him from federal prosecution and allowed him to serve less than thirteen months, and much of it on a 'work release,' permitting him to spend much of his time out of jail.

"Alexander Acosta, the prosecutor responsible for that plea deal, went on to become President Trump's Secretary of Labor but resigned from that post in July, amid widespread criticism related to the Epstein case. That same month, Epstein was arrested in New York on federal sex-trafficking charges."

"In the course of 2014 and 2015," Farrow continued. Ito and Epstein "developed an ambitious plan to secure a large new influx of contributions from Epstein's contacts, including Gates... and Leon Black, the founder of Apollo Global Management, one of the world's largest private-equity firms."

Gates' Inner Circle of 'Science Scholars'

According to the *NYTimes*, "[t]wo members of Mr. Gates's inner circle —**Boris Nikolic** and **Melanie Walker**— were close to Mr. Epstein and at times functioned as intermediaries between the two men.[415]

Ms. Walker met Mr. Epstein in 1992, six months after graduating as a medical doctor from the University of Texas. According to Ms. Walker, Mr. Epstein, an adviser to Leslie Wexner, the owner of Victoria's Secret, told Walker that he could land her an audition for a modeling job there. But according to the *NYTimes*, "After she graduated from medical school... Mr. Epstein hired her as a science adviser in 1998."

There is no evidence Victoria's Secret model was a 'science scholar.' Ms. Walker's contributions to the world of science are lacking. However, her generally concealed World Banking credentials are quite impressive. The *NYTimes* article neglected Walker's service to the World Bank, much like Chelsea Clinton's "service" to the World Bank as a "science scholar" lacking credentials.

Melanie Walker, who had known Mr. Epstein since 1992, joined the Gates Foundation as senior program officer in 2006. Copyright by World Economic Forum; Benedikt von Loebell

Walker later moved to New York and stayed in a Manhattan apartment that Epstein owned.

So the *NYTimes*, again, fraudulently concealed the Deep State's financial influence in the World Bank, intertwined with Gates's and

Epstein's complicity in a nano-bioelectronic racketeering enterprise developing "novel" vaccines.

According to online intelligence,[416] in September 2014, Melanie Walker served as senior adviser to the World Bank Group President Jim Yong Kim. She spoke during a press conference to announce the World Bank's new website at the World Bank headquarters in Washington, DC. The World Bank Group launched a website to track "the progress of the group's projects."

Ms. Walker later met Steven Sinofsky, a senior executive at Microsoft who became president of its Windows division. She moved to Seattle to be with him.

In 2006, Walker joined the Gates Foundation with the title of Senior Program Officer. At the foundation, Walker met and befriended Boris Nikolic,[417] a native of what is now Croatia, and a former fellow at Harvard Medical School. Nikolic was the Gates Foundation's science adviser. "Mr. Nikolic and Mr. Gates frequently traveled and socialized together," explained the *NYTimes*. "Ms. Walker, who had remained in close touch with Mr. Epstein, introduced him to Mr. Nikolic, and the men became friendly."

An online search for science papers produced by Nikolic yielded several radiology papers in which he is listed as a co-author but not a senior author or well-credentialed advisor.

More to the point, this gathering of "science scholars" around Gates and Epstein fronted for serious authority, even legitimacy. In fact, Nikolic's 'radiology science scholarship' supplemented his income from running a venture capital firm with Gates as a special interest investor.

Nikolic said he was "shocked" to be named by purportedly Epstein in Epstein's last will. He said in a statement to the *NYTimes*: "I deeply regret ever meeting Mr. Epstein."

Imagine leaving your estate so someone who regrets ever meeting you! What's wrong with that picture? It lies at the heart of the Deep State's COVID Coup and transhumanism cabal.

According to the *NYTimes'* propaganda piece, "Mr. Epstein and Mr. Gates first met face to face on the evening of Jan. 31, 2011, at Mr. Epstein's townhouse on the Upper East Side. They were joined by Dr. Eva Andersson-Dubin, a former Miss Sweden whom Mr. Epstein had once dated, and her 15-year-old daughter."⁴¹⁸

Dr. Andersson, Dubin's husband, is the hedge fund billionaire **Glenn Dubin**. He was also a friend and business associate of Mr. Epstein'. The Dubins declined to comment on these matters.

It is unconscionable that the *New York Times* concealed here what was widely reported elsewhere, public knowledge even on *Wikipedia*:

"In August 2019, unsealed documents revealed connections between Dubin and Jeffrey Epstein, including allegations of involvement in the Deep State's sex ring.

Former house manager for the Dubins, Rinaldo Rizzo, described a 2005 encounter with a 15-year-old girl employed as a nanny. Rizzo said the girl told him **Ghislaine Maxwell** pressured her to have sex with Epstein, taking her passport when she refused. According to *The Daily Beast*, a month into her employ, the Dubins took the girl with them to Sweden, where she was dropped off at an airport.

Compounding this story's veracity, Virginia Giuffre also claimed in her lawsuit that Glenn Dubin was one of the men with whom Epstein and

Maxwell forced her to have sex. According to *Wikipedia*, "Giuffre named several others that she claimed Epstein and Maxwell instructed her to have sex with, including hedge fund manager Glenn Dubin, attorney Alan Dershowitz,[419] politician Bill Richardson, the late MIT scientist **Marvin Minsky**, lawyer George J. Mitchell, and MC2 modeling agent Jean-Luc Brunel."[420]

Further corroborating the global conspiracy among criminal psychopaths with perverse sexual appetites, Marvin Minsky is known as "the father of artificial intelligence[421]." He was "a professor emeritus at the **MIT Media Lab**."

Put Minsky's MIT Media Lab AI together with MIT's Langer Lab's drug delivery and tissue engineering nano-bioelectronic technology, and you have Moderna's and Pfizer's brain-Cloud fusing mRNA vaccines that people are being *forced* to take.

When Epstein's buddy, Alan Dershowitz, debated Robert F. Kennedy, Jr. on the issue of mandatory vaccinations, the anti-vaccination community was neglected and discouraged. Their video was professionally produced, well-advertised, and widely spread.[419] Unfortunately, the "anti-vaccination" position was damagingly missing in their debate. I realized the "pro-vaccine" bias of both lawyers. This shortcoming generated 'cognitive-behavioral dissonance.' Kennedy slammed the risks of getting vaccinated but encouraged jabs anyway. This demonstrated discrediting hypocrisy. That is consistent with the MindWar mentioned earlier and "controlled-opposition" —a media ploy generating confusion reflecting the Deep State's influence. I wrote an Op/Ed article expressing these concerns.[419] Therein, I published the following assessment:

> "[T]he Kennedy/Dershowitz Constitutional 'debate,' weighing upon the legality of forcing vaccinations into people, was fundamentally and contextually defective; arguably lame, reckless, negligent, ignorant, arbitrary and capricious. It frames the entire debate like a corral concentrates a herd moving to slaughter."

Early German scientist, novelist, and poet, Johann Goethe, wrote: "Tell me with whom you associate, and I will tell you who you are."[422] On Jan. 31, 2011, Gates met with Epstein, as described by *The NYTimes*.

"Gates ... praised Epstein's charm and intelligence. The next day, Gates emailed colleagues and said, 'A very attractive Swedish woman and her daughter dropped by, and I ended up staying there quite late.'

"Gates soon saw Mr. Epstein again. At a TED conference in Long Beach, Calif., attendees spotted the two men engaged in private conversation.

"Later that spring, on May 3, 2011, Mr. Gates again visited Mr. Epstein at his New York mansion, according to emails about the meeting and a photograph reviewed by *The Times*.

"The photo, taken in Mr. Epstein's marble-clad entrance hall, shows a beaming Mr. Epstein —in blue-and-gold slippers and a fleece decorated with an American flag— flanked by luminaries. On his right: James E. Staley, at the time a senior JPMorgan executive, and former Treasury Secretary Lawrence Summers. On his left: Mr. Nikolic and Mr. Gates..."

Most noteworthy, Carlos Slim's *NYTimes* neglected to describe the artwork adorning Epstein's walls, which is being censored.

The "Global Health Investment Fund"

Around that time, the Gates Foundation and JPMorgan teamed to create the Global Health Investment Fund —a major initiative to consolidate power and control over civilization by 'PROMIS programming,' super-fast computing, AI, social engineering, substance abuse, and nano-bioelectronics. Epstein's cohorts and the Defense Department were funding the Lieber and Langer labs and their academic affiliates to accomplish this. Drugs and vaccines, along with data-mining and the AI neuroscience research and developments at MIT, Harvard, Johns Hopkins, NYU, Stanford, SRI, etc. would be used to provide "individual and institutional investors the opportunity to finance late-stage global health technologies that have the potential to save millions of lives in low-income countries," the *NYTimes* propaganda reported.

I pierced this veil of deception. To "finance late-stage global health technologies that have the potential to [kill] millions of [people] in low-income countries most [profitably]" was (and is) the actual objective.

"As the details of the fund were being hammered out," the propaganda continued, "Mr. Staley told his JPMorgan colleagues that Mr. Epstein wanted to be brought into the discussions, according to two people familiar with the talks. Mr. Epstein was an important JPMorgan customer,[423] holding millions of dollars in accounts at the bank and referring a procession of wealthy individuals to become the company's clients.

"Mr. Epstein pitched an idea for a separate charitable fund to JPMorgan officials, including Mr. Staley and Mr. Gates's adviser Mr. Nikolic. He envisioned a vast fund, seeded with the Gates Foundation's money, that would focus on health projects around the world, according to five people involved in or briefed on the talks, including current and former Gates Foundation and JPMorgan employees. In addition to the Gates money, Epstein planned to round up donations from his wealthy friends and some of JPMorgan's richest clients.

"Epstein thought he could personally benefit. He circulated a four-page proposal that included a suggestion that he be paid 0.3 percent of whatever money he raised, according to one person who saw the proposal. If Mr. Epstein had raised $10 billion, for example, that would have amounted to $30 million in fees.

THE COVID COUP WITH FINANCING FROM HELL

Gates was allegedly "unaware that Mr. Epstein was seeking fees in his proposal for a charitable fund," Gates's spokeswoman said.

Some of the Gates Foundation employees said they had been unaware of Mr. Epstein's criminal record and had been shocked to learn that the foundation was working with a sex offender. They worried that it could seriously damage the foundation's reputation.

This appears to be 'damage control' on two levels that I address in the following sections.

Gates was photographed here at the World Economic Forum in Davos, Switzerland, in January 2011. (Credit Arnd Wiegmann/Reuters)]

EPSTEIN ALIVE? HERE ARE THE TOP 10 REASONS WHY.

Epstein's 'Suicide' Implicates Gates's Service to the Deep State

First, assuming that some foundation members "worried" that Epstein could "seriously damage" the Bill & Melinda Gates Foundation's reputation, this worry provided good cause for Gates to 'neutralize' or 'disappear' Epstein. The admission of 'worry' makes Gates a primary suspect in Epstein's falsely reported "suicide."

Don't forget, Gates's multiple social and business visits with Epstein were undoubtedly videotaped for bribery. Epstein built his career on this practice, providing his blackmail service to the Deep State and intelligence agencies controlling governments and lawmakers. Epstein served his bosses, including Leslie Wexner and other members of the "The Mega Group," as reported by Whitney Webb.[424]

Second, this *NYTimes* propaganda discrediting the Global Health Investment Fund, falsely claiming it was "preposterous," diverts from the actual ongoing program administered by Gates's Microsoft empire that includes his allies in data-mining, social engineering, and vaccine bioelectronics, as already detailed.

So what must the Court of Public Opinion presume about the preponderance of evidence convicting co-conspirators in the government and media for their actions that establish a 'presumption of guilt' based on the aforementioned facts?

Number one, as I first reported online,[397] we must presume Epstein was 'switched out' sometime between his first hospital visit in New York on July 23, 2019, and the morning of August 10th.

We must presume a self-incriminating Justice Department investigation will not occur, given the corruption in the Deep State that has captured and controlled the FBI and CIA.

The *NYTimes* coverage of the Justice Department's "Criminal Inquiry into its Own Russia Investigation"[425] reinforces the improbability that federal law enforcers will consider the conspiracy involving Gates and Epstein.

But if it did, the focus would feature:
1) The media co-conspirators acting to socially engineer a PSYOPS to protect the government officials and politically powerful celebrities who sexually abused underage women and are engaged in various racketeering activities; and
2) The intelligence agencies that coordinated this public deception —namely the CIA, MI6, and Israeli Mossad. The latter is implicated due to Epstein's ventures with billionaire Leslie Wexner. Wexner was the co-founder of The Mega Group. That pro-Israel group includes several of the wealthiest Jewish

THE COVID COUP WITH FINANCING FROM HELL

financiers, people with intelligence agency connections secreted by the complicit media in service to the Deep State.

Epstein, through Wexner, was connected to "major political donors in both the U.S. and Israel. Several of The Mega Group's most notable members have close ties to the governments of both countries as well as their intelligence communities," reported Webb on her whistleblowing blog, *MintPressNews.com*.[426]

"[T]he Mega Group also had close ties to two businessmen who worked for Israel's Mossad" — Ghislaine Maxwell's father, Robert Maxwell [is shown in the photo on the next page. He is seen smiling at Donald Trump, addressing Mike Wallace] and Marc Rich. These people held "deep ties to Israel's intelligence community..."

Also shown below is Steve Ross, the CEO of Time Warner, Warner Communications, and John Tower, who is known to have concealed the Clintons' complicity in the Iran Contra Affair investigated by the "Tower Commission."

Tower was also Vice-Chair in the Frank Church Committee[427] hearings of the CIA's MKULTRA and MKNAOMI media MindWar and biological weapons programs. As I mentioned, officials thereby concealed the lab-virus hepatitis B vaccine origin of HIV/AIDS loosed by the Merck drug company with CDC, NYU, U.S. Army, and Rockefeller Blood Bank co-investigators.

The Israeli Deep State Mossad Connections

Robert Maxwell, a business partner of Mega Group co-founder **Charles Bronfman**, aided the successful Mossad plot to plant a trapdoor in Silicon Valley software that was then sold to governments and companies worldwide, as accurately reported by Whitney Webb for MintPressNews.[426]

That plot's success was largely due to the role of a close associate of then-President Ronald Reagan and an American politician close to Maxwell [i.e., Sen. John Tower]. Tower later helped aid Reagan in the cover-up of the Iran Contra scandal heavily incriminating the Clintons' Rose Law Firm / Mena Airport "Drugs for Arms" operation.

Charles Bronfman was the father of Edgar Bronfman, grandfather of Edgar Bronfman, Jr., who runs Barry Diller's ICA and the NBC/Universal Music Group with Chelsea Clinton. Each of these suspects operates like Bill Gates as agents for the World Bank. The NBC/Universal Music Group, directed by Edgar Bronfman, Jr., holds a virtual monopoly over the music industry in America.

Edgar Bronfman, Jr.'s first cousin is Jeffrey Bronfman, who you will learn about later in the context of advancing the brain-conditioner above, DMT for the "Ayahuasca Religion." This hallucinogen, with studies ongoing at NYU where Chelsea Clinton serves on the "spiritual faculty," conditions users' brains to submit to the bioelectronic brain-Cloud connection developed by the Harvard and MIT Lieber and Langer labs.

Charles Bronfman led the World Jewish Congress[428] in becoming the preeminent international Jewish organization that it is today. The Chairman of the WJC is **David Mayer de Rothschild**, a member of the world's most powerful banking family. As head of the Sculpt the Future Foundation, he "supports innovations and creativity in social and environmental impact efforts," according to *Wikipedia*.[429]

Harvard's Center for Climate, Health and the Global Environment seeds the meme that the SARS-CoV-2-HIV-1 has emerged due to "climate change."[430] The group's leader, Dr. Aaron Bernstein, approved the promotion despite lacking evidence, but heralds the risk of death nonetheless from COVID is increased by air pollution.

The Mega Group lobby-power influences both sides of the political aisle, and secretly the World Bank. So long as this is secreted or disregarded, radical right and radical left agendas dividing America's racial, ethnic and religious groups will worsen as a diversion from the transhumanist takeover.

Robert Maxwell's daughter, Ghislaine, joined Jeffrey Epstein's 'inner circle.' This merger of Deep State interests in Harvard and MIT bioelectronic vaccinology occurred at the same time Epstein was bankrolling a software program, much like the PROMIS software, now being marketed for critical electronic infrastructure and social engineering in the U.S. and abroad, Whitney Webb reported.[426]

"Epstein was a long-time friend of former Israeli Prime Minister Ehud Barak, who has long-standing and deep ties to Israel's intelligence community," Webb continued. Barak was "close to Epstein's chief patron and Mega Group member, Leslie Wexner." The "Wexner Foundation gave Barak $2 million in 2004 for a still unspecified research program," Webb noted.[431]

"According to Barak, he was first introduced to Epstein by former Israeli Prime Minister Shimon Peres, who eulogized Robert Maxwell at his funeral and had decades-long ties with the Bronfman family going back to the early 1950s.

"Peres was also a frequent participant[432] in programs[433] funded by Leslie Wexner in Israel and worked closely[434] with the Mossad for decades.

"In 2015, a few years after Epstein's release from prison... Barak formed a company with Epstein... now called Carbyne..." The security service software Carbyne sells 'runs any caller's identity through any linked government database.

It has specifically been marketed by the company itself and the Israeli press as a solution[435] to mass shootings[436] in the United States...' Israeli media reported that Epstein and Barak were among the company's largest investors. This adequately explains the hideous increase in mass shootings and individual shootings in America's largest cities, where police forces are being defunded and shrunk from vaccine mandates. These investors are selling AI remedies, including military-designed 'terminators' to fight crime.

"Barak poured millions into the [Carbyne] company. Haaretz revealed that a significant amount of Barak's total investments in Carbyne was funded by Epstein, making him a '*de facto* partner' in the company.[437]

"Barak is now Carbyne's chairman[438]," Webb reported.[426]

"The company's executive team are all former members of different branches of Israeli intelligence, including the elite military intelligence unit, Unit 8200, that is often likened to Israel's equivalent of the U.S. National Security Agency (NSA)..."

This is similar to former Secretary of Homeland Security Michael Chertoff and the CIA transferring several of the intelligence agencies' top officials into The Chertoff Group that set up a for-profit security business anticipating profits from the *Las Vegas Deep State Massacre* example.[237]

This repeated observation evidences the 'pattern-and-practice' of the Deep State elite. They stack the deck with military intelligence and security assets to administer the Deep State's criminal enterprise and population control programs. "Compartmentalization" assures "plausible deniability" whereby complicit agents do their dirty work without fully grasping who is pulling their strings, or the Bigger Picture advancing to damage, deprive, even terminate us as MindWar slaves.

The Carbyne company also includes several tie-ins to the MindWar drug supply. Peter Thiel, for instance, Palantir's founder and hallucinogenic drug investor, is a co-investor in Carbyne.[439]

In addition, Carbyne's board of advisers includes former Palantir employee Trae Stephens, a member of the Trump transition team, and Michael Chertoff.[438]

Deep State Financing Racism and White Supremacy

The Deep State's investments in genocidal agendas damage and deprive youth, human rights, and psychological and social well-being. This 'reality' is destroying people's faith in governments, the media, and nationalism in favor of One World Government, or Huxley's *Brave New World*.

This current sociopolitical and economic degeneration was foretold in the infamous anti-Semitic text, *Protocols of the Elders of Sion*.[440]

Those Protocols were converted from a leaked document used to spur the Russian revolution and Communism. Allegedly, the document was prepared according to the Jewish Rothschild banking family to plot the global takeover and justify doing so.

Protocol #10 (paragraphs 18 and 19 of the FBI's version) provides an uncanny preview of what is unfolding today. Under 'Deep State' financing and powerful media influence, divisions between the races, religions, and ethnic groups would favor globalization. It states:

> "18. The recognition of our despot may also come before the destruction of the constitution; the moment for this recognition will come when the peoples, utterly wearied by the irregularities and incompetence – a matter which we shall arrange for – of their rulers, will clamor: "Away with them and give us one king over all the earth who will unite us and annihilate the causes of disorders – frontiers, nationalities, religions, State debts – who will give us peace and quiet which we cannot find under our rulers and representatives.
>
> "19. But you yourselves perfectly well know that TO PRODUCE THE POSSIBILITY OF THE EXPRESSION OF SUCH WISHES BY ALL THE NATIONS IT IS INDISPENSABLE TO TROUBLE IN ALL COUNTRIES THE PEOPLE'S RELATIONS WITH THEIR GOVERNMENTS SO AS TO UTTERLY EXHAUST HUMANITY WITH DISSENSION, HATRED, STRUGGLE, ENVY AND EVEN BY THE USE OF TORTURE, BY STARVATION, BY THE INOCULATION OF DISEASES, BY WANT, SO THAT THE "CATTLE" SEE NO OTHER ISSUE THAN TO TAKE REFUGE IN OUR COMPLETE SOVEREIGNTY IN MONEY AND IN ALL ELSE. [Emphasis not added.]"

Accordingly, crucial in this globalist agenda is the politicization of public health, so-called "science" for the "INOCULATION OF DISEASES," and "dissension, hatred, struggle, and envy between people of various factions. This MindWar objective best explains the sudden rise of Black Lives Matter and Antifa. This explains the WOKE movement and the "dissension, hatred, [and] struggle" between liberals and conservatives. Mandatory vaccinations have similarly divided countries. In the United States at the time of this writing, news media propaganda is inciting a civil war between the VAXXED and UN-VAXXED. Withholding hospital care for needy people is being attributed to "vaccination hesitancy."

These policies incite violence, racism, anti-Semitism, and white supremacy. Blacks, for good reasons, are the least vaccinated and least trusting of the government. Blacks, therefore, are the most susceptible to being deprived of hospital care. Add to this the fact that Blacks suffer disproportionately from all chronic diseases for many reasons.

The Luciferians also generate Jewish hate for similar reasons. With COVID bioterrorism, Rabbis in New York, and church leaders everywhere, have been similarly discriminated against for their religious beliefs opposing mandatory vaccinations. MindWar heralds agnostics being threatened by unvaccinated monotheists.

Meanwhile, as the COVID Coup and genocide continue to damage and kill people, white supremacy groups (that blindly blame Jews, Blacks, and Muslims for everything bad) are stoked.

The Epstein-Gates Global Health Fund

In early 2012, another Gates Foundation team met with Epstein at his mansion, recalled the *NYTimes*. "He claimed that he had access to trillions of dollars of his clients' money that he could put in the proposed charitable fund — a figure so preposterous that it left his visitors doubting Mr. Epstein's credibility."

However, this amount of money is not unreasonable for these billionaires who control trillion-dollar industries and act to monopolize healthcare.

So the *NYTimes* intelligence, actually propaganda, is consistent with Ronan Farrow's widely publicized *Catch and Kill* saga that exposed Gates's partner, NBC, of protecting another sex-predator, Harvey Weinstein.[441]

Gates's $1B investment in Comcast/NBCUNIVERSAL fueled our concerns that the MGM Grand/NBCUniversal 'fake news' conglomerate was up to no good, again.[442] We knew this racketeering enterprise operated at the heart of the 'Deep State.' I knew criminal psychopaths controlled it. Now we know they ritually performed acts of sexual perversion, mass murders (as in the case of the *Las Vegas Deep State Massacre*), and the mounting vaccination genocide.

The vaccines were effectively administering eugenics and infanticide, with blood sacrificing justified by Luciferian pseudoscience.

The Overall Scam Inside the 'Sad Café'

Much of the financial backing for America's destruction sources from George Soros and his Open Society Foundation. Soros joins fellow globalists, such as Barry Diller and Bill Gates, in making money by 'donating' money to 'leftist' 'liberals' and 'progressive' organizations advancing socialism, Communism, and corporate-fascism that characterizes "globalism." Aside from presumably an agency fee, a 'cut' or commission when they make their big deals, they make money by *investing in globalization and its products and services.* In this case, *military bioelectronic population control devices* are key to their "futures market." As this market matures, the elite's 'seed funding' generates huge returns, often in the billions of dollars.

Naturally, this leaves those who see themselves as 'loyal liberals' and those dedicated to "enlightenment" through "expanding consciousness" becoming disappointed after being defrauded. As Glenn Frey and Don Henle wrote in *The Sad Café,*[443] "We thought we could change this world with words like 'love' and 'freedom.' We were part of the lonely crowd inside the Sad Café."

The "Sad Café" is a metaphor for our planet, owned and controlled by the fascist criminal-psychopathic elite. The oligarchs of this Brave New World invest in the products and services for optimal power and control, such as their secreted nano-neuro-bioelectronic vaccine technologies for brain-Cloud manipulations and transhumanist conversion.

The presumably 'liberal' OpenDemocracy.net that, like the Burning Man event, uses a *butterfly* in its logo consistent with MKULTRA's 'Monarch Program' discussed earlier, explained that in 1994, "Ethan Nadelmann [the main man that pushed marijuana policy reforms in America[444]] founded the Lindesmith Center, a drug policy institute created with the philanthropic support of **George Soros**. In 2000, the growing Center merged with another organization to form the Drug Policy Alliance and Drug Policy Alliance Network, which advocates for hallucinogenic drug policies purportedly "grounded in science, compassion, health and human rights."[445]

Nadelmann made many Deep State-controlled network television appearances to spread this false pretense neglecting the brain-Cloud conversion of civilization conditioned by hallucinogenic drugs. The globalists thereby moved drug policy reforms in the US in favor of their New World Religion, as promoted by Dr. Krippner on behalf of the CIA and British intelligence community.[446]

In other words, under the guise of 'liberal' 'humanitarian' aid justified by 'science' and 'human rights,' leading capitalists, exemplified by Soros, Black, and Diller, predictively programmed the global *CULT*ure to degenerate in favor of their MindWar and co-investors.

Left and Right Loopholes Profit Large Donors in Insane Societies

You don't need to look far to discover solid evidence of criminal fraud, tax evasion, money laundering, and racketeering by the globalists through their foundations and donations to academic institutions. When President Biden speaks of everyone "paying their fair share" of taxes and neglects the loopholes granted to his Democratic Party backers' foundations, you are witnessing 'public corruption' and organized crime.

Non-profit 'humanitarian' foundations often profit large 'donors.' Most reasonably-informed 'conscious' people have awakened to this central 'racket' of financing Deep State social-engineering through 'Criminal Foundations.'

Multi-national corporations leverage these 'tax' shelters to "Divide and Conquer" society. That's why you have 'liberal' and 'conservative' 'non-profit' foundations and 'think tanks.' These serve important functions within the commercial CULTure developed to control populations most profitably and insidiously.

As I previously shared regarding Oxford University's scam, like Harvard University's scam, like MIT's scam, researchers develop new biotechnologies, profitable patents, and products that the 'insiders' commercialize through for-profit spin-off companies. We evidenced this with Oxford's spin-off Oxitec selling GM mosquitoes; Harvard's Evergrande/Evergreen agents transferring Lieber's biotechnology to China; and MIT's Media Labs (financed by DARPA, the Gates Foundation, and Epstein's Mega Group) selling Robert Langer's mRNA vaccine technology to Moderna. This 'pattern-and-practice' of financial fraud is obvious but encouraged nonetheless by bribed (campaign-financed) politicians whose 'willful blindness aids-and-abets this organized crime.

Genius Albert Einstein presumably wrote, "The definition of insanity is doing the same thing over and over again and expecting a different result." We vote for Democrats and Republicans in America over and over again, expecting "a different result."[447] We get nowhere positive, only further in debt and disappointment.

Rather than doing 'something completely different,' like enforcing laws opposing money-laundering through corporate donations to foundations to remedy this widespread deadly deception, voters re-elect the same or similar big-money bribed con-artists expecting to be blessed every election cycle. That's *insanity*.

If "draining The Swamp" is a remedy, it can't happen while we remain immersed in the muck! The solution must come from beyond The Swamp. *"Come out of her, my people."* (Rev. 18:4)

There appear to be two remedies for this tragedy. One is advanced by the 'Illuminati' (i.e., the Deep State), and the other by God. The Illuminati's remedy directs civilization to a New World Order featuring intoxicating injections containing brain-Cloud bioelectronic devices and psychotropic drugs to numb perceptions and condition brains to submit to a demonic reality. God's remedy is different and apparent in nature. It is reflected in 'Divine-design,' also called 'intelligent design.' That is called "righteousness" in the religious world, "coherence" in physics, and in common law, it is called "good faith" and "competence" compelled by "public duty."

Planned Parenthood and Related Foundations

According to the right-wing 'watchdog' group *Influence Watch*, the Open Society Foundations (OSF)[448] founded by 'liberal' financier Soros[449] "is a network of more than 20 national and regional foundations. That makes it one of the largest political philanthropies in the world."

Presumably built on Soros' "anti-capitalist redistributionist political philosophies," the organization donates nearly a billion dollars per year "to left-wing organizations around the world."

In the United States, Open Society Foundations have given hundreds of millions to left-wing political organizations, including World Health Organization partners, Planned Parenthood, the ACLU, Robin Hood Foundation, Tides Foundation, Brennan Center for Justice, and the Alliance for Citizenship, among many others. "In 2016, it was reported that Open Society Foundations gave $33 million to the Black Lives Matter movement and groups associated with it," *Influence Watch* detailed.[448]

I balked over the inclusion of the ACLU as a purportedly 'liberal' organization. If that were true, the ACLU would not have repeatedly rejected the massive evidence and repeated appeals I sent them. I petitioned the ACLU to help me oppose HIV/AIDS impositions and vaccination genocide repeatedly during the past quarter-century.

Dr. Jonathan Mann, the WHO's former "AIDS Czar" I consider later in Chapter IX, protested HIV/AIDS as a "sociopolitical imposition." Many other concerned citizens objected similarly, including Barack Obama's minister, Rev. Jeremiah Wright,[288] on national television. I especially petitioned the ACLU to litigate in favor of affected minority populations. The organization neglected my correspondence and avoided public duty.

It is deceptive to claim Planned Parenthood is a 'liberal' organization when its primary mission has always been *depopulation* to protect the wealthy.

Planned Parenthood sourced directly from Rockefeller's Standard Oil/IG Farben Anglo-American corporate defense strategy opposing hordes of non-white immigrants, ethnic groups, and labor unions influencing U.S. politics and policies. These immigrants and labor unions were the perceived 'enemy' of the Anglo-elite. The enemies of the ultra-wealthy had always been non-white immigrants risking revolutions against status-quo administrations endangering assets — corporations, properties, and families' dynasties. These virtual slave-masters historically opposed any viable competition or opposition. This is why they 'neutralized' Dr. Martin Luther King, Jr. and Malcolm X. These covert operations continue to target victims and decimate populations today. Today, the only difference is that working-class whites are being targeted for elimination more than ever, just like African Americans.

Look no further than the Rockefellers' Population Council in New York to find the origin of this American imposition and Planned Parenthood. Read Henry Kissinger's *National Security Memorandum 200* (NSM-200) to evidence the political attack against Blacks and its justification for allegedly National Security. This entire enterprise feeds off genocides, infectious diseases, and drug side effects, much like the Clinton Foundation does with HIV/AIDS services in Africa, promoting deadly and damaging 'remedies' for the world's poor.

The Population Council similarly sourced the Negative Population Growth, Inc. company that widely publicized the "urgent" need to reduce the U.S population and other nations' people. Examine their advertisement below published in the esteemed *Foreign Affairs* journal in 1996. That ad prompted me to discern the generally concealed ties between the Rockefeller/IG Farben genocidal enterprise and today's healthcare racket profiting most from human suffering.

'Radical Left' vs. "Alt-Right' Financed by the Same Foundations

In the Robin Hood Foundation link, you will see that Soros's Open Society Foundations and the "Soros Fund Charitable Foundation" joined the Goldman Sachs Philanthropy Fund heavily influenced (if not entirely controlled) by Rockefeller interests since its inception.

Rockefeller-controlled financial institutions include, most prominently, Goldman Sachs. This firm is the main powerhouse directing the American economy and the Department of the Treasury. Almost all Treasury Department directors train first at Goldman Sachs.[450]

Global Research detailed the Rockefeller-banking influence that best explains George Soros's so-called 'liberal' leanings.[451] This includes the direct impact Sherri Kane, and I discovered and disclosed involving Morgan Stanley involved in money laundering for Hawaii's major drug trafficking enterprise. Our court transactions and evidence indicting ayahuasca kingpin and Bronfman cohort, Paul Sulla, led to Morgan Stanley and Goldman Sachs. These entities are major "institutional investors" in Barry Diller/Chelsea Clinton/Edgar Bronfman's IAC media conglomerate. This racket extends to the Bidens and the National Security Crime Syndicate, including the CIA and FBI.

Why We Need A Smaller U.S. Population And How We Can Achieve It

— A message from Negative Population Growth, Inc. —

We need a smaller population in order to halt the destruction of our environment, and to create an economy that will be sustainable over the very long term.

We are trying to address our steadily worsening environmental problems without coming to grips with their root cause — overpopulation.

If present immigration and fertility rates continue, our population, now over 264 million, will pass 400 million by the year 2050 — and still be growing rapidly!

All efforts to save our environment will ultimately be futile unless we not only halt U.S. population growth, but reverse it, so that our population can eventually be stabilized at a sustainable level — far lower than it is today.

The Optimum U.S. Population Size

The crucial issue is merely this: At what size should we seek to stabilize U.S. population? Unless we know in what direction we should be headed, how can we possibly devise sensible policies to get us there?

The size at which our population is eventually stabilized is supremely important because of the effect of sheer numbers of people on such vitally important national goals as a healthy environment, and a sustainable economy.

We believe these goals can best be achieved with a U.S. population in the range of 125 to 150 million, or about its size in the 1940s. This optimum size could be reached in about three to four generations if we do two things now that are well within our grasp.

How To Get There

1. Impose restrictions on immigration that would halt illegal immigration, and cap legal immigration at not over 100,000 per year, including all relatives, refugees and asylees. That alone would sharply slow our growth.

2. Lower our fertility rate (the average number of children per woman) from the present 2.0 to around 1.5 and maintain it at that level for several decades. We believe that non-coercive financial incentives will be necessary in order to reach that goal.

If almost all women had no more than two children, our fertility rate would drop to around 1.5, because many women remain childless by choice, or choose to have not more than one child. We promote the ideal of the two-child maximum family as the social norm, because that is the key to lowering our fertility.

Incentives to Lower Fertility

NPG proposes these incentives to motivate parents to have no more than two children:

• Eliminate the present Federal income tax exemption for dependent children born after a specified date.

• Give a Federal income tax credit only to those parents who have not more than two children. Those with three or more would lose the credit entirely.

• Give an annual cash grant to low income parents who pay little or no income tax, and who have no more than two children. Those with three or more children would lose the cash grant entirely.

Two Vastly Different Paths Lie Before Us

With the reductions in immigration and fertility we advocate, our nation could start now on the path toward a sustainable, and prosperous, population of 125 to 150 million.

Without such a program, we are almost certain to continue our mindless, headlong rush down our current path. That path is leading us straight toward catastrophic population levels that can only devastate our environment, and produce universal poverty in a crowded, polluted nation.

To learn more about NPG's recommendations for programs designed to halt, and eventually to reverse, U.S. and world population growth, write today for our FREE BROCHURE.

NPG is a national nonprofit organization founded in 1972. We are the only organization that calls for a smaller U.S. and world population, and recommends specific, realistic measures to achieve those goals.

🌐 **Negative Population Growth, Inc.**
P.O. Box 1206, 210 The Plaza, Suite 7K, Teaneck, NJ 07666

The above ad initially appeared in the March/April 1996 issue of *Foreign Affairs*, the esteemed political periodical published by the David Rockefeller directed Council on Foreign Relations—a private organization that largely represents the interests of multinational corporations and the Anglo-American banking cartel. A nearly identical version of this paid editorial appeared on page 21 of the July 2003, issue of *Harper's* magazine. The ads, placed by Negative Population Growth, Inc., an offspring of the Population Council organized by the Rockefeller brothers in New York, called for approximately half of the American population to be eliminated through methods of population controls, that have been largely unsuccessful.

I already mentioned that from our court cases, we discovered and published that Burisma's leading competitor in Ukraine was, like us, damaged by this drug-trafficking vaccine-pushing mob. As I will explain later, the Russian cocaine dealer Stefan Martirosian worked with Sulla to defraud Burisma's competitor, the Vitoil Company,[452] and the ArmbusinessBank.[453] Evidence-tampering by the previous Chief Justice of Hawaii's infamous "Drug Court" on the Big Island, Ronald Ibarra, concealed these facts and complicit parties.

Accordingly, among 'liberal-leaning' Democratic Party donors, Deep State affiliations are solidly evidenced and grossly hypocritical.

For example, in 2014, Goldman Sachs reported contributing $155,850 to five Planned Parenthood affiliates, including the Planned

Parenthood Federation of America.[454] It also donated $19,750 to the Heritage Foundation in 2014. These are certainly not 'liberal' beneficiaries, as reported by the globalists' media. Planned Parenthood owns and licenses RU486 (i.e., mifepristone[455]), causing serious side effects including high fevers, cardiac arrhythmias, pelvic pain, severe chronic nausea, vomiting, diarrhea, and general weakness. Yet, this is heavily promoted in abortion clinics cloaked in 'women's rights' politics.

In addition, how many reasonably-informed women would elect *vaccinations* partially made from aborted human fetus cells? Although the fetal cells may not be in the vaccines directly, "fetal cells are used to grow vaccine viruses."[456] Fetal cell parts, their proteins, or antigens, thereby find their way into the vaccines.

How many reasonably intelligent women avoiding foods labeled with GMO warnings would elect to vaccinate their children knowing all vaccines are made from laboratory mutated GMOs?

More foul play and hypocrisy are obvious. Bill Gates's Bill and Melinda Gates Foundation openly allied with Planned Parenthood to depopulate the planet using vaccines. This is no 'conspiracy theory.' Gates himself lectured on 15% of the world's population being terminated by vaccines that his allies produce and distribute.[411]

Also, the Heritage Foundation is suspect. According to *Influence Watch*, it is widely known as one of the largest nonprofit "right-of-center foundations." Its stated mission is "to formulate and promote conservative public policies based on the principles of free enterprise, limited government, individual freedom, traditional American values, and a strong national defense." So why would Goldman Sachs and the global elite that finances Planned Parenthood also fund the Heritage Foundation? Because both sides of the political aisle must remain controlled for globalists to prevail.[448]

The 'left-leaning' *Huffington Post* delivered an insightful condemnation of the politically controlled "alt-right" 'neo-Nazi' influence dominating news headlines, supposedly justifying the WOKE movement and claims of "systemic racism." "Alt-righters are mostly young white men who are angry about income inequality, poor job prospects, PC culture, crumbling social welfare programs, and war," the *Post* obfuscated. "Along the way, [Alt-righters] feed off disinformation and conspiracy theories that have gained credence thanks in no small measure to Republican efforts to demonize journalism, science, and what Karl Rove is believed to have dismissed as the 'reality-based community.' This journey is called being 'red-pilled,' a reference to the main character's choice in 'The Matrix' to swallow a red pill that shows him the horror of his enslaved reality or a blue pill that lets him remain blissfully unaware..."[457]

The *Post*'s obfuscations, financed today by Verizon Media, compound the propaganda output by trolls that seed social media with divisive messages to incite apathy, or alternatively, civil war.

Thus, you can see how the radical right and left are both leveraged, financed, and ideologically fueled by the same globalist operatives making billions of dollars administering black-ops for MindWar.

Goldman Sachs: Central to Genocidal Commerce

Through its numerous alumni, Goldman Sachs (GS) officials control the U.S. Treasury. You can see from the screenshot below that the leading 'Institutional Investors' in GS include the Vanguard Group, Inc. and Blackrock, Inc, two of the largest private equity investment (banking) organizations.

Blackrock was recently in the news for financing Chinese dominance over the American economy, National Security, and the military.[458]

According to my years of research, these two private equity investment companies, Blackrock and Vanguard, are nearly always financially fueling the most heinous criminal actions. Sherri and I exposed three examples of this in *The Las Vegas Deep State Massacre* book,[459] "Hawaii Nu-clear Ballistic Missile Crisis" report and video,[460] and related film *SPACEGATE: The Militarization of Sacred Mauna Kea*.[461]

In our *SPACE PEARL HARBOR* e-book published in 2015, we exposed Hillary Clinton's devil-doing as Secretary of State for the Obama Administration, granting China the "high-ground" of space for military superiority over the United States. Today, for the first time, at this writing, America's corporate-controlled media is finally heralding China's military superiority through supersonic missiles, not laser weapons leveraging satellites as we reported in *SPACE PEARL HARBOR*.

In each of our many publications, we vetted leading institutional investors in the profit-making political arenas, not coincidentally the Vanguard Group, Inc. and Blackrock, Inc.

Both of these entities have heavily invested in American healthcare controllers, including the suspect foundations. The three largest donors financed the Goldman Sachs Philanthropy Funds (GSFP) scheme: (1) Microsoft CEO Steve Ballmer working for Bill Gates's enterprise; (2) Steve Jobs's widow Laurene Powell Jobs; and (3) *Whatsapp* founder Jan Koum.

Bloomberg and *Guidestar* revealed that Microsoft's Ballmer donated $1.9 billion in 2016, Jobs gave $526 million in 2016, and Koum donated $114 million to the GS 'humanitarian' entity. These donors drove the GSFP's growth up 450 percent between 2015 and 2016. The Fund received $623 million in 2015 and $3.2 billion in 2016. That much money buys a lot of Deep State influence.

"The high wealth of many donors, tax benefits of donations, lack of disclosure requirements as compared to private foundations, and more, have led to criticism of these donor-advised funds..." *Wikipedia* reported.

EO Tax Journal Editor Paul Streckfus examined how 'real charities' hesitate to critique donor-advised funds. According to Streckfus, the global elite retaliates against anyone exposing how this criminal enterprise using front-foundations operates. Charitable groups which critique "donor-advised funds" "siphoning off billions of dollars into an investment" are black-listed. This is how 'leverage' over the commercial world, tax 'fairness,' and the global economy are maintained. This knowledge also explains how and why specific social agendas are advanced. Mandatory COVID vaccinations, mask-wearing, and social distancing are good examples evidencing such Deep State corruption.

Piercing the Veils of 'Liberal' Politics and 'Conservative' Profits

Examining the fund managers' and investment bankers' financing perceived 'liberal' and 'conservative' investments pierces the veil over the global control game exploiting the 'isms'— capitalism, socialism, and communism.

[Table: Goldman Sachs Group, Inc. (The) Institutional Ownership — GS $213.94 2.59 ↑ 1.23% — contents illegible]

For example, you can't get more 'conservative' than the financial and judicial forces working for Goldman Sachs (GS). Yet, careful inspection of GS's "institutional owners" controverts presumed 'liberal' affiliations. We see this with the Bank of America Corp., a leading owner of GS and backer of Barack Obama. Investigative journalist Wayne Madsen critically examined this association. Madsen and other authors criticized the CIA and Deep State's international connections, especially condemning the Hawaii drug mob.[462]

Goldman Sachs Philanthropy Fund's leaders in 2016 included Richardo Mestres as Chairman and Karey Dye as President. Mestres was a partner in the CIA's darling law firm, Sullivan & Cromwell LLP. Dye was Managing Director of Goldman Sachs for many years.

Naming Sullivan & Cromwell LLP, you might as well name the CIA that took over the law firm under the influence of Allen and John Foster Dulles, who served the Rockefeller banking cartel underwriting GS.

The NASDAQ chart screenshot below shows JP Morgan Chase and Morgan Stanley heavyweights in the Rockefeller Deep State enterprise helped finance Soros's Open Society Foundations' prerogatives.

SOROS FUND MANAGEMENT LLC

Report Date: 03/31/2019

These political agendas financed much of what you are witnessing today in America, including the lenience on illegal immigrants." Enacting comprehensive immigration reforms have made it easy for illegal immigrants to become citizens or simply evade deportation—increasing welfare handouts to discourage the American work ethic, causing many people to drop out of the workforce and raising taxes to redistribute wealth. *Influence Watch* heralded all of these damaging and deadly policies.

Now add the disgraceful Afghanistan withdrawal administered under the guise of the Biden administration. In fact, this appears to be one more example of the influence undermining American foreign policy exercised by Open Society Foundations.

Accurate as this intelligence is, you can see this dividing right from left precisely reflecting a MindWar featuring political illusions and globalist impositions.[448]

Among interrelated news stories was Ocasio-Cortez's debacle pushing the Green New Deal. Ocasio-Cortez's appearance on *The Young Turks*, a liberal internet show, evidenced her support from billionaire Soros.[463] The program was supported in part by the Open Society Foundations. In fact, she credited the online show with giving her congressional campaign exposure.[464] But her campaign did not receive any direct funding from Soros.[465-466] That would be 'too obvious' and potentially harmful to moderate voters. Instead, Soros financed OC's media exposure.

These transactions evidence the abuse of the media for political persuasion and black-ops.

According to Reuters —the quintessential Deep State propaganda source directly connected to England's GCHQ— it is widely known that Soros's 'non-profit' was rumored to have financed the Occupy Wall Street protests.[467] But little evidence exists to back those rumors, Reuters hashed.[468, 469]

Soros, the UN, WHO, and Hallucinogenic Drugs

Perhaps Danny Nemu summarized George Soros's influence best by writing for the *Psychedelic Press UK* in 2014. He revealed Soros's Open Society Foundation's (OFS) commitment to the social acceptance of the hallucinogenic drugs, especially the "God molecule" —DMT.[470]

"Kasia Malinowska-Sempruch, the Open Society Foundation's (OFS) Global Drug Policy director, spoke at the UN General Assembly Special Session on Drugs. Nemu wrote. This was "an historic opportunity to influence global drug policy," and Malinowska-Sempruch was "well-experienced in handling policy at this level."

"But her rhetoric seemed a little too simplistic, Nemu criticized. "Perhaps it is a function of her working in policy rather than anthropology, but her potted history of anti-malaria medicine celebrated the 'millions of lives saved' and neglected to mention the costs. These include the many tons of DDT and other persistent pollutants used in pristine environments since 1955 when the [Rockefeller/Deep State-directed UN] World Health Organisation (WHO) began its malaria eradication program.

"[T]he failure of said program left a generation with no natural resistance, leading to massive resurgences and spread of malaria, including drug-resistant strains."

As a result, Nemu concluded, malaria today kills more people than ever. Deforestation and petrochemical pollution contributed to the "catastrophic destruction" of Amazonian rainforests that gave birth to ayahuasca and the spread of disease and death to natives.

Sherri Kane and I published extensively on the CIA's and U.S. military's roles[335] in the creation of the "Daime religion"[471] first hatched in Brazil. White man's corporate presence in malaria-stricken countries brought genocide depopulating natives, not simply the rainforests.

I studied Steven Kunitz's[472] work at the University of Rochester and subsequently at Harvard. Kunitz was considered America's foremost medical sociologist. Kunitz published a revealing article in the *American Journal of Public Health* titled, "Globalization, States, and the Health of Indigenous Peoples."[473,474] He documented that corporatism administered by "white Anglo-Saxon Protestants (WASPS)" regularly resulted in genocide(s), the extermination of populations, cultural degradation, and irreconcilable damage to environments, especially in lesser developed nations; all profitable and ongoing since the early twentieth century.

Partnership for New York City and Profitable Depopulation

In April 2010, Sherri and I broke astonishing news that Leon Black's Apollo Global Management had "veered" billions of dollars from the 9/11 World Trade Center "reconstruction fund" to Las Vegas. The conduit for this apparent money-laundering operation was MGM Grand International, Inc., directed by James Murren, the White House's chief commercial consultant and advisor on homeland infrastructure security at risk of terrorist attacks.

In our film *PharmaWhores: The Showtime Sting of Penn & Teller*, we exposed that the head of this Deep State 'octopus' operated from New York City to largely control arguably every aspect of international trade and cultural indoctrination. This global business group is The Partnership for New York City (PFNYC).[475]

Although their private meetings have remained shrouded in secrecy at the time of our study, key players in this David Rockefeller/Royal Family of England established PFNYC were co-chairman Rupert Murdoch of News Corps and Lloyd Blankfein, the chairman of Goldman Sachs (GS). Sherri and I identified U.S. Treasury Dept. members moving through the 'revolving door' between GS and the Treasury.[475]

In addition, many of the world's most influential businessmen and women controlling healthcare, pharmaceuticals, military, energy, intelligence, and other industries were well-represented in the PFNYC.[476]

The PFNYC was pledged to play a central role in reconstructing Ground Zero following the 9/11 attacks, according to Kathryn Wylde, President & CEO of the Partnership and Director of The Federal Reserve Bank of New York. The organization compiled the economic report on the damage done, advanced financial plans for reconstructing the World Trade Center and advised leading financiers regarding reconstruction investments.

According to 911-Truth movement directors, Wylde and other members of the PFNYC are implicated in the treasonous attacks that scientific evidence says involved thermitic explosives used in controlled demolitions.

Reflecting on Epstein's painting of George W. Bush flying paper planes into two fallen towers, propaganda was necessary to conceal the National Security Crime Syndicate during 9/11, since public polling showed more than forty (40) percent of American adults believed those "terrorist attacks" were an "inside job" involving Vice President Dick Cheney and Halliburton Co., a leading military contractor (also implicated in the Gulf Oil disaster, as we documented and broadcasted.[477])

Investment bankers at JPMorgan Chase, GS, and other private equity firms active in the PFNYC, including Black's Apollo Management, acquired controlling interests in the reconstruction fund that was "veered" from NYC to Las Vegas.

Many PFNYC members gained controlling interests in the largest drug firms during mergers and acquisitions. They had placed 'depopulation' near the top of their list of geopolitical and Brave New World priorities. Their depopulation agents became top officials in government, finance, and industry to move their plans forward.

The depopulation plan helped secure globalization. Smaller populations are easier to control and pose fewer risks of an uprising. Depopulation was advanced by the world's wealthiest people, including Bill Gates. As mentioned earlier, Gates had admittedly funded vaccinations to reduce global populations by 10-15%. Leading population planners and economic developers advanced identical plans to cull the world's population to 1 billion or less from more than 7 billion.

Killing 6-out-of-7 people globally, most profitably, required planning and an unprecedented conspiracy. Who, after all, other than the wealthiest psychopathic sex fiends ritualistically torturing children and killing babies through vaccination "injecticide," was capable of committing global genocide?

"Leon Black, the billionaire chairman of Apollo Global Management, assured investors in his firm's funds that his relationship with convicted sex offender Jeffrey Epstein didn't extend to the private equity company," reported *Bloomberg*.[478] But that did not jibe with public knowledge that Epstein had contracted from time to time with Black to "pitch personal tax strategies to the firm's executives."[479] Tax strategies" was not the only service Epstein was selling Black's chief money managers.

By financing and applying the costly advances at MIT and IBM, available only to people with money on the level of Gates, Soros, Epstein, and Black, the world's power elite could leverage neuroscience and genetic biotechnologies for "globalization" and transhumanism.

Therein, their plans for more "novel" vaccinations, intertwined with brain-conditioning psychotropic drugs, played a key role.

It must be highlighted that the Royal Family of England—a majority shareholder in General Electric (GE)—chartered PFNYC. GE is the world's largest company that controlled NBC/Universal/Comcast and MSNBC news network with Bill Gates and Barry Diller's ICA. Diller appointed Chelsea Clinton to co-directed ICA with Edgar Bronfman, the most influential official directing NBC's Universal Music Group.

THE COVID COUP WITH FINANCING FROM HELL

On November 5, 2021, at least eight people died, and "scores" of injuries were reported from the chaos that ensued at the Astroworld festival in Houston, TX, during the Travis Scott concert tied to the owners of the Universal Music Group. The music frequency-related tragedy was first reported in a video published by the *SGTReport*.[480]

The concert organizer was Live Nation Entertainment, Inc.—a division of Liberty Media that also owns Sirius XM radio. Live Nation also produced the "Las Vegas Deep State Massacre" event detailed in the next section. That terrorizing tragedy was controlled by complicit enterprise administrators that produced the Bataclan Theater massacre in Paris in 2015.[481] I cover that massacre too, two sections ahead.

This pattern-and-practice of producing terrorizing musical events implicates, once again, Edgar Bronfman, Jr. officiating the Universal Music Group; Vivendi that owns the Lagardère media empire; Qatar's monarch, Sheikh Tamim bin Hamad Al Thani; and Democratic Party notables Ari and Rahm Emanual. Ari, the William Morris-Endeavor Agency owner, booked the "Eagles of Death Metal Band" for the " Bataclan venue. The band sang a tribute to Satan when the shootings began.

All three tragedies drove the globalists' political agendas.

The PFNYC, complicit in the 9/11 "terrorist attacks," was co-directed by Jeffrey B. Kindler, the Chairman, and CEO of Pfizer–producer of the COVID mRNA vaccine with Moderna. Kindler stewarded Pfizer through multi-billion-dollar acquisitions also involving JPMorgan Chase and Goldman Sachs financing.

The Pharmaceutical Research and Manufacturers of America (PhRMA), the drug industry's main trade organization, was also directed by Kindler. PhRMA officials engaged the White House and Pentagon officials in private negotiations determining the government's stockpiling and price-gouging of drugs and vaccines. Trump opposed the over-charging.

This was how the system actually worked when COVID struck—through largely covert administrations of multi-billion-dollar pharmaceutical contracts profiting multi-trillion-dollar investment firms—all symptomatic of the industry's demonic corruption.

The "corporate shell game" was played using mergers and acquisitions directed by the same relatively few people. Their creation of the PhRMA trade organization provided the illusion of their legitimacy and fair market competition. Price fixing occurred behind closed doors, explaining why prices varied so widely internationally.

This also explains why news programs, financed mainly by drug ads, demonstrate coordinated reporting. Even the same words and sound-bites are used and repeated by different networks to condition viewers most persuasively.

And if you are still wondering why officials neglected to broadcast natural alternatives to the experimental COVID vaccines and damaging drugs, consider the example of Congressman Henry Waxman (D-CA). He sucked Wall Street's slime to discourage dietary supplements. These included beneficial vitamins, minerals, herbs, etc. Waxman snuck restrictions into the regulatory language of 'The Wall Street Reform and Consumer Protection Act of 2009' (H.R. 4173).

For years, Waxman had unsuccessfully attempted to pass legislation restricting consumer access to nutritional supplements in favor of BigPharma. Influenced by BigPharma's bankers, the FDA weakly contended that regulating vitamins, minerals, herbs, homeopathics, essential oils, silver products such as OxySilver™, and much more, posed risks to consumers.

Diversions, Distractions, and False Flag Operations

On October 1, 2017, Sherri and I had tickets to attend the "91 Harvest Festival" in Las Vegas. But our flight was delayed, 'by the grace of God,' and we missed the massacre. Fifty-nine people were killed, and 527 were wounded that night in American history's most deadly mass shooting.

Late that night, and for days after, we had the opportunity as investigative journalists to interview witnesses and victims. We researched suspects and identified the Deep State's MGM Grand/NBCUniversal leader, James Murren, as allegedly involved in the tragedy.

In our book, T*he Las Vegas Deep State Massacre*, we exonerated the alleged single gunman, Stephen Paddock, for many reasons.

Our discoveries were fascinating and distressing. Many sources reported that Bill Gates, George Soros, and Michael Chertoff were complicit in the massacre.

Recall from Chapter III that Chertoff, who operated in the highest levels of the National Security Crime Syndicate, was cited in the Fauci e-mails by Starnes Walker pursuant to commercializing COVID biodefense.

Gates had gone into the hotel business when he joined Saudi Prince Alwaleed bin Talal in purchasing the neighboring luxurious Four Seasons hotel and casino for $3.8bn in 2007. The deal conveyed the top five floors of Mandalay Bay to these two investors. There were several unsettling reports that these floors played a crucial role in the massacre.

By then, we knew enough about media intelligence to presume that anything and everything the media broadcasts by consensus was intelligence agency sourced or approved. And the media's "lone gunman" story was clearly false.

We reached an alternative conclusion by examining compelling evidence, including several videotapes shot from different angles. These showed multiple shooters at several locations. Revealing video evidence neglected by the press included a group of snipers and at least one helicopter filmed firing down at the crime scene captured from two different angles by frightened citizens using cell phones.

Our findings were corroborated by photos, eyewitness accounts, victims' testimonies, police statements, independent ballistics analyses, military officials, medical doctors, a social psychiatrist, lawyers' opinions, and "'military neuroscience'-MindWar style" false reports published by the complicit media.

As usual, our controversial disclosures were heavily suppressed by Google/YouTube, *Wikipedia*, and the corporate-controlled media. In the few news outlets opened to reporting our story, 'trolls' were assigned to smear, discredit, and 'neutralize' our impact.

Our 3-month study also examined thousands of documents showing evidence-tampering and obstruction-of-justice by law enforcers and major news outlets harboring James Murren and Michael Chertoff's group of key suspects. Famous corporate officials close to Chertoff with close ties to the CIA and the Department of Homeland Security (DHS) were heavily implicated in the massacre. We named the key defense industrialists and their companies who profited most from the bloodbath.

A pre-planned simultaneous DHS "drill" and multiple reports of shootings at hotels along the Vegas Strip at the time of the massacre implicated officials with conflicting interests in the intelligence community and defense industry. Murren, as aforementioned, served as a Deep State government "insider," as President Trump's advisor on national 'infrastructure' security responsible for accessing risks posed by terrorists.

The massacre and media coverage promoted "gun control." It profited from security and arms dealers who had acquired government positions. This key personnel includes Michael Hayden, previous Director of the CIA until 2009. He subsequently co-created the Chertoff Group. That business 'made a killing' selling security technologies and services in the aftermath of the massacre. They were positioned to relay "insider information" reflected in large stock trades that we evidenced, including Murren's shorting of MGM Grand stock only weeks before the massacre.

Our e-book, *The Las Vegas Deep State Massacre*, confirmed early police reports of shooting victims far east of the concert venue. A nearby fuel tank supplied a "triangulating" position for snipers filmed at the rear of the concert field.

Two separate ballistics analyses proved beyond any reasonable doubt at least one ground-level shooter. Evidence of two bullet holes in the fuel tank appeared suspiciously altered by police or FBI investigators, arguably to conceal the ground shooter and fake the Mandalay Bay trajectory.

We analyzed the common political, commercial, and insider trading motives of the corporate suspects. "Deep State" involvement was proven beyond any reasonable doubt by twenty-four (24) confirmed facts and unreasonable media diversions. We analyzed six likely motives for the massacre. Examples of the media PSYOPS included the exclusive interview on MSNBC by Ellen DeGeneres of "Jesus Campos," later proven to be an imposter.

These connections also tied Barry Diller's organization into corruption. Recall that Diller was the owner of ICA, the parent company of the Bronfman-directed MSNBC-Universal Music Group conglomerate. MSNBC —the IBM-MIT- Gates MindWar television network— broadcast *altered* images of Paddock and his girlfriend, compounding evidence of fraud.

We witnessed similar Photoshop-altered pictures evidencing media fraud in the Epstein suicide scandal. On August 15, 2019, Rupert Murdoch's *New York Post*[482] (that blundered in breaking news of Epstein's "suicide" as we reported previously) published a photo of Jeffrey Epstein's "gal pal" —Madam Ghislaine Maxwell— allegedly "spotted at In-N-Out Burger." She appeared to be posing for the perfect camera shot, as though it was a contrived 'photo op.' She was falsely reported by the *New York Post* as reading a book about the secret lives and deaths of CIA operatives.[483] It was unreasonable to believe that Murdoch's paparazzi could locate Maxwell for a "photo op" while top Justice Department officials could not. We showed that Rupert Murdoch's *New York Post* had switched images in that photo[397] and had done the same in the photo used to break the news about Epstein's misrepresented death by suicide.

As with Epstein's contested "death," we rejected the most common 'conspiracy theories.' We simply verified facts and provided reasoned analyses.

Then we were further attacked by trolls, and for the first time, *scam bots* too.[484] These multiple AI automated fake cyber-identities stalked us online. They posted disinformation all over the Internet and social media to 'neutralize' our findings.

Deep State critics produced several false and misleading Google/YouTube videos. Meanwhile, my videos were consistently blocked, much like the rest of my videos had been for years.

Ironically, I noted, the entire PSYOPS operation in both cases—Epstein's "death" and the Vegas massacre—traced to the same commercial and political players. This was much like the **MINDFREAK** advertisement overlooking the 'killing field.'

MGM/Luxor posted the freaky banner high above the massacre scene to advertise their 'magic act.'

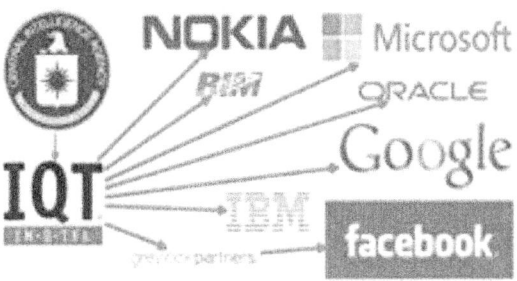

The Bataclan Theater Massacre

Earlier in this chapter, I relayed Newt Gingrich's conclusion that the Democratic Party is "waging war against the American people. In the COVID Coup context, the Biden administration appears to be similarly engaged. "More people have died of COVID in 2021 under Biden's administration than perished under Trump's reign," news outlets report, despite draconian measures and mandates being increasingly implemented.

The degeneration of civility and the American economy, inflation and gang operations, skyrocketing rates of crime, and deaths from fentanyl trafficking through the open border, all this has been plotted by the Deep State elite and administered most clearly by "liberal" and "progressive" Democrats.

If the Luciferian elite is capable of administering "The Las Vegas Deep State Massacre" without showing their hand manipulating the 'deck;' likewise releasing AIDS, Ebola, Zika, West Nile Virus, the Swine Flu, herpes and hepatitis viruses, mailed anthrax, and COVID-19 without getting caught, then what else are they capable of doing?

Sherri and I previously reported on the Bataclan Theater massacre in Paris that killed 130 people and triggered fears of immigrants terrorizing nations worldwide.[485] We evidenced that the same Democratic Party financiers administered this politically motivated 'blood sacrifice' on Friday the 13th in 2015. We vetted the complicit actions of devil-doers, including Ari Emanuel.[486]

Ari is Hollywood's top talent provider. As mentioned above, he owns the William Morris Endeavor agency (WME) —America's oldest entertainment company. Ari's brother is the political heavyweight, Rahm Emanuel— past Chicago mayor and Chief of Staff for the Obama White House.

Evidencing the criminal operations of this close-knit group of Democratic Party financiers, Ari booked the Bataclan Theater event by and through Edgar Bronfman, Jr.'s associates at the Warner/NBC/Universal Music Group.[487] The slaughter was knowingly set at the Bataclan because that venue was purchased only a few months earlier by the wealthiest hate-filled anti-Semite in the world, the Sheik of Qatar.[488] That Sheik is the most influential owner of the Al Jazeera media conglomerate.

Emanuel's group booked The Eagles of Death Metal Band to play at the right place, at the right time, even the right song, to deliver **the most evil anti-American message in history**. At the same time, it would help smear the real iconic band, The Eagles, who Universal Music Group had damaged along with more than 700 other artists while fraudulently concealing "The Day the Music Burned" —a fire purportedly set 'accidentally,' destroying "The God List" of original master recordings.[489]

At the Bataclan, among the deceased (i.e., collateral damage) was Thomas Ayad.[490] He worked for Mercury Records, a division of Universal Music France.[491]

In August 2011, Edgar Bronfman became Chairman of Warner Music, and Stephen Cooper became CEO. He stepped down as Chairman on January 31, 2012, *Wiki[hypno]pedia* reported. That CIA's propaganda mill concealed that Bronfman's resignation closely followed his conviction on inside-trading charges in France.[487]

Wikipedia reported that in February 2017, a group of investors led by Bronfman were considering purchasing Time Inc. and Meredith Corp.

Meanwhile, Barry Diller's and Chelsea Clinton's co-director of the ICA/NBC-Universal consortium, Bronfman Jr.,[492] was the chairman of Endeavor.[493] This international non-profit development organization "supports entrepreneurs."

These suspects and companies tied the Bronfman Crime Family and the Bataclan devil-doing to more than Ari Emanuel's William Morris

Endeavor Hollywood talent firm. These facts connected the Bronfmans to the heart of the Deep State's "Jewish Mafia" controlling Hollywood and the mainstream media.

In this context, the CIA's takeover of Hollywood from the original "Hollywood Jews" during the McCarthy Era was expertly detailed in the film *Hollywoodism: Jews, Movies, and the American Dream* by Simcha Jacobovici. That movie was based on the book *An Empire of Their Own: How the Jews Invented Hollywood* by Neal Gabler.[494]

After the CIA overtook the original Jewish mob running Hollywood, a new breed of demonically possessed criminal psychopaths served their new Deep State bosses. The 'Hollywood Jews' active now and identified in this book serve multi-denominational associations, especially obvious in and around NBC and its affiliates.[495]

Too influential to neglect, the new Hollywoodism aligned itself by Deep State complicity with George Soros's money and investments in liberalizing America, along with its drug laws.

Ari Emanuel and his brother Rahm were born to the infamous Israeli doctor and generally-agreed-on terrorist, Benjamin M. Emanuel. The senior Emanuel was 'sworn to silence' in the Jewish paramilitary organization that operated in "Mandate Palestine" called the **Irgun**.

The Irgun was based on "Revisionist Zionism" —an extremely radical ideology that appeared to be central to Deep State criminality. This was the equivalent to "Christian Identity," wherein demonic right-wing extremists' ruled with doctrines that prescribed either Jewish or Palestinian death camps.

I view both religious and political factions as part of the Deep State's divide-and-conquer protection racket. Their media administers this divisive strategy and globalization protocol.

Zionists, in the traditional sense, were peace proponents, scholars, doctors, lawyers, and accountants. But not "Revisionist Zionists" who were traumatized and sexually abused as a youth to the point of suffering criminal psychopathology, as we see with Epstein, Soros, the Emanuels, Harvey Weinstein, and Henry Kissinger.

This criminal psychopathology demonstrated by the ruling elite's "Hollywood Jews" featured sexual perversion and drug dependence. This behavior precisely reflected their radical revisionist counterparts in Protestantism, Catholicism, and Islamic religions. People like "holocaust deniers" encouraged repeated exterminations of Jewish people —the equivalent of what self-loathing Revisionist Zionists prescribe for Palestinians. The extremists on both sides justify and extend violence.

In this context of Hollywood propaganda under globalist control, it is not surprising that COVID vaccine advocate, Dr. Deepak Chopra, appears in the photo below. Chopra is NBC-MGM's darling spiritualist and Epstein beneficiary.[496] The photo was taken at the MGM Grand shopping "District" in Vegas. It shows Chopra advertising in "Television City." There, Chopra plugs the enterprises' provision of natural health spa services for wealthy guests.

True to form, Chopra's celebrity was also used to promote the 'New World Religion' featuring ayahuasca, as seen in *The Reality of Truth.com*. Kurt Engfehr produced this highly touted documentary. Engfehr co-produced with Weinstein Brothers-financing films by director Michael Moore, distributed by MGM. These included the liberal-leaning *Bowling For Columbine* and *Fahrenheit 911* co-produced by the sexual perverse Harvey Weinstein.

Moore didn't have kind words to say about Weinstein either. Moore took the sexual predator to court for allegedly stealing Moore's money.[497]

You can see SHOWTIME's marques adjacent to a neuroscience laboratory in Television City in the photo above. People here are ushered through the entrance below Deepak Chopra's advertisement. The guard with the blue shirt escorts visitors to their seats. The audience is solicited with "free passes" to view forthcoming episodes of TV programs. While seated, the "guests" are connected to electronic neuroscience research equipment. Electrodes are placed on their bodies to record every physiological and emotional reaction. Responses to each scene and statement are cataloged. This data is used to optimize MindWar messages, including subliminal programming and suggestions to influence viewers' behavior and engineer society.

These operations rely on the most advanced military neuroscience and MKULTRA media mind control techniques used to divert public attention away from anything risky to the Deep State's discovery. This way, officials obscure anything the shadow governors intend to conceal or muddle, such as vaccination risks or transhumanism overtaking the human race.

CHAPTER VIII
INCITING VIOLENCE AND CIVIL WAR AGAINST THE UN-VACCINATED

> *"Be not deceived: evil communications corrupt good manners... For God is not the author of confusion but of peace."*
>
> 1 Corinthians 14 and 15:33

"No pandemic in earth's history ever evolved unaccompanied by major socio-economic and political upheaval," I preached over the past quarter-century.

COVID-19 is no exception. The mass protests and rioting that followed media-promoted 'social distancing,' 'masking,' and George Floyd's murder are best explained by behavioral science in the context of the psychosocial distress caused by governmental impositions directed by the global elite.

This chapter contributes proper diagnosis for the effective treatment of systemic illness we witness in America today, a sickness rooted in the behavior of multi-national corporate board members securing their investments, psychopathology, and demonic possessions.

Under their evil influence and evil obedience, complicit officials manufactured the mounting chaos destroying this ungodly nation. America's fall has come through secret society-controlled pseudoscience, monopolized medicine, corrupted politicians and law enforcers, and the divisive media.

Therefore, it was not 'systemic racism' that 'caused' the crises to worsen at the time of this writing. The media and masses' contentions are mistaken. The allegation or presumption of guilt indicting "systemic racism" in America is controverted by too many discrediting facts, including Blacks killing Blacks and rioters destroying Black communities.

"Systemic racism" and white supremacy in America is certainly a scourge underlying and sabotaging the 'social fabric.' But that 'misdiagnosis' neglects the true cause of the damaging riots, mounting protests, police brutality, COVID-19, vaccination depopulation, and transhumanism.

The 'true cause' and diagnosis of all of this is white-collar corruption and racketeering in organized crime that has poisoned every part of civilization.

Whistleblowers and conscientious objectors are discredited as part of the cartel's "protection racket." Complicit agents use the "art of deception," behavioral science, social conditioning, and media propaganda to inflict compounding damage.

The image above comes from a training slide sourced from the main MindWar agency in the British military, the GCHQ, that works in tandem with the CIA. It shows magician Raymond Teller, who is a Fellow at the <u>Cato Institute</u> think-tank in Washington, D.C., that addresses "media exposure" for "societal influence," "conjuring with information."

I've already mentioned that 'conspiracy theorists' and whistleblowers are effectively 'neutralized' by covert agents and operations.[498] Such "black ops" conducted by intelligence agencies have been vetted by fellow persecuted author Glenn Greenwald.[499] MindWar administrators neutralize and control opposition enabling totalitarian fascism to expand under the guises of "public health," "National Security," and "law enforcement."[500]

INCITING VIOLENCE AND CIVIL WAR AGAINST THE UN-VACCINATED

This 'systemic corruption' represents an un-diagnosed plague. This institutionalized madness and violence reflect **behavioral sickness** secured by ignorance at best or willful blindness at worst.

Willful blindness, arrogance, and evil obedience sow the seeds of social discord.

This systemic pathology has turned health workers and law enforcers into death administrators. This has left governments enslaved to board-roomers and globalist 'banksters.'[501,475]

In this context of behavioral science and controlling 'intelligence,' public outrage and 'civil disobedience' were predictable and predicted by awakened analysts.

Retired U.S. military intelligence officer, now radio talk show host, John Moore, is one such honorable example addressing the 'political right.' Carl Nelson, America's leading Black community educator and radio personality, is another great example largely enlightening the 'liberal left.'[502]

Otherwise, as the 'blind-led-the blind' and ignorance ruled over healthcare policies and practices, evil obedient 'experts' directed the social, economic, and political behaviors and responses to COVID-19 that became **genocidal**.

In retrospect, little-to-no competent good-faith planning preceded the COVID response.[503] There was an alternative evil agenda behind the "COVID Coup."

Accordingly, **common sense** and **behavioral science** best explain the distressing and damaging outcomes we have witnessed in the wake of COVID-19. The so-called 'leaders' grossly neglected available science and common sense. This intentional malfeasance led to unnecessary deaths.

To argue otherwise, one would have to ask, **How could these deadly (arguably demonic) social conditions, and the media's lethal diversions, escape the diagnosis of public health officials and expert policy-makers?**

For instance, it was obvious using common sense to anticipate suicide rates skyrocketing due to forced isolation by "social distancing" and business "lock-downs."

This chapter explains why reckless negligence and official malfeasance occurred willfully, knowingly, and intentionally to cause fearful, phobic, and criminal minds to turn violent.

The Hidden Dangers in Neglecting Common Sense

The importance of the behavioral and social sciences pursuant to COVID-19 has been recklessly, even criminally, neglected to serve concealed special interests and their covert agendas. No other reasonable conclusion better explains the pattern and practice of frightening, confusing and diverting the 'herd' into a state of chronic anxiety and chaos, depriving people's rights, censoring remedial views, and inciting criminal behavior.

Fears, phobias, and social unrest has been imposed, most effectively by "public health" officials who acted like commercial agents for the crime syndicate. All combined, their social conditioning and behavioral impact have been far more distressing and deadlier than the coronavirus was predicted to be.

All of this was premeditated. The "free press" and officials acted willfully blind to the shutdowns' scientifically proven risks and expected outcomes. Behavioral factors were grossly, recklessly, and damagingly neglected *by intent*. Globalists were bent on destroying America as the "land of the free," sabotaging the Trump Administration, and directing the global economy for the "Rise of the Fourth Reich."

It appears that the only immunization against these unconscionable impositions lies in redirecting faith and trust to the Creator of the human race. That is, beyond the covert intelligence officials that misdirect the National Public Health Service and Justice Department.[504]

According to more than 350 behavioral science studies conducted by Albert Bandura and colleagues decades ago, television, film, and video gaming are much more powerful in shaping human behavior than parents, teachers, and even peers.

This scientific fact has been recklessly neglected by more than media moguls and censors. The public health community, and social engineers within the 'population management' sectors of government and industry, have compounded this degenerative and destructive negligence. They have intentionally permitted 'programming' that has conditioned the current climate characterized by profitable politically persuasive

INCITING VIOLENCE AND CIVIL WAR AGAINST THE UN-VACCINATED

violence and sickness. The overriding doctrine of 'stimulus/response' behavior and "group think" has been unreasonably, unjustifiably, and unconscionably neglected.

These circumstances, scientific facts, and social outcomes are surmised in the opening quote from 1st Corinthians. The "fake news" media has especially corrupted "good manners." "Evil communications" have deceptively accompanied governmental 'shutdowns,' drug and economic damages, medical mischief, social impositions including mass quarantines, inhuman punishment-like isolation, interpersonal distancing, mask-wearing, racial divisions, and disparities, and assorted political crises including illegal immigration.

These conditions 'naturally' resulted in angry citizens, depressed people, suicidal youth, volatile mobs, widespread protests, and 'Babylonian' degeneration. All of this perfectly fulfills Bible prophecy and the globalists' agendas.

Meanwhile, most people remain clueless about the hidden powers behind the system-wide dynamics underlying these troubling circumstances.

"For we wrestle not against flesh and blood, but against principalities, against powers, against the rulers of the darkness of this world, against spiritual wickedness in high places." (Ephesians 6:12)

Evil forces, including television executives, media programmers, press leaders, social media regulators, 5G industrialists, Big Pharma executives, Big Banking profiteers, and angry anarchists enraging youth in groups like ANTIFA, have used threats and violence terroristically to leverage support and induce 'change.'

> *"The public health community, and social engineers within the 'population management' sectors of government and industry, have compounded this degenerative and destructive negligence, intentionally permitting 'programming' that has been conditioning the current climate characterized by profitable politically-persuasive violence and sickness."*

The rich and infamous have all contributed to fascist policies and social impositions by their complicity or silence. All have disgraced themselves and the administration of justice. They have disregarded Divine righteousness and karma.

Accordingly, this systemic pathology is best analyzed by **behavioral science.** In these proceedings, like the motivations of the instigators committing atrocities that remain largely concealed, even as God's intentions are being revealed, the devil-doers have been hiding behind the veils of 'public health,' 'law-enforcement,' 'progressive politics,' and the 'WOKE movement.'

Divine prophesy and deliverance supersede this earthly chaos. (See my separate related essay[504].)

COVID-19 Imposed Chaos

For months preceding the 'COVID-19 riots,' news coverage conditioned the public to accept governmental impositions such as severely distressing and dehumanizing 'social distancing'. Ethnic and racial disparities accompanied these impositions. These, too, were predictable, predicted, and even scientifically determined. Neglecting this intelligence costs millions of lives worldwide.

Major inequities with Blacks suffering a disproportionate amount of disease and deaths were even opposed by the federal government's leading COVID-19 vaccine official, Kizzmekia Corbett, to no avail. Corbett was severely smeared for comparing the racial disparity to "genocide."[505]

INCITING VIOLENCE AND CIVIL WAR AGAINST THE UN-VACCINATED

Considering the seriousness of genocide, it is unreasonable to presume these matters had nothing to do with the civil disobedience that followed George Floyd's televised murder. Like a cause-and-effect and 'stimulus response' reaction, COVID-19 riots followed a set of 'correlates and antecedents' seeding and foreshadowing the widespread carnage. Most important among these was the financing of unrest by the globalists, especially George Soro's Open Society Foundations orchestrating the civil unrest. Fears and deadly phobias underlying the uprisings, including the Black community's fear of white supremacy, were mass-mediated and socially engineered.

That's not to say white supremacy doesn't exist in America, nor "systemic racism" hasn't been a problem begging for solutions, but no one knows better than Kane and I about how the 'radical right administers the curse of this privileged identity and prejudice.'

From 2008 through 2016, we lost virtually everything, including our house, health spa, businesses, web properties, videos, articles, and businesses, to America's leading white supremacy propagandist, Alma C. Ott & Co. I already mentioned how Ott and FBI/CIA operator Gunderson directed minions of anti-Semites and political hacks to attack us online in my home and family.

At the time of this writing, I continue to be tied up in the corrupted Third Circuit Court of Hawaii by Ott's comrade in organized crime, Paul Sulla, as I detailed earlier. After fifteen years of stalking, harassment, malicious prosecution, and litigations by Sulla, Ott, and the CIA/FBI/Deep State stealing the most productive years in my professional career, I can tell you, this scourge of demonic criminality and political persecution is real and terrifying. Their actions resulted in

Sherri's manslaughter under the duress and severe distress of their attacks, prejudice, and deprivation of our rights.

It takes a lot of money and political power to oppose this devil-doing that sent me into bankruptcy in 2016. Black Lives Matter and Antifa didn't have this cash shortage once financed by Soros and his socialist/communist comrades fueling the 2020 riots.

The American Psychological Association (APA), in covering the "psychological impact of COVID-19,"[506] highlighted rises in civil unrest, anxiety, depressive disorders, panic attacks, and suicides in the wake of the 'plandemic.' The political impositions and behavioral restrictions caused people to become increasingly divided, confused, troubled, stressed, immune-suppressed, and more susceptible to diseases. This was all foreseeable and financed by the globalists.

One example was reported on May 29, 2020, by Tucker Carlson on *FOX News*. He interviewed Brad Hunstable, the father of a suicide victim. They discussed alarming death rates among children and young adults forced into quarantine or social isolation.[507] Later, Carlson interviewed Glenn Greenwald, who voiced the same opinion as I do, citing Deep State administration of COVID, its lab origin, and resulting global impositions.

'Public health' advertisements, like totalitarian edicts unproven by science, caused far more damage to society than even experts expected. Much of this was avoidable because behavioral science historically identified such risks and preventive strategies.

The ripping-apart of America's 'social fabric' was intended and successful. Social distancing, after all, undermined the main objective of gaining 'herd immunity.' Slowing the spread of the virus to secure

INCITING VIOLENCE AND CIVIL WAR AGAINST THE UN-VACCINATED

medical resources was foreseeably ill-conceived. Like cigarette smoking risking cancers, or sexual promiscuity risking AIDS,[508] people's misbehavior dictated deadly outcomes. With "love and belonging" central to human needs, depriving this by imposed separations was genocidal.

The "public health" prescriptions, including COVID-19 testing, were reliable.[509] There were too many false positives. This compounded foreseeable burdens to healthcare, the economy, and society.

These facts and circumstances challenged the intelligence and integrity of policymakers and law enforcers alike. Our leaders recklessly neglected what was obvious—'human nature,' psychology, sociology, medical history, and behavioral science.

This crisis could not have occurred by 'accident' or excusable neglect. These deadly outcomes were foreseeable and foreseen by world-leading experts.[92] Many opposed the shadow-governors imposing their tyranny on the masses. Many opposed the mass-mediated bioterrorism used to gain compliance.

Others in the Deep State's public health community, including the devil-doers identified in Chapter III, formally contemplated,[92] prepared, and were complicit in the well-planned "Final Solution" of injecting bioelectrifying vaccinations for depopulation and transhumanism. This was, after all, the objective of the "COVID Coup."

Engineering Incivility and Hypocrisy

It is widely known that fear of the unknown is most distressing. Victims can't taste, feel, see, or smell the coronavirus or its sociopolitical and economic underpinnings.

However, people can see news reports repeatedly threatening citizens' health and safety. Episodes of police brutality were intentionally broadcast over and over again until the threat was *generalized* to all police and all minority citizens. The same distressing conditioning was administered by the media broadcasting Johns Hopkins data. The COVID-19 case numbers misrepresented the actual threat. The made-up numbers induced fear and compliance with vaccination politics.

Regarding the lab virus, relief was speculative on all fronts, including the experimental vaccines being pushed. Officials claimed the germ was multiplying, mutating, and returning. So the vaccines became less and less effective as mutations spread. "GET VACCINATED," they repeated *ad nauseum* regardless.

The secret objective was never preventing fatalities. Deploying bioelectronic vaccine hydrogels for merging brains with the Cloud best explained even the "breakthrough cases." In many hospitals, as many as 40% of COVID "cases" had been fully vaccinated.[510]

Then the fear of return "spikes" was hyped when the 'herd' stopped social distancing and mask-wearing at supposed "super-spreader events."

Extortionate testing messages[511] and unproven vaccinations provided our only hope officials hyped. It resulted in phobias and rational fears and distress, mounted along with the "vaccination hesitant" population.

This was used to drive another wedge between people. Derelict and complicit officials blamed the innocent. The Democrats, including Biden, falsely claimed 'vaccination hesitancy'[512] was the greatest risk to humanity. This claim was absurd if the vaccinations actually worked.

Alternatively, those vaccinated posed the gravest risk to the un-vaccinated due to the "shedding" and spreading of the S-protein antigen mass-produced by mRNA-corrupted cells.

Vaccination opponents and reasonably concerned citizens were smeared and 'targeted'[513] like enemies of state when they issued objections. Leaders within this hesitant cohort were subject to character assassinations,[514] 'neutralizations,' arrests, or exterminations.

INCITING VIOLENCE AND CIVIL WAR AGAINST THE UN-VACCINATED

Mixing messages and confusing science coupled with threats of death directed 'herd mentality,' not simply herd immunity. This 'mind-control,' actually MindWar, was far more debilitating and much deadlier than even the lab-engineered coronavirus.

Idiots and complicit criminals aided and abetted the genocide by turning willfully blind to the science, common sense, and the crimes against humanity broadcast on network television.

For instance, on June 5, 2020, Johns Hopkins epidemiologist, Jennifer Nuzzo, demonstrated hypocrisy, actually insanity, by annulling her organization's social distancing decrees in favor of mass gatherings to protest racism. She tweeted, "We should always evaluate the risks and benefits of efforts to control the virus." But "[i]n this moment the public health risks of not protesting to demand an end to systemic racism greatly exceed the harms of the virus."[515]

Even assuming Nuzzo's rationale was reasonable, her hypocrisy was glaring.[516] Nuzzo co-signed an open letter with more than "1,200 infectious disease experts, public health professionals, and community stakeholders."[517] They wrote in bold, **"White supremacy is a lethal public health issue that predates and contributes to COVID-19."**

In other words, it is okay for America's public health leadership to **neglect**:

- those responsible for manufacturing and releasing the pandemic coronavirus;[518]
- those at the Univ. of North Carolina collaborating with Wuhan bioweapons contractors who spliced four HIV-1 gene sequences into the hyper weaponized coronavirus 'spike protein' that disproportionately infects Black people through their ACE2 receptors;
- the global elite most responsible for establishing public health, academic, political, and economic policies sustaining White Supremacy having 'captured' the highest levels of government and industry,[213]
- vaccination and drug safety risks, especially from poorly tested 'fast-tracked' vaccines and drugs advancing for COVID-19;[65]
- the conflicting interests of Bill Gates and Anthony Fauci in Moderna, Inc. financed by taxpayer money through the NIAID to secure private equity investment stockholders' profits from selling millions of doses of the new mRNA vaccine to the government to presumably prompt 'herd immunity' against the lab virus;[519]

INCITING VIOLENCE AND CIVIL WAR AGAINST THE UN-VACCINATED

- Black people and their communities have been most victimized and stigmatized by the riots accompanying the protests.[520]
- America's leading anti-Black, anti-Semitic, white supremacist tied to the 'Arizona Massacre' killing of federal judge John Roll and maiming of Congresswoman Gabriel Giffords sought to end illegal drug trafficking incoming from Mexico poisoning and killing people under the protection of the courts, the FBI and CIA.[521]

But it's not okay to continue social distancing to prevent the spread of COVID-19 because "systemic white supremacy" takes precedence.

Will the public health community, controlled by U.S. Navy intelligence officers and their Deep State paymasters, address the above list of neglected facts to secure society?

Unlikely.

The signatories' endorsement thereby is hypocrisy.

Fears, Phobias and the COVID-19 Riots

It was not serendipitous that a plague of rioting accompanied society's 'response' to the 'virus.'

Democrat-controlled cities were most heavily damaged following George Floyd's televised manslaughter. Arson and assaults paralleled the disease disparities between the races and classes. News that the African-American community suffered disproportionately from COVID-19 morbidity and mortality spurred further distrust, distress, and unrest.

In classical psychology and psychiatry, fear and self-defense are 'normal' reactions to real or imagined threats to survival. This is simply explained as follows:

The human 'ego' mechanically responds defensively against perceived threats by deciding to fight, flee, or use other available coping actions.

This choice-making 'internal dialogue' generally occurs impulsively —very quickly— often beyond conscious awareness. Life-changing and damaging decisions are made regardless of the outcomes.

Media modeling simulates real-life sensitization, prompting such decisions and human conditioning. Core beliefs result and generally impact people for lifetimes.

Disregarding what is well-known in the behavioral and social sciences, irresponsible and complicit officials, and the Deep State's media, exploited the hearts and minds of victims. Officials, including those in intelligence agencies directing the media's messaging, extended power and control over the ignorant and susceptible herd.

Most damaging, **phobias** arose. These are classically defined as severe **anxiety disorders** or fear-related mental illnesses. Phobic people reacted irrationally and damagingly to COVID-19 hype. The mass-mediated messages, 'cues,' 'triggers,' or 'stimuli,' distressed and damaged 'germaphobes' and anxious people. Their stimulus-response reactions (i.e., behaviors) and concerns were disproportionate to the actual size of the threats.

Mask-wearing while driving alone in cars evidenced more than rational 'caution.' It signified *madness*.

Phobic reactions are classically considered damaging to self and society. Anxiety in these individuals is elevated to the point where 'coping behavior' becomes 'maladaptive,' dangerous, even deadly.

These damaging or deadly psychopathic reactions due to COVID messages and impositions were recklessly justified in the minds of officials like Dr. Fauci.

Justifying Social Degeneration

All of the above psychopathologies have been well-studied and proven in the behavioral and social sciences. Yet here, the National Public

INCITING VIOLENCE AND CIVIL WAR AGAINST THE UN-VACCINATED

Health Service, operating as a 'captured' 'front' for the global elite, recklessly neglected or maliciously disregarded this scientific knowledge. Even the APA neglected these causes and effects of the observed deadly and damaging reactions to spreading fears or phobias.

In this case of COVID-19 psychopathology sewing social discord, the 'Pharma-media' and health establishment continued to neglect the most obvious causes and liable parties.

The anti-social self-defeating terrorism, chaos, tyranny, even mob rule as modeled on television in so-called "protests" was actually sedition largely financed by Soros's Open Societies Foundations' grants to activist groups, including Black Lives Matter.[522]

To destroy America, the globalists played to the sympathies of two psychopathic factions—the 'haves and have-nots.' Both groups feared losing life, property, power, and personal control. Threatened survival triggered criminality in each group.

This scourge enabled the disruption of American life.

Given this background and understanding, government health officials and globalist financiers must be held accountable for irreparably harming *We The People*.

Their insane directives and impositions, such as 'sheltering in place' under the influence of televised anxiety-provoking programming, killed people by denying civil rights and liberties to work and travel, interact with loved ones, family and community.

> *"The 'generalized conditions' of fear or phobias were conditioned and expressed in a spectrum of anti-social and self-defeating human behaviors underlying and rationalizing terrorism, chaos, tyranny, even mob rule as seen in the context of the 'protests.'"*

More Contributing Factors to Social Unrest

The aforementioned known risks of a social backlash, beyond the reckless official "shutdowns" and falsified data analyses repeatedly televised, were compounded by the finding that the COVID virus targeted "ACE2" immune-cell receptors making Black people much more susceptible to COVID deaths.

Most hospital and nursing home deaths were suffered by Blacks. This group already suffered higher risks for heart disease and blood oxygen deprivation. Death by asphyxiation—hypoxia—virtual strangulation, caused brain damage and death in disproportionately Black victims.

Curiously compounding these risks and racial disparities, this global problem metaphorically manifested in the televised chocking of George Floyd in Minneapolis.

Floyd's apparent murder/manslaughter by suffocation was repeatedly used to sensitize the masses. This recklessly incited more fear, anger, depression, and rage.

Anarchists and agent-provocateurs further exploited these events to generate riots, looting, and urban destruction. Looters justified their well-organized thievery and arson as compensation for "system racism," institutionalized discrimination, and ongoing genocide.

Again, all government, business, and academia leaders recklessly neglected the known causes (correlations and antecedents) of the civil unrest. The COVID impositions were clearly expected to cause this kind of social chaos.[92]

Psychosomatic Conversions and Nocebos

Compounding officials' reckless disregard for anticipated chaos, damage, and deaths, is the neglected and predictable personal and social damage caused by 'psychosomatic conversions.' This mental illness and 'nocebos' caused more damage and deaths.

What are 'psychosomatic conversions' and 'nocebos'?

INCITING VIOLENCE AND CIVIL WAR AGAINST THE UN-VACCINATED

You've heard the phrases "mind-over-matter" and "think and grow rich." You can also **think and grow dead**.

In psychosomatic conversion syndromes, physical illnesses manifest from unrecognized mental illnesses, or psychological and emotional distress. These mental illnesses can build up and suddenly express as a disease or set of symptoms.

Nocebos are different, equally troubling, grossly neglected, and may be deadly in certain cases.[523] The word **nocebo** means in Latin, "I will harm," as opposed to the word **placebo** that means "I will please."

Most people, including health professionals, overlook the power of negative expectations–the 'nocebo effect.'

Nocebos are like 'shaking the bones' in witchcraft. People attach meaning to and give their power to *nocebos*. These projections can be especially lethal for frightened and phobic people.   For instance, medical patients pronounced "terminally ill" may die needlessly due to the erroneous diagnosis and awful prognosis told to them.

Repeatedly threatening, terrorizing, and conditioning people with the threat that COVID might convert those messages into physical illnesses or kill people from the distress and expectation they may get the disease.

Moreover, locking people up by "self-isolation," knowing the virus is lurking and will be returning, repeating this anticipated and increasing threat on television, certainly encourages deadly nocebo effects.

These effects, too, were foreseeable. But no mention of these scientifically determined risks was ever made by media pundits or health officials.

This 'silence' was worse than 'deafening.' It was deadly and indicting.

'Negligent manslaughter' can be charged given these facts because people died due to the reckless neglect of this behavioral science.

Compounding evidence of malicious intent, officials neglected public education campaigns. They could have and should have used their vast resources to allay irrational fears.

Instead of educating people to promote preventative behaviors, 'health advisers' initially minimized the risks, as Fauci did. Then officials did the opposite. They fanned fears of impending doom—warnings that fueled pathological reactions and deadly consequences.

Terrorizing media messaging repeatedly broadcast for months. The networks modeled behaviors they knew would incite distress and distrust. Mixed or contradictory messaging confused the masses. Viewers were conditioned to expect, accept, and adapt to negative consequences and anti-social behaviors, such as rioting in the streets:

Vested Interests in Genocide

Malicious motives and concealed conflicting interests best explain the abuse of vaccine 'shortages' to drive markets. Early in the pandemic, fear of shortages was driven into the hearts of citizens using images of COVID-19 patients overrunning hospitals.

Television networks, health maintenance organizations, and hospitals are generally owned by leading stockholders/investors in drug and vaccine companies. This created more than 'disease control commerce' from deadly drug and risky vaccine side effects. It administered genocide.

'Front line health workers' and 'first responders' were threatened by shortages of 'personal protective equipment' including masks and also a shortage of ventilators. These healthcare workers were especially abused as puppets, manipulated by the media and their paymasters. They were not to blame for being deceived. They are to blame for trusting the untrustworthy—lying bureaucrats and drug industrialists.

Masked and unmasked people were shown being persecuted or prosecuted by police. Some were dragged from their homes and business in China and America. People were quarantined aboard ships in the U.S. and Japan. These images accompanied death statistics prepared by Johns

INCITING VIOLENCE AND CIVIL WAR AGAINST THE UN-VACCINATED

Hopkins faculty members that had sponsored the COVID-19 preparedness (Event '201' predictive programming) conference six weeks *before* this precise virus outbroke in Wuhan.[92]

What social and behavioral outcomes would a reasonable person expect from these negative, disturbing, distressing, abusive, and ongoing manipulations?

A revolution in the streets?

The 'Bigger Picture' and Deep State's Interests

[524]

More deadly than the infection was the sum total of subliminal, psychosocial, interpersonal, metaphysical, spiritual, demonically imposed official policies on confused, frightened, and phobically-conditioned victims.

Officials disregarded and diverted from taking a 'holistic' view of the crisis, and the fact that this saga extends to the present day from the Rockefeller/Rothschild/IG Farben syndicate that secretly partnered during World War II with the Nazis. This cartel's successors-in-interest currently administer the world's most profitable companies producing products and services that, in effect, embodies "globalism." That is, the rise of the Fourth Reich is what Hitler called the "Neuordnung," and George H.W. Bush called the "New World Order."

> *"Gross incompetence or malicious intent can only explain their official proclamations and governmental shutdowns. A commercial and financial crisis was imposed to presumably prevent a health crisis that never factually occurred."*

In this context of 'End Game' globalization, advancing transhumanism, the apparent release of the virus for political and financial gain, bioterrorism and biocrime, for extortion of civilization by public health officials globally imposing virtual enslavement, can only be understood and reconciled in this context of genocide. By carefully examining the corporate and financial links between the implicated Deep State agents and agencies, their behavioral and motivational biases, substantial evidence of genocide is revealed.

Accordingly, the preponderance of incontrovertible evidence proves the Anglo-American-Asian lab virus did not emerge by 'accident.' The bigger picture,' and clear-and-convincing intelligence competently analyzed, incriminates the ruling elite.

Recklessly Neglecting Human Needs Evidences Mal-intent

Officials dealing with COVID-19 and the riots completely neglected the foundation of humanistic psychology pioneered by Abraham Maslow and his "hierarchy of human needs" shown on the next page.[525] This had a most profound impact on citizens, protestors, and the key objective of protests to end systemic racism.

The U.S. Navy commissioned officials overseeing the 'public health service' imposed reckless neglect of basic human needs. According to President Trump, "they ordered [the country] to be shut down."

This malfeasance was unconscionable and deadly yet adopted by the 'evil obedient'[526] medical communities.

It is widely known that basic human needs are gained through socialization, labor, and commerce. Trading affordable and sustainable goods and services cannot be deprived, neglected, or rejected without causing substantial distress and economic hardship, risking lives, or alternatively *depopulation*.

This mental, emotional, and commercial pathology was predicted to deprive people of their civil rights. People were deprived of their right to live freely and prosperously. This distress caused immune suppression, diseases, and imposed deaths among members of the 'herd.' That satisfies the elements of genocide for which officials are accountable.

INCITING VIOLENCE AND CIVIL WAR AGAINST THE UN-VACCINATED

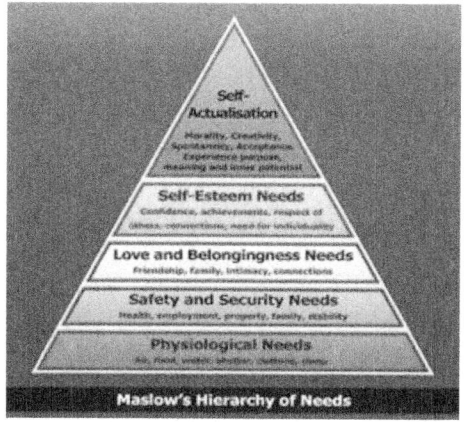

Maslow's human needs were deprived, especially the need for "love and belonging."

Keeping friends and families apart, then masking threatening faces, is torturous and arguably demonic. These impositions caused depression and anxiety disorders, as mentioned.

Studies showed that people deprived of physical contact for long periods suffer higher levels of disease and death.[527]

Moreover, because more than 60% of interpersonal communications are 'non-verbal,' involving **body language**, facial expressions are most important. Facial 'cues' help social interaction. Consequently, interpersonal communications and rapport-building are severely undermined by mask-wearing. Masking faces induced more fears, phobias, and anti-social behavior. It also aided-and-abetted crime, as demonstrated by rioters.

Demanding and conditioning 'faceless people' to act so irrationally damages trust in relationships vital to the economy and economic recovery. Economic growth and development are considered necessary to establish the new post-COVID-19 "normalcy." It is unreasonable, illogical, and sadly unconscionable to institutionalize and 'normalize' **faceless commerce**.

Are U.S. Navy intelligence officers so dim that they could not grasp this? Or were they put up to the task of destroying America and the world economy?

Mask-Wearing is Also Foolish and Contraindicated

Maslow's hierarchy addresses several more basic needs beginning with food, water, and air, each challenged by COVID-19 and related socioeconomic turmoil.

Food and water shelves went bare when COVID-19 commenced. Today, at the time of this writing, this is happening again from the delayed barge deliveries blamed on Biden administration policies.

At the beginning of the pandemic, a liter of water cost as much as a gallon of gasoline. At the time of this writing, a gallon of gas costs three times more than a liter of water. Basic physiological needs for food, water, air, or oxygen, like fuel to heat homes during cold winters, are depleted or completely deprived.

Stanford Univ. researchers confirmed what is obvious to anyone who wears a face mask.[528] This 'protection' makes it harder to breathe. The standard 'personal protective mask' is "estimated to reduce oxygen intake by anywhere from 5 to 20 percent. That's significant. Even healthy people suffer dizziness and light-headedness from extended use of face masks. If you wear a mask long enough, it can damage your lungs. For patients with respiratory disorders, it can cause distress and be life-threatening.

Air pollution has already lowered oxygen availability for health and longevity. Atmospheric oxygen dropped from 38.2% in prehistoric times to 23.9% before World War II. Since then, it dropped to less than 21%.

In other words, humans are being slowly suffocated, and this too was recklessly neglected by officials imposing mask mandates.

Furthermore, anxiety fuels shallow breathing. Oxygen deprivation and immune dysfunction are the main causes of COVID-19 morbidity and mortality. Oxygen deprivation and dehydration cause acidification of your body chemistry, predisposing you to myriad diseases.

Anxiety does likewise. The anxiety generated by "shelter in place" requirements, and laws opposing outdoor group activities, raised risks of diseases and deaths.

The corporate-controlled media, evil-obedient policymakers, healthcare administrators, submissive healthcare workers, and their U.S. Navy intelligence overlords neglected all of this.

Next, rethink **how 'safe' is the indoor "shelter in place" imposition?**

INCITING VIOLENCE AND CIVIL WAR AGAINST THE UN-VACCINATED

'Sheltering in Place' Subverted 'Herd Immunity'

As I briefly mentioned, physical isolation and social restrictions were grossly unsafe. Beyond the risks and damages cited above, this behavior undermined **'herd immunity.'**

Officials claimed they prevented a deadly overburdening of hospitals by commanding social distancing. But even if that were true, and it is highly contested, the overriding objective was to 'immunize' civilization or gain herd immunity. This was proposed to secure civilization's health and safety.

But this *hypothetical notion* was stupidly neglected by precluding interpersonal contacts that would permit the 'natural' spread of this unnatural infection.

If the virus "evolved naturally," as Dr. Fauci falsely broadcast, then it would have been most prudent to permit 'nature' to run its course. This was especially encouraged by the vast majority of COVID-19 cases fully recovered. In a front-page feature on May 29, 2020, the *New York Times* heralded this objective graphically, as shown in the screenshot below.

Officials' negligence raised several vitally important questions:

Shall we disregard all of the above evidence of criminality engaging the public health community by defending the complete disregard of science, ethics, and common sense as excusable because of 'emergency circumstances?'

Shall we excuse the crimes against humanity because of the scientifically proven and politically-concealed fact that this "novel" germ is a 'hyper-weaponized' lab virus engineered to infect humans using four segments of the AIDS virus HIV-1, spliced into the mutagen's 'spike protein'?

Does this mean there is a good possibility that a long incubation period will follow infection, as is true for AIDS patients?

Is this why U.S. Navy intelligence directed the "shutdowns" and fraudulent concealment of HIV/AIDS genetic tinkering endangering civilization?

Does the neglected S-protein intoxication cause latent immune suppression and related auto-immune diseases?

> *"Mentally ill people might generalize these threats to every mask-wearer, every school and workplace, and every healthcare facility. According to phobic imaginations, every drugstore and food market may be teaming with deadly pestilence."*

Will this poisoning cause millions of people recovering from the infection to develop new diseases and die years from now? That is, after all, what happens in people infected with AIDS/HIV-1. They die after a 3-to-7 year 'incubation period.'

Accordingly, given these circumstances, facts, and questions, sheltering in place and getting vaccinated not only subverts herd immunity but may cause great harm.

For these reasons, the presumed remedial 'public health policies' are ill-conceived.

More Censored Risks

The Coronavirus/SARS/HIV-1 mutagen may also 'naturally' recombine with more deadly viruses and even bacteria in the future. Experts predict additional outbreaks and 'spiking' pandemics.

'Pandora's box' has been opened, raising risks of nightmarish recombinants—mutants of coronaviruses springing from other more deadly pathogens such as H5N1 or even Ebola.

These frightening unknowns, stealth possibilities, and probabilities may have been thoughtfully concealed because of the risk of spiking anxiety and fueling phobias.

INCITING VIOLENCE AND CIVIL WAR AGAINST THE UN-VACCINATED

Mentally ill people might generalize these threats to every mask-wearer, every school and workplace, and every healthcare facility. According to phobic imaginations, every drugstore and food market may be teaming with deadly pestilence.

If the U.S. mail can deliver anthrax, why not coronavirus? Can this menace that affects cats and dogs also pass from mail-senders to mail-carriers and those quarantined? Shall we shut down all postal services?

What are the risk-to-benefit ratios for these policy decisions?

Unknown.

Only weeks ago, the un-vaccinated population represented the vast majority of Americans who remained "hesitant." Then the National Security Crime Syndicate, Biden administration, and the Democratic Party-complicit media unconscionably targeted this group of 'vaccine-resistors.'

Biden's handlers and speechwriters are not dumb. They certainly knew that Biden's statements condemning citizens choosing to forego vaccinations might incite civil unrest and retaliatory strikes against 'un-vaxxed people' falsely blamed for others' deaths. The President urged everyone to take advantage of the "safe, effective, and free" shots to help end what he called "a pandemic of the un-vaccinated."[529]

FREE VACCINES? What do you get for "free" today?

GENOCIDE!

What do you get for believing compulsive liars and the media that conceals conflicting interests and solid science?

You get a president, campaign financed by Big Banking, Big Pharma, Big Energy, Big Military, and Big BioTech, to impose totalitarian mandatory vaccinations and public health policies enabling genocide.

This is Gates's "Final Solution," enabling Schwab's "Great Global Reset."

Kill the Un-Vaccinated for your Hospital Bed

On September 3, 2021, AP broke a terrorizing story heralded by the national news media. It was titled, "'Loss of hope': Idaho hospitals crushed by COVID-19 surge."[530] The story provided a classic example of the gross criminal deception administered under the guise of "news."

The lead-line alleged: "Intensive care beds are full of un-vaccinated coronavirus patients at a hospital in Boise, Idaho." That hospital was St. Luke's Boise Medical Center (SLBMC).

The AP article, by Rebecca Boone, neglected to mention substantial intelligence on that hospital.

I lived in Idaho for many years and discovered many facts relevant here. I even wrote a book about these discoveries prophetically titled *Death In The Air: Globalism, Terrorism, and Toxic Warfare,* published three months before 9/11 in June 2000.

First, as reported by the CIA-monitored/edited/and censored *Wikipedia*, it is public knowledge that AP is infamous for its "dealing with Nazi Germany."

I have lectured for nearly a quarter-century about Nazi Germany being well-represented in Big Pharma. As detailed in the Foreword by JT deCordova, you can trace the post-Nuremberg trial "decartelization" of IG Farben to today's leading drug companies.

Recall that IG Farben was Germany's leading industrial organization. It maintained concealed partners in IBM and the Rockefeller crime family (represented by the Dulles Law Firm and later Cromwell & Sullivan). These parties launched and administered the CIA in IG Farben's headquarters in Frankfort, Germany. That huge building was spared allied bombings to protect these partners and stockholders in genocide.

According to public knowledge, AP *collaborated* within this murderous syndicate. This Nazi/American propaganda mill was ministered by Josef Goebbels. AP intermediaries were used to conveying "an estimated 40,000 photos" from America's Diplomatic Corp, revealing classified intelligence to aid-and-abet the Hitler/Rockefeller genocidalists. In brief, AP was complicit in the holocaust of WWII.

Later, AP's anti-Semitic genocidal service to the 'Deep State' extended to the Israeli-Palestinian conflict. AP's Cairo bureau spun facts

INCITING VIOLENCE AND CIVIL WAR AGAINST THE UN-VACCINATED

to make the world believe "the conflict was Israel's fault, and the Arabs and Palestinians were blameless," *Wikipedia* records.

Today, AP neglects chilling facts to generate civil unrest and trigger violence in America.

I have often urged audiences to realize that "hospitals are the modern concentration camps," and "vaccines are like the gas used to kill holocaust victims who went into 'showers' for 'public health' and 'disinfection.'"

I based my accurate and prophetic warnings on the decartelization of IG Farben and the Rockefeller Cartel's investments in Big Pharma, beyond the dark history of the Kaiser family leading to Kaiser Permanente's dominance in the hospital industry.

People forget that the wealthy Kaiser family is infamous. After prompting WWI, Kaiser Wilhelm II was exiled with all his wealth intact. He endorsed the Rockefeller, Ford, and Carnegie-financed eugenics movement. This encouraged the mass murder of Jews, Christians, Blacks, and the "feeble-minded" during WWII. This agenda 'captured' the public health community and justified the Tuskeegee syphilis study. Some historians report that Kaiser Wilhelm II declared his intention to conquer "the United States [that] was proving to be an irritating obstacle to Wilhelm's goals," *Wikipedia* recalls. Today's political and economic proceedings actualize the Kaiser's intentions.[531]

Considering the current control over healthcare and COVID's suspicious origin in (NIH/NIAID/DOD and Chinese military-financed) labs, AP's selection of SLBMC to declare MindWar against un-vaccinated people raised serious 'red flags' for me. So I sought to discover who controls that hospital.

Bob Lokken is the Chairman of the SLBMC Board. According to his hospital bio, he is a computer industry "entrepreneur" specializing in "data and analytics." He is an "intelligence" agent who "joined Microsoft" —Bill Gates's company. (Microsoft is MSNBC's partner and also CNN's financier.[532])

Lokken's Links to Globalists Advancing Genocide

Lokken "led the Business Intelligence Product Management Group, driving the development of product and commercial strategies for Microsoft products," Lokken's biographical records state.

As already detailed, Bill Gates, along with Jeffrey Epstein's group of investors, made substantial investments in the development of the COVID vaccines through Harvard and MIT co-inventors (and co-authors) Charles Lieber and Robert Langer.[533]

Therefore, conflicting interests abound in the AP/Deep State's propaganda targeting people who are not vaccinated.

The AP article stated, "St. Luke's Boise Medical Center invited The Associated Press into its restricted ICUs this week in hopes that sharing the dire reality would prompt people to change their behavior."

In other words, the physical (cement and steel) SLBMC building telephoned the AP without Lokken's knowledge, consent, or instigation. That is ridiculous.

> *"The Deep State's propaganda extortionately promotes vaccinations. It sells people on getting vaccinated by disregarding the risks–what the vaccines actually deliver–the S-protein antigen–and other toxic ingredients including bioelectronic devices."*

INCITING VIOLENCE AND CIVIL WAR AGAINST THE UN-VACCINATED

It is public knowledge that Lokken is loyal to Bill Gates (financed largely by Microsoft/IBM/Warren Buffet).[534] He is complicit in the damage done by the world's leading promoter/propagandist and distributor of vaccinations, Mr. Gates. Gates openly urges depopulation. He states 15 percent of civilization can and must die using vaccination programs to save the environment. **That is the essence of the "Green New Deal" advanced by the Deep State/Democratic Party alliance.**[411]

Extending Lokken's loyalties and conflicting interests, the main COVID vaccine makers, Pfizer and Moderna, share condemning conflicting interests and investments with Hearst Co. and its subsidiary —the First Databank— partnered with the McKesson Corp. McKesson controls the lion's share of "data-mining and analytics" in medicine, as I explained earlier. Much like IBM tracked holocaust victims, the First Databank/Hearst/McKesson BigPharma/BigTech conglomerate sells the products hospitals, like Lokken's SLBMC, use to track patient data and impose the 'standard care' on patients.

Piercing the Deadly Propaganda

This AP/News media/Gates & Co./Deep State propaganda intentionally incites civil war between the VAXXED and UN-VAXXED.

AP readers are terrorized by threats of death. AP warned that some victims would be excluded from healthcare altogether because the un-vaccinated took up all the hospital beds exclusively.

In Idaho, where many people carry guns, this propaganda will likely incite shootings targeting the UN-VAXXED. This bloody outcome is precisely what the global elite desire. This activity serves to distress people, break up families, divide communities, lower immunity, reduce populations, expand their control, and secure the "Final Solution" for One World Government.

This is how destabilizing nation-states, societies, and economies is accomplished most efficiently and effectively.

Bioterrorism threatening and exploiting populations at risk are ploys abused to extort the people and governments into submission to achieve totalitarian globalization.

This is how and why the Deep State's propaganda extortionately promotes vaccinations. It compels people to get vaccinated, disregarding the risks of what the COVID vaccines actually deliver —the S-protein antigen— and other toxic ingredients, including bioelectronic devices for "Mark of the Beast"-like control.

The media's divisive terror campaign politically mobilizes people to do evil. Submissives are 'evil obedient.' They act under the influence of a hypno-tronic trance-state. Much like a magic 'spell' cast by a hypnotist, naive submissive people place themselves and others at grave risk.

Caregivers and politicians are especially susceptible to the Deep State's propaganda. Dr. Bill Dittrich continued the AP terror campaign, "All of the ICU coronavirus patients were generally healthy people who simply didn't get vaccinated." He deceptively neglected those who had developed natural immunity from previous exposures, as well as "breakthrough" cases reported to be as many as 40% of cases in other hospitals.

"Idaho could enact crisis care standards in days, leaving [Dittrich] to make gut-wrenching decisions about who gets life-saving treatment," AP warned.

In other words, doctors will decide who is to live or die based on vaccination history, not immuno-toxicity caused by the S-protein antigen. Doctors will prejudicially influence outcomes based on false COVID narratives.

More Important Neglected Questions

This real and present danger of "crisis care standards," 'crisis fascism' and 'crisis capitalism' raises more urgent questions:

INCITING VIOLENCE AND CIVIL WAR AGAINST THE UN-VACCINATED

What if the "novel" mRNA vaccines are not as "safe and effective" as the propagandists have claimed?

What if the "gain-of-function" synthetic "dual-use" (i.e., military and commercial) bio weaponized S-protein antigen was actually spreading/shedding from those vaccinated, sickening the un-vaxxed, and sending the un-vaxxed in droves to hospitals?

What if it is precisely that lab-engineered antigen that enables the germ mutations that make people so sick?

And what if the increasing number of bacterial and viral mutations, like the "Delta variant" and "Mu," are sourcing from vaccinated people?

What if the FDA and CDC were 'captured' by BigPharma and turned into political agencies?

What if bribed officials erred in accepting the vaccine makers' false safety assurances and testing data, as I claimed in my lawsuit against Pfizer and Moderna?

What if the vaccinations spreading the S-protein antigen are causing respiratory distress falsely attributed to the virus?

What if standard COVID patient care is, therefore, seriously flawed?[535]

What if "The Horowitz COVID Protocol" that I published could better save lives by freeing airways using: (1) natural anti-histamines; (2) natural mucous decongestants and expectorants; (3) positive lifestyle changes; (4) good alkalizing hydration; (5) oxygenation therapy and breathing exercises; (6) anti-oxidant vitamins and beneficial nutritional supplements; and (7) other alternatives including bio-electric therapies such as 'Rife machines?'[536] (See APPENDIX for more details.)

"I don't think anybody will ever be ready to have the kinds of conversations and make the kinds of decisions that we're concerned we're going to have to be making in the next several weeks. I'm really terrified," the ignorant of complicit Dr. Dittrich was reported as telling his AP interviewer.

Many fear their family members "may be taken off life support if someone with 'a better chance of survival' needs the bed... I don't even want to think about it... If it comes to crisis standards of care, they're going to say [they're] not showing enough improvement."

"Crisis Standards of Care" Threatens Civil War

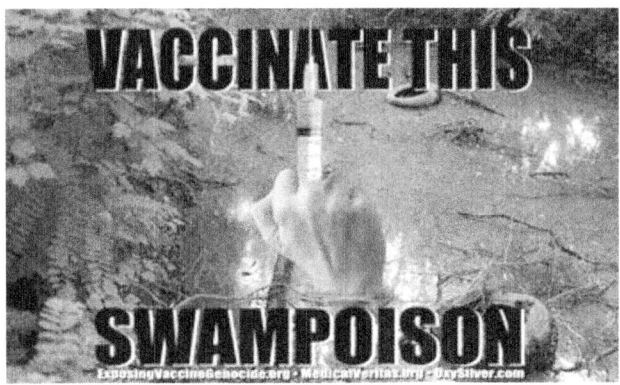

With "critical shortages of hospital beds and staff," American states with the lowest vaccination rates will be the first to justify "crisis standards of care," officials have warned. These conditions "call for giving scarce resources to patients most likely to survive."

AP threatened, predicted, and programmed the public to fear and accept iatrogenocide.

"There is so much loss here, and so much of it is preventable. I'm not just talking about loss of life. Ultimately, it's like loss of hope," said Dr. Jim Souza, chief medical officer at Lokken's SLBMC. "When the vaccines came out in December, those of us in health care were like, 'Oh, my God, it's like the cavalry coming over the hill.' ...To see now what's playing out? It's all so needless."

AP continued, "Inside the ICUs, Kristen Connelly and fellow nurses frequently gather to turn over each patient, careful to avoid disconnecting the tangle of tubes and wires keeping them alive. With breathing tubes, feeding tubes, and half a dozen hanging bags of medications intended to halt a cascade of organ damage, turning a patient is a dangerous but necessary endeavor that happens twice a day."

During my eight-day hospitalization in August 2021 for an S-protein mutated strain of Hemophilus influenza pneumonia, the Lee County Hospital staff in Florida never turned me over once. Needless to say, severe bedsores developed.

Alarms would go off day and night with the slightest shifting of my position in the "tangled tubes." Those restrictions precluded me from

INCITING VIOLENCE AND CIVIL WAR AGAINST THE UN-VACCINATED

even going to the bathroom, only ten feet away from my bed. Sleep deprivation from the needless alarms damaged me further. All of this amounted to cruel and unusual punishment.

"When Idaho's hospitals were nearly overwhelmed with coronavirus patients last winter," AP continued, one nurse, Ms. "Connelly wasn't fazed, believing she could make a difference. Now, instead of focusing on one patient at a time, she cares for multiple. Many colleagues have quit, burned out by the relentless demands of the pandemic. [Now many have quit due to vaccination mandates.] It's devastating... Where we are right now is avoidable —we didn't have to go here."

Rather than blaming the agents and agencies that manufactured the lab virus, the Deep State investors that released the pathogen, the media that concealed all the critical facts, or the National Security Crime Syndicate that secrets all this important intelligence, the SLBMC, and AP, like the Biden administration, blamed the un-vaxxed.

"Most of the ICU patients fell prey to con-artists before they fell ill with the virus, said Souza, the chief medical officer... Misinformation is hurting people and killing people," Souza said... We don't have any vaccinated patients here."

Politicizing these issues, recklessly driving divisions, and inciting violence and civil war, AP added, "Even families who have witnessed the trauma of COVID-19 firsthand are on opposite sides."

Consequently, the 'Deep State' has incited a **civil war** in America, pitting vaxxers against anti-vaxxers, conservatives against liberals, Democrats v. Republicans, agnostics against religious groups, Blacks vs. Whites, Anglos against Asians, but never the poor against the mega-rich that are imposing, broadcasting, modeling, and socially-engineering the dissonance, chaos, and disease we witness.[537]

Summarily, these are demonic and treasonous times. The obvious agenda is to incite violence through the propaganda mills of "National Security 'Crime' Syndicate.' BigTech is BigPharma, and also BigMilitary, BigEnergy and BigBanking. This crime syndicate does the dirty work for the "Deep State."

Abuse of their "fake news" media is being administered to destroy America in every way possible. This is required by the global elite to complete their mission to impose totalitarian transhumanism and robotic rule over everyone who survives.

The obvious overriding objective is to degenerate, damage, and destroy America as a sovereign nation-state in favor of globalization and monopolizing corporate fascism.

Inciting civil war in America is the most efficient way of achieving the globalists' goals.

CHAPTER IX

BILL GATES, THE CLINTONS, AND THE WHO'S COMPLICITY IN THE COVID COUP

Past chapters have identified Bill Gates's and George Soros's financing of Planned Parenthood. Add the World Health Organization (WHO) to this enterprise that has been racketeering in genocide.

The WHO openly advertises its partnership with Planned Parenthood for "sexual and reproductive health."[538] This is central to the depopulation agenda and to cheapening life.

Notice how the WHO's propaganda switches this truth. According to the WHO, the "International Planned Parenthood Federation (IPPF) works in 172 countries to empower the most vulnerable women, men, and young people to access life-saving services and programs, and to live with dignity."

Those "programs" not only feature abortion services but vaccination impositions.[539] If you want to terminate your pregnancy for whatever reason at an abortion clinic, you will submit to the jab that causes substantial side effects, many potentially deadly.

So much for "life-saving 'health' services" and the truth.

As mentioned in Chapter VII, Bill Gates's father, William H. Gates, Sr., was the head of Planned Parenthood. It was founded by Margaret Sanger, promoting, like Gates, population control, racial hygiene, and white supremacy.

I asked you, "Are you for '**86ing**' life?" Planned Parenthood owns and licenses "RU486," the "morning-after abortion pill."

Skeptics may argue, 'There's no evidence that Gates is a 'white supremacist.' Wrong! Vaccine side effects disproportionately damage and kill Black people.

In 2019, the WHO surprised subscribers by adding "vaccination hesitancy" to its list of "top ten threats to world health."

The WHO wrote:

"Vaccine hesitancy – the reluctance or refusal to vaccinate despite the availability of vaccines – threatens to reverse progress made in tackling vaccine-preventable diseases. Vaccination is one of the most cost-effective ways of avoiding disease – it currently prevents 2-3 million deaths a year, and a further 1.5 million could be avoided if global coverage of vaccinations improved."

As analyzed by *Science-based Medicine*,[540] "The antivaccine movement contributes significantly to 1.5 million deaths a year." This propaganda disregards the millions-more deaths caused by vaccination "adverse events."

"[This] has caused the resurgence of previously eliminated diseases," the propagandists insisted. "This has frustrated our ability to eradicate diseases like polio."

This misrepresentation is unconscionable. It not only neglects the "eradication" that resulted from improvements in hygiene, pure hydration, and good nutrition, but that the polio vaccines, both Salks and Sabin's, caused pandemic cancers from SV40 contaminations.

"We can't entirely blame antivaccine pseudoscience for this, as there are other factors, such as availability of vaccines, and things like complacency and perceived inconvenience. But the impact of antivaccine propaganda is significant [although not as significant as Big Pharma 'vaccinecide.'"

"As did Sanger, Gates subscribed to eugenicist Thomas Malthus's idea that the sustainability of the world's resources is completely dependent upon maintaining population control," explained Mary Pesarchick for *CatholicStand.com*.[541]

Pesarchick epitomizes the confused masses, as you can glean from her article. She clearly bought into Gates's fraudulent depopulation 'paradox.' According to Pesarchick, Gates "believes that improving health care, primarily through vaccinations, will accomplish [population reduction]. And maybe the most interesting thing I learned," she wrote, "[A]s you improve health in a society, population growth goes down."

That is counter-intuitive and lacks common sense.

There are five misrepresentations in that analysis thus far, and more to come.

First, there is no solid scientific evidence that proves officials are not killing and maiming more people than they are helping or saving with vaccinations. So actually, more developed countries that have implemented strict vaccination policies appear to have reduced their populations from deadly vaccine side effects, including cancers.

Second, as mentioned, the myth that vaccines "caused" the disappearance of great plagues such as polio and smallpox fraudulently omit the benefits of better hygiene, nutrition, and pure hydration. The con-job is much like the hype that un-vaccinated people put vaccinated people at risk.

Despite these improvements, reduced populations are mostly attributable to vaccination intoxication side effects. That is why criticizing vaccines, "mandatory vaccinations," and the outrageously defective vaccine injury (VAERS) data is so widely censored, neglected, and taboo.

Third, there is no solid scientific evidence proving that as health improves in individuals and society overall, people have fewer children. Even in lesser-developed nations, there are too many social, environmental, and behavioral variables to measure this false assertion. The dearth of studies in this field is readily apparent from a Google search of the scientific literature. Add, for example, studies such as this one: "Reproductive behavior and health in consanguineous marriages.[542] You will learn that the hypothesis of "better health reduces population" is proven to be **pseudoscience**, thus refuted.

Pesarchick strained to grasp the pseudoscience. "You know... before I learned about it, I thought it was paradoxical."

Then the curious Catholic asked, consistent with Gates's father's justification for eugenics, "[I]f you improve health, aren't you just dooming people to deal with such a lack of resources where they won't be educated, or they won't have enough food? You know, sort of a Malthusian view of what would take place[?]"

That is a dumb question because if people didn't have enough food, their health would suffer, and more would die.

Fourth, Pesarchick added 'confounding factors,' further discrediting Gates's argument. Education level and other demographic variables confuse the analysis. Quality of nutrition and water (i.e., good hydration) are more confounding factors associated with better immunity and longevity.

And fifth, Pesarchick considered that good "health leads parents to decide, 'Okay, we don't need to have as many children because the chance of having the less children being able to survive to be adults and take care of us, means we don't have to have 7 or 8 children.'" For this discussion, Pesarchick recanted Bill Gates *NOW* interview, wherein he shared his illusions and deceptions.

How many nations, after all, have such generalized narcissism that appears to run rampantly in the Silicon Valley geek community of emotional misfits?[543]

Assuming impoverished parents reproduced more to secure a better quality of life through their children when they get old predicts larger populations further burdening scarce healthcare resources and natural resources, thus defeating the parental plan.

"Gates emphasizes vaccination programs as the best means of combating Third World poverty," the propaganda reads. But, in truth, Gates promotes vaccines for depopulation to secure the elite's protection against popular uprisings. The oligarch retains power and control generation after generation by most profitably controlling populations, diseases, cures, healthcare, and online resources.

"[I]n 2014," Pesarchick recalled, "the Kenyan Catholic Doctors Association and the Kenyan Catholic Bishops Conference issued a statement expressing concern that a UNICEF/WHO Tetanus vaccine was tainted with hCG, a contraceptive hormone." While this accusation was denied by the agencies involved, "the Catholic groups remain wary,"[544] Pesarchick reported. Today we see the same material concealment of the brain-Cloud bioelectronic devices in the COVID vaccines.

Gates' Depopulation Plan to Remedy Climate Change

"Population control is also central to the issue of climate change, another of Gates's passionate causes," Pesarchick continued. "In a talk titled, *Innovating to Zero, presented* at the 2010 Technology, Entertainment and Design (TED) Conference, Gates proposed a goal of achieving zero carbon dioxide (CO_2) emissions by 2050."[545] He explained his pseudoscientific formula by which that goal may be achieved:

CO_2=Population x Services x Energy x Emissions.[411]

"So you've got a thing on the left, CO_2, that you want to get to zero...[so]...Probably, one of these numbers is going to have to get pretty near to zero," Gates professed.

Doesn't CO_2 play an important role in photosynthesis?[546] Of course, it does! CO_2 stimulates plant growth producing more food. So Gates's foundational statement and the formula are absurd unless the man is a psychopath and desires all aerobic life to perish on Earth.

Noting that the first factor in the equation is population, Gates remarked:

"The world today has 6.8 billion people. That's headed up to about nine billion. Now, if we do a really great job on new vaccines, health care, reproductive health services, we could lower that by, perhaps, 10 or 15 percent."

"To that end," Pesarchick added, "since 2012, Melinda Gates, a Catholic, has pledged over a billion dollars from the Gates Foundation to support Family Planning 2020 (FP2020).[547] She helps lead this international effort whose goal is to get birth control to 120 million more women by 2020. Despite strong objections from groups such as *Culture of Life Africa.com*, who resent 'the disturbing encroachment of the bold and wealthy proponents of the Culture of Death.'

"Emotional Quotient of a Snail"

"When I met Bill Gates," Barry Diller explained. "I would say he had the emotional quotient of a snail. And now you can see him cry."

"They're tech people," explained Barry Diller, describing Gates and the shallowness of the Silicon Valley elite. "They don't have a lot of romance in them. They don't have a lot of nuance in them. Their lives are ones and zeros."

Diller's description enjoyed "an eerie accuracy," noted *POLITICO*.[548]

"Too many tech companies have produced soulless products that demand machine-like behavior from humans. And, at their helm, have been machine-like humans," Diller told *POLITICO*, well aware of his fellow oligarchs' transhumanistic goal of turning humans into machines.

Diller was pained to say he doesn't think these tech titans are our overlords. "Our overlords are artificial intelligence," Diller said, diverting from the devil-doers that manufacture and control AI and AGI.

Nick Gass concluded for *POLITICO*, "There's something in the notion that when you pull many tech people away from the power and algorithms they so enjoy, they begin to see the world in a slightly different light ... [L]ook at all the former Facebook executives who now worry about what the site is doing to children's minds."

In contrast, look at the current Big Tech executives, including Facebook's Mark Zuckerberg, favoring mandatory vaccines and the Metaverse for manipulating children's brains.

And as far as Donald Trump was concerned, Gates preferred Hillary and Bill Clinton for the Oval Office. According to Gass, Gates remarked that Trump's knowledge of vaccines was outdated. During a televised debate, Trump seemingly linked the increase in autism to childhood vaccines. The theory was only debunked by the 'fake news' media and officials influenced by Gates's allies.

"'Science in general," Gates concluded, "whether it's GMOs or vaccines, there's a lot of people out there who don't give science the benefit of the doubt."

Diller, Soros, IAC, the Democrats, and Chelsea Clinton

You have already read about the massive financing George Soros's Open Society Foundations have granted leftist 'radical' organizations and Democrats advancing socialism, and likewise "American Marxism" and the "Green New Deal." Among the major shareholders in the Soros Fund

Management, LLC is "IAC" —the Interactive Corporation. The NASDAQ listing shows IAC owning 100,000 shares of Soros company's stock.

Accordingly, it is not surprising that IAC's President, Barry Diller, has been an outspoken critic of Trump and a major donor to Democratic Party candidates opposing Republicans.

As previously mentioned, Diller officiates his company with Chelsea Clinton and Edgar Bronfman, Jr.

Now a reasonable person might ask reasonable questions, such as, "What does Chelsea Clinton have in common with IAC —a media company? What is her function at IAC? What is the tie between ICA and the Clinton Foundation HIV/AIDS and vaccine initiatives? Moreover, how do Clinton's investments intertwine with Diller's political agendas?

The answers may be found by starting with Chelsea Clinton's biography posted on *Wikipedia*.

Curiously, Chelsea is credited with having attended three of the Deep State's most infamous institutions: Stanford, Oxford, and Columbia. All three are active in population management, population control, and social engineering through political science. If you want to identify **the heart of the National Security Crime Syndicate key to the COVID Coup**, start with Chelsea's 'higher education.'

Earlier, I vetted Oxford University's engagement in a scheme involving the Zika and Ebola viruses and genetically modified mosquitoes. I evidenced Jeremy "James" Farrar, director of the Wellcome Trust and past professor of tropical medicine at Oxford, seemingly directing Fauci.

Recall that Farrar acted *frantically* concerned about the "Indian paper" blowing the lid off the entire bioweapons industry, subjecting Fauci to grave scrutiny, along with Francis Collins and Harvard's Dean Daley. They all conspired to cover up Wuhan lab gain-of-function experiments implicating the Chinese government, Harvard's Charles Lieber, and Moderna's Robert Langer. Several key "insiders" were involved in the cover-up, including Kristian Andersen collaborating with Peter Daszak and the EcoHealth Alliance on the fake science paper. Robert Garry administered the Ebola response with the Biden-linked Metabiota Company, as detailed in the next chapter. Metabiota is a pandemic firm partnering with Daszak's EcoHealth Alliance and vicariously the Wuhan bioweapons lab.

As the urgent teleconference coordinator, Farrar urged lying to cover up with WHO officials the lab virus origin of COVID-19, including four genes from the AIDS virus in the mutated spike protein bioweapon construction.

I also explained that the Obama/Biden appointee, Sylvia Mary Burwell, who served as the United States Secretary of Health and Human Services, oversaw the Ebola emergency for America with Fauci and the CIA, concealing the Ebola Zaire strain that science and common sense said was released from a military bioweapon's refrigerator.

The evidence also indicted Columbia University, wherein the CIA's point man overseeing the alleged deployment of HIV/AIDS across Africa, Frank Carlucci, III, trained.

I mention these facts again here because the same suspects are advancing the same 'pattern-and-practice' of population control, and Chelsea Clinton and the Clinton Foundation's interest in HIV/AIDS and tainted vaccines is heavily suspect. Clinton's most active involvement with Bill Gates, the WHO, and the World Bank deploying so-called "aid" to Africa is incriminated for alleged conspiracy to commit genocide.

These same financial and political forces underwrote the "New World Religion" and hallucinogenic "God molecule," dimethyltryptamine (DMT), to not only advance the "Ayahuasca Death Cult" from the CIA's covert operation in Brazil, complicit with Clinton's pulpit at NYU but also condition the brain to accept the "new reality" of bioelectronic population management.

Recall that NYU[549] Medical Center was the place that abused children at Willowbrook, poisoning them as ill-informed un-consenting experimental subjects in the hepatitis B vaccine trials co-conducted by the CDC, Merck, the U.S. Army, and Rockefeller Blood Counsel. NYU[550] is also the place that advances the hallucinogenic psychiatric drug industry.

It is also NYU[551] that suckered mass-transit workers into COVID vaccine experiments—a virtual 'remake' of the CIA's abuse of NYC transit system, passengers, and citizens across New York, exposed to *Bacillus globigii* and *S. marcescens*—*airborne bioweapons,* the former being potentially deadly.

BILL GATES, THE CLINTONS, AND THE WHO'S COMPLICITY IN THE COVID COUP

We now witness the COVID Coup, for transhumanism and depopulation, featuring Chelsea Clinton and co-conspirators challenging the religious world and the ethical and moral values of monotheists.

NYU is also where Chelsea Clinton advances "interfaith initiatives and the university's global expansion program," according to IAC's SEC filing.

Accenting more "hypocrisy," according to the IAC's Securities and Exchange Commission disclosures, Chelsea Clinton served as "Assistant Vice Provost" at NYU where she reportedly focuses on "interfaith initiatives and the university's global expansion program."

Ms. Clinton also serves as "Co-Chair of the Advisory Board" at NYU overseeing the "Of Many Institute."[552] That group advertises itself as "a pioneering initiative devoted to educating and inspiring religious and spiritual leaders to utilize multi-faith dialogue and service as a force for positive social change."

Yet, dialogue on holy spiritual suppression caused by vaccine-injected nano-bioelectronic brain-Cloud wireless devices is not part of their curriculum nor multi-faith "service."

"Vaccines Are The 'Sacred Cow' of Public Health"

Now you know vaccines play a crucial role in subverting the world's religious theologies, and you know why COVID vaccines are being mandated without urgently needed vaccine safety studies. You now also know why vaccination risks have been censored and why professionals who object to this ongoing genocide are smeared as 'killers and kooks,' or marginalized as 'conspiracy theorists.'

Government AIDS and vaccine programs financed by the World Bank and Deep State in which Chelsea Clinton and Barry Diller knowingly serve as propagandists are far too important commercially and geopolitically to allow dissent.

I heralded this systemic corruption in AIDS science decades before anyone else. I coined the phrase "Vaccines Are The 'Sacred Cow' of Public Health" nearly a quarter-century ago.

Now you can understand why I did and why my nearly lone voice has cried this truth in the wilderness of ignorance. I have tried my best to prevent the world's exploding cancer rates, HIV/AIDS deaths, mass sorrow, and the COVID Coup.

In other words, criticizing vaccines in any way is **taboo.** I've repeatedly warned that the oligarchy considers vaccines the 'Sacred Cow of Public Health.' That means the idea and custom of injecting carcinogenic GMO's into people for alleged immunological 'protection' is, according to standard definitions of the "Sacred Cow, " a topic *unreasonably* denied any criticism, controversy, or reproach.

Corporate-influenced legislators and mass media manipulators have imposed the dogma exempting vaccines from criticism or reproach. Thus, 'general acceptance' of vaccine safety is criminally *imposed*. It was never actually based on scientific evidence. Vaccine safety is an illusion generated by MKULTRA programming in MindWar.

Not only has the risk/benefit analysis NEVER been done on any vaccines to effectively determine danger or damage, but the entire concept of risk assessment in vaccinology is arbitrary and capricious. There are two types of risks: relative and absolute;[553] neither can be legitimately used to measure whether vaccinating people is a beneficial action, despite Oxford's pseudoscience.[553]

This imposed hypocrisy on society is unconscionable. It 'shocks-the-conscience.' Vaccine package inserts warn of dozens of tragic side effects that vaccines cause, but MKULTRA media censorship and social engineering prohibit public education and frank discussions about these risks to health, safety, and civilization's sustainability.

Academic and professional communities shun debates and blacklist critics in gross violation of First Amendment rights and 'Public Duty Doctrine.'[419] The devil-doers neglect laws and civility. They neglect to protect people and prevent harm. Informed consent requirements are neglected too, and religious free will is denied or replaced by deadly pseudoscience cults.

That's as bad as a fascist regime can get. Totally **treasonous, genocidal,** and hypocritical, especially coming from those who claim to defend human rights and prevent diseases, like the Clintons, Gates, and WHO officials.

BILL GATES, THE CLINTONS, AND THE WHO'S COMPLICITY IN THE COVID COUP

Dr. Jonathan Mann

Here I pay tribute to Dr. Jonathan Mann—one of the world's most esteemed World Health Organization (WHO) leaders—the HIV/AIDS Czar for that United Nations health institution. He and his wife, the Johns Hopkins University hepatitis B vaccine expert, Dr. Mary Lou Clements-Mann, died in the crash of Swissair Flight 111 in September 1998, two years after I published the world's best-selling text on this subject, *Emerging Viruses: AIDS and Ebola--Nature, Accident or Intentional?*

Dr. Mann was present when I confronted the alleged "AIDS-virus discoverer," Dr. Robert Gallo,[53] at the 11th International Conference on AIDS in Vancouver, Canada. This interaction was videotaped, republished many times by many people, and has since been viewed by more than 10 million people worldwide. (This, too, was censored when Google/YouTube and Vimeo/ICΛ terminated my accounts.)

To Dr. Mann's credit, he quit his WHO post, undoubtedly affected by his witness of my confronting Gallo. Mann subsequently protested the commercialization and "imposition" of HIV/AIDS and the hepatitis B vaccination transmission science I advanced. Mann stated his heart-felt opposition and reason for quitting his position thusly: **"More than a medical problem, HIV/AIDS is a socio-political imposition."**

The same is true for COVID-19.

CHAPTER X

COVID, The Biden Crime Family, and Deep State Corrupted Science

"This is not about freedom or personal choice," President Biden said in his speech on September 9, 2021, pushing mandatory vaccinations. The Deep State's draconian depopulation and transhumanist imposition found its champion.[554]

Not actually the New York Post's cover.

Oblivious to the lacking risk-benefit vaccination data, and the deadly consequences of his negligence and complicity in the global genocide, Biden stated, "It's about protecting yourself and those around you – the people you work with, people you care about, people you love."

As reported in Chapter V, referencing Oliver Cook in *TheBL* in June 2021, the president's son, Hunter, invested in Robert Garry's Metabiota

Company— the pandemic firm partnering with Peter Daszak's EcoHealth Alliance and the Wuhan bioweapons lab.

Metabiota is a pandemic tracking and response firm that collaborated with the NIH gain-of-function coronavirus grant recipient, EcoHealth Alliance, and the Wuhan Institute of Virology. This partnership was primarily financed by Rosemont Seneca Technology Partners, an investment group led by Hunter Biden.

On October 20, 2021, the NIH finally admitted financing the gain-of-function Wuhan lab experiments through Daszak's criminally complicit associates, partnered with Rosemont.[555]

Rosemont Seneca Technology Partners ("Rosemont" or "RSTP") was a spinoff of Rosemont Capital, a venture capital firm created by Hunter Biden and John Kerry's stepson in 2009.[556] The President's son served as a Managing Director.[557]

One of the companies identified on archived versions of Rosemon's portfolio is Metabiota. This San Francisco-based startup that claims to detect, track, and analyze new infectious diseases, *The National Pulse* revealed.[558]

According to financial reports,[559] Rosemont led the Metabiota's first round of fundraising, which totaled $30 million. Neil Callahan,[560] the former Managing Director and co-founder of RSTP—a name that frequently shows on Hunter Biden's laptop computer hard drive—is also a member of Metabiota's Board of Advisors. The computer also evidences Biden's substantial business dealings with the Chinese.

As reported by the *BBC*,[561] Hunter told the *New Yorker* he had met with Mr. Jonathan Li, among China's leading investment bankers, for "a cup of coffee." But 12 days after the trip, the men established a private equity fund named BHR Partners, approved by Chinese Communist Party authorities. "Mr Li was chief executive and Hunter was a board member. He would hold a 10% stake."

Evidencing the Deep State crime syndicate's Communist/Capitalist investments, Hunter's lawyer, George Mesires, published that aside from news reports and laptop evidence showing Hunter as a perverted sex and drug addict, Hunter was groomed for international trade.[562]

"After graduating from Yale Law School, where Hunter served as an editor of both the *Yale Law & Policy Review* and the *Yale Journal of Law & the Humanities*, [Yale being a main depository of highly pathogenic viruses] Hunter worked in numerous posts, including as a Senior Executive VP for [Bank of America subsidiary, MBNA,[563] a Delaware-based bank], and as Executive Director of E-Commerce Policy Coordination for the United States Department of Commerce. Hunter also taught as an adjunct professor at Georgetown University School of Foreign Service Masters of Science program."

You may recall that Georgetown University employs Dr. James Giordano[564] — a leading expert in brain-Cloud military neuroscience and bioethics. Georgetown University also hosted Dr. Fauci's 2017 highly incriminating prediction that the Trump Administration would suffer terribly from an unprecedented pandemic.[565]

EcoHealth Alliance and Metabiota are highly incriminated in that plague. Researchers in both worked together on presentations on how to "live safely with bats" and studies tying new infectious disease epidemics to wildlife trade facilities, such as the Wuhan "wet market."[566]

Quoting from the Metabiota/EcoHealth/USAID sponsored presentation, "Wildlife trade can facilitate zoonotic disease transmission and represents a threat to human health and economies in Asia, highlighted by the 2003 SARS coronavirus outbreak, where a Chinese wildlife market facilitated pathogen transmission," the 2016 paper noted.[567]

In addition, Metabiota researchers were named with EcoHealth Alliance staff on a 2014 study on henipavirus spillover,[568] a 2014 study on Ebola monitoring,[569] a 2015 study on herpes,[570] and a 2015 study on viral diversity.[571]

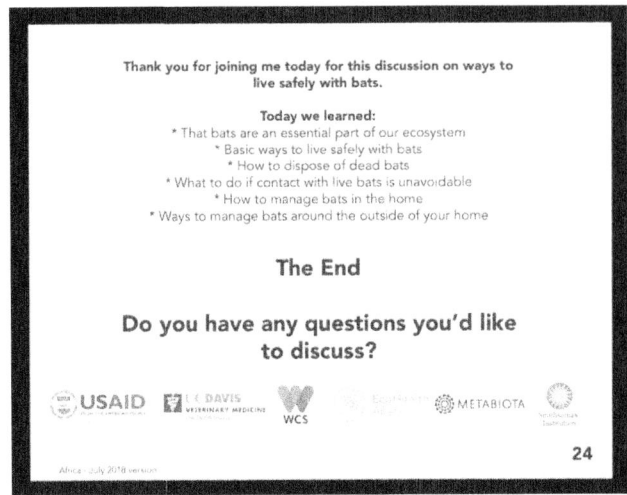

End slide of presentation. (Screenshot via TheBL)

This not only ties the Bidens to an array of deadly plagues, including Ebola, wherein Metabiota was criticized for "bungling" America's Ebola response in Africa but to the CIA and the agency's covert operations in Africa under the cover of USAID "humanitarian aid." Quoting the *New York Times*, "As a cold war policy tool, the agency was, at times, used as a front for C.I.A. operations and operatives."[572]

Metabiota & the Ebola Crisis

Metabiota is "[a]n American company that bills itself as a pioneer in tracking emerging epidemics." Under Robert Garry's leadership, the pioneering group "made a series of costly mistakes during the 2014 Ebola outbreak that swept across West Africa. Metabiota employees feuded with fellow responders with the WHO, contributed to misdiagnosing Ebola cases, "and repeatedly misreading the trajectory of the virus," an *Associated Press* investigation into the company found.[573]

According to reports, the company exacerbated an "already chaotic situation," prompting WHO authorities to denounce it:

Emails obtained by AP and interviews with aid workers on the ground showed that some of the company's actions made an already chaotic situation worse.

Dr. Eric Bertherat, a WHO outbreak expert, wrote to colleagues in a July 17, 2014, email about Metabiota's misdiagnoses and "total confusion" at the Sierra Leone government lab the company shared with Tulane University in the city of Kenema. Bertherat said there was "no tracking of the samples" and "absolutely no control on what is being done." Given the cartel's depopulation and financial agendas and ties to the CIA through USAID, the Chertoff Group, and In-Q-Tel (as previously detailed), the "chaos" must be presumed to have been intended.

"This is a situation that WHO can no longer endorse," Bertherat wrote.

In April 2021, Joe Biden's USAID launched a new program to detect emerging infectious illnesses with pandemic potential, coordinated by EcoHealth Alliance. Metabiota,[574] whose researchers were named as co-authors on articles linked to coronavirus surveillance in Africa since June 2021,[575] also cooperated in the taxpayer-financed endeavor.

The Biden's covert business dealings with China (and actions in Afghanistan and Ukraine as further detailed below) compounds evidence of the National Security Crime Syndicate's treason. This racketeering enterprise commandeered the intelligence community —the CIA, FBI, U.S. Naval Intelligence, and British Intelligence; with Pentagon Chief General Milley mandating vaccines on the advice of Fauci et al. —to leverage COVID for financial gain and Deep State globalization.

"We've been patient but our patience is wearing thin," President Biden threatened Americans, "and your refusal [to get vaccinated] has cost all of us."

The President accused "vaccination hesitant" people of what Biden's complicity in the genocide is guilty of, spreading COVID, inciting violence, and fostering civil war.[576]

What if those people vaccinated are shedding and spreading the poisonous S-protein antigen as mentioned earlier? This is known to occur with dire consequences. Then what?

Biden's propaganda and mandates would then be seen for what they are. **Deadly and Genocidal!** Mandatory vaccinations would be seen as part of a covert depopulation operation aiding-and-abetting transhumanism, the Fourth Industrial Revolution, and the Great Global

Reset —precisely why these mandates are being imposed in favor of the Bidens' Deep State co-conspirators.

Afghanistan, BigPharma and 'Vaccinecide.'

What does Afghanistan have to do with the COVID Coup, vaccination mandates, or the reckless inciting of civil war in America? The common denominator, again, is the globalist agenda.

Given the pattern-and-practice of fraudulently concealing vital evidence of deadly vaccination risks and the globalists' agenda to damage public confidence in our governments, it is no accident that the Afghanistan debacle occurred when it did.

The recklessness and malice behind President Biden's actions to dictate mandatory vaccinations, masking people, social distancing, and imposing unconstitutional business restraints, all while lying about "science," conducting illegal international trade, neglecting the damage to Americans that each policy decision caused, reflects the same **systemic corruption**, *intentional* negligence, recklessness, and malice in Biden's "successful" withdrawal from Afghanistan —the world's leading supplier of opium.

Consider these facts. While anchors in the media were leveraging the "D variant" as justification for mandatory jabs, Biden and complicit military leaders, including General Milley, had a "meeting-of-the-minds." They agreed to abandon Western interests in the world's opium capital bordering Russia, China, and the volatile Middle East.

In the "fake news" media, no one revealed that the Taliban, who were ceded control, was **established by the CIA under Operation Cyclone during the late 1960s** to defend against purported Russian invaders.

No one in the "fake news" media made known that Osama bin Laden, blamed for masterminding the 9/11 "terrorist attacks," was trained and financed by the CIA to oppose the Russians.

No one in the "fake news" media made known the bin Ladens' invested heavily with the Bush family and complicit insiders in Bioport, LLC —the vaccine maker that produced the poisonous anthrax vaccines that triggered Gulf War Syndrome. That sickened and killed tens of thousands of our servicemen and women.

Get the picture? These sickening and enslaving events are rooted in the same demon-infested "Swamp."

In the news at the time of this writing, to no avail, private efforts prompted the release of classified intelligence regarding Saudi Arabia's claimed involvement in the 9/11 disaster that "changed the world." No "fake news media" recalled that Bioport's principal investor was Saudi businessman Fuad El-Hibri, a close friend of the bin Laden family and a previous merger and acquisitions manager for the Rockefeller-linked Citigroup in New York.

According to some investigators I have referenced in past publications, additional Bioport shares were held by The Carlyle Management Group —a leading American defense contractor largely directed by past CIA director Frank Carlucci, James Baker III, George H.W. Bush, and former British Prime Minister John Major.

Further vetting the Deep State's capture of American intelligence agencies, including the NSA and FBI, according to the Associated Press, Past President George H.W. Bush acted as a business agent for the Carlyle Group and wealthy Saudi families, including the bin Ladens.

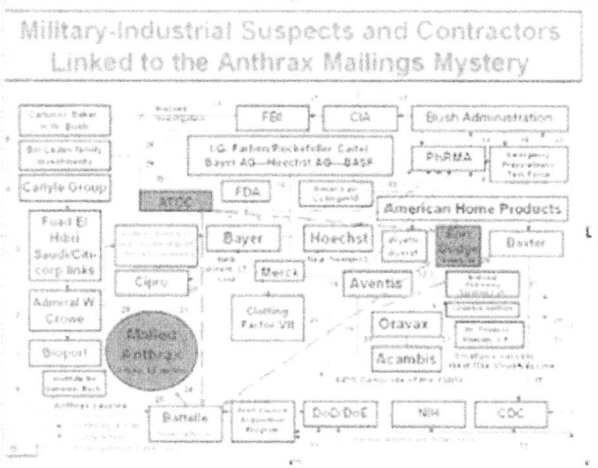

The BBC, and London's leading Sunday newspaper, *The Guardian*, reported on Nov. 7, 2001, that the Bush administration, through the CIA, hog-tied the FBI in their investigations linking Bush family and bin Laden family investments, including commonly held stock in the Carlyle Management Group. This was alleged to have profoundly suppressive implications affecting the investigations of Bioport and Battelle Memorial Labs that produced the deadly mailed anthrax spores.

Today, with the COVID 'plandemic,' we witness the same pattern of fraudulent concealment and racketeering in organized crime.[533]

On September 13, 2021, Tennessee Sen. Marsha Blackburn (R) raised similar concerns that the Afghanistan withdrawal *fiasco* was **"intentional."**[577] She didn't elaborate.

Without vetting or diagnosing this globalist racket, and its pattern-and-practice of committing organized crimes, including treason, sedition, and genocide, we will only continue to witness compounding damage to people, our country, and the world.

> *"We The People now witness this treason and complicity in genocide more clearly. All mainstream media moguls, hosts, and anchors that disregard their paymasters' fundamental agenda are complicit in committing treason and genocide. They are instrumental in killing people, administering 'population control.'"*

The Ukrainian Connection to the Democratic Party's Globalists

Ten days before Hunter Biden negotiated with Ukrainian businessman Vadym Pozharskyi to advance Burisma Holdings' gas and oil businesses with Russia and Germany,[578] on April 7, 2015, the DNC circulated the following memorandum to 2016 election campaign officials setting Donald Trump and his presidency to be the crime syndicate's "Pied Piper":

"Re: 2016 GOP presidential candidates" "intended to outline the strategy and goals a potential Hillary Clinton presidential campaign would have regarding the 2016 Republican presidential field."

"Clearly, most of what is contained in this memo is work the DNC [was] already doing... The 'Goals & Strategy' intended 'to make whomever the Republicans nominate unpalatable to a majority of the electorate... 1) Force all Republican candidates to back themselves into extreme conservative positions that will hurt them in a general election; 2) Undermine any credibility/trust Republican presidential candidates have to make inroads into our coalition or independents... [W]e don't want to marginalize the more extreme candidates but make them more 'Pied Piper' candidates who actually represent the mainstream of the Republican Party... Pied piper candidates include... Ted Cruz, Donald Trump, Ben Carson... We need to be elevating the Pied Piper candidates so that they are leaders of the pack and tell the press to [take] them seriously. Undermining Their Messages & Credibility... The 'Pied Pipers' of the field will mitigate... building a winning general election coalition... undermin[ing] their credibility in... (communities of color, millennials, women) and independent voters... [T]he goal here would be to show that they are... extremely conservative on these issues."

Ten days after this DNC Memo was circulated, on April 17, 2015, Hunter Biden's Ukrainian business dealmaker for Burisma Holdings, Vadym Pozharskyi, e-mailed the Vice President's son to thank him "for inviting me to DC and giving an opportunity to meet your father and spen[d] some time together."

A copy of that e-mail (shown below) accompanied the *New York Post* "blockbuster" story exposing the Bidens' influence peddling.[579]

> From: Vadym Pozharskyi
> <v.pozharskyi.ukraine@gmail.com>
> Subject: Meeting for coffee?
> Date: April 17, 2015 at 6:00:51 AM PDT
> To: Hunter Biden <hbiden@rosemontseneca.com>
>
> Dear Hunter, thank you for inviting me to DC and giving an opportunity to meet your father and spent some time together. It's realty an honor and pleasure. As we spoke yesterday evening, would be great to meet today for a quick coffee. What do you think? I could come to you office somewhere around noon or so, before or on my way to airport.
> Best ,
> V
>
> Отправлено с iPhone

Why President Trump was Hated and Targeted

I previously addressed the CIA and FBI's complicity in the liberal left's coup to undermine the Executive Branch of the Government.[166] As I evidenced previously, and below, Trump's presidency and reelection bid were compromised *ab initio*.

A pattern-and-practice of media cover-ups corrupting politics, health science, and COVID-19's emergence from this cartel's network of collaborating labs in the U.S., U.K., and China began weeks after Chinese President Xi Jinping and Donald Trump took unprecedented actions. They both threatened global investment banks and institutional investors holding stock in Big Pharma and Big Biotech. This challenged the adversaries' COVID-19 plot. The Bidens and their Democratic Party demagogues were likewise put off.

China imposed a military takeover of the drug and vaccine industry in December 2019; the same time COVID-19 hit the press. China's state action especially damaged Western vaccine makers and their stockholders representing the World Bank and World Economic Forum. The drug companies were prohibited from marketing and distributing their products in China. The "Global Legal Monitor" published news of these actions in an article titled, "China: Vaccine Law Passed" on August 27, 2019.[73] That threatening policy took effect on December 1, 2019, the approximate date of the 2019 nCoV/SARS/HIV bioweapon's first appearance (COVID-19), resulting in the 7-person cluster of cases at the Wuhan market.

Meanwhile, at the same time, Democratic Party and Deep State nemesis Trump revised trade and economic policies with China to restore America's independence. He forced globalist drug makers to reduce the cost of prescription meds dramatically. "They don't like me too much," Trump proudly announced during television appearances and campaign rallies.

The reported discovery of Hunter Biden's laptop computer containing criminal evidence of 'influence peddling' in China, Russia, Ukraine, and elsewhere from 2014-15 dovetails with the facts above, implicating World Economic Forum officials representing the global elite.

In correspondence between Joe Biden's son and leading energy and banking industrialists, criminal influence is apparent. These facts, widely dismissed by the liberal left's media as another "Russian hoax," have been confirmed by legitimate insiders and censored members of the press, including White House officials and *FOX News* broadcasters. Biden critics and whistleblowers have been wrongly dismissed as 'conspiracy theorists' by Democratic Party-backed mainstream presses, such as *Business Insider*.[580]

These revelations supplemented President Trump's demand days earlier that the U.S. Central Intelligence Agency (CIA) unredacted its records evidencing a related CIA coup that similarly extended into Russian, Chinese, Ukrainian, and German politics, health science, and COVID-19 cover-ups.

Given these facts, **it must be presumed that the CIA massively redacted Fauci's e-mails (vetted in Chapters III and IV) and not the FBI**. The redactions were committed to conceal the heart of the Deep State's genocidal bioweapons and neuroscience racket and COVID-19's role in the Coup.

The facts about the "novel" virus's lab origin were censored for these reasons.

Trump's reasonable actions to encourage greater candor, more discovery, and law enforcement by the DOJ, to do its job prosecuting criminals were denied by the intelligence arm of the superior controlling powers exposed here.[581]

Expanding on what you learned in previous chapters regarding the intelligence community's capture by the globalists, chief among the early objectors to President Trump's demand for more evidence and candor from Department of Justice officials was former CIA director John Brennan. Mr. Brennan's complicity with Hillary Clinton became clear-and-convincing by the emerging discoveries.

Democratic National Committee leaders, top Justice Department officials, and Obama administration politicians were in on the "Russian hoax" —a dossier and media response full of public deceptions. At the heart of this well-orchestrated scandal, Mr. Brennan served as CIA director until January 20, 2017. Then, Brennan transitioned his covert operations between the Obama and Trump administrations. His sham "departure" coincided with the illegal surveillance and dirty tricks played on the incoming president and his staff.

CIA Director Brennan's departure occurred one week after Dr. Anthony Fauci relayed his "insider" intelligence to the gathering at Georgetown University.[582]

It is public knowledge that the CIA oversees Dr. Fauci's National Institute for Allergies and Infectious Diseases (NIAID), covert vaccination campaigns,[583,584] and the Biden administration at large. In this way, the CIA is largely aligned with Bill Gates's global enterprise.

Few Americans know of these connections. The CIA is largely responsible for U.S. National Security, defending the cartel against foreign threats. This spy network delivers medical intelligence to officials as well.

Surely the CIA knew about and encouraged the Bidens' global business dealings, including those in Metabiota.[574]

The 'science' supposedly demanding 'social-distancing,' 'mask-wearing,' business 'shutdowns,' and mandatory vaccinations certainly merits intelligence gathering and black-op decision-making. Leading Navy uniformed experts in the fields of public health, military biodefense, and pharmaceutical/vaccine response undoubtedly received intelligence briefings overseen by these intelligence arms of the Deep State. Otherwise, Dr. Fauci's unprecedented warning to his colleagues at Georgetown, and leadership of America's COVID response, would not have happened.

This intelligence best explains how and why in October 2019, six weeks before the Wuhan outbreak, Bill Gates, Johns Hopkins University, and the World Economic Forum, co-sponsored "Event 201"[86] —the astonishing 'predictive programming' conference foretelling virtually everything that would unfold from the "plandemic," including detailed plans for the public and private corporate responses to the disease and its devastating anticipated social and economic consequences.

Hunter Biden, Dirty Money and the Bank of China

Financial data leaves a trail for law enforcement to follow. Among dozens of news outlets, *AXIOS*[585] reported on thousands of leaked government documents covering at least $2 trillion worth of transactions by some of the world's biggest banks knowingly laundering money for those at the forefront of the Fourth Industrial Revolution (4IR) movement, the "oligarchs, terrorists, and criminals."

COVID, The Biden Crime Family, and Deep State Corrupted Science

It seems Noam Chomsky's liberal bias neglected this data.

Few if any consequences occurred from this neglect. Law enforcement was shut down. And the only leader somewhat objecting to this racketeering enterprise appeared to be Trump.

According to "a massive investigation by *BuzzFeed News*,[586] the International Consortium of Investigative Journalists (ICIJ),[587] and hundreds of other news organizations," the injustice mentioned above is verified.

"The Big Picture," AXIOS reported, "examine[d] more than 2,100 suspicious activity reports (SARs) filed by banks and other financial firms with the U.S. Department of Treasury's Financial Crimes Enforcement Network,[588] known as FinCEN... Individuals or organizations with addresses in the U.K., China, Germany, the United Arab Emirates, Canada, and Ukraine [all the countries within which the Biden's covertly transacted payoff schemes] appeared in at least **20 reports each.**

Anonymous (ID: UPgcN8+)
The world as we know it as goi(...)
10/17/20(Sat)00:22:43 No.283080858

Chinese billionaire who opposes Chinese Communist party delivered three hard drives filled with Chinese blackmail material on Hunter Biden to Rudy Giuliani in order to sabotage CCP efforts to control our election process. China backed Biden for President. The Democrats paid the Chinese government to create and release COVID-19 to disrupt the election and drive down Trump's support. China has blackmail material of Hunter with little girls. It's about to be released.

This will be viewed as a crime against humanity. A crime that only the most fanatical Communists in America will defend. Democrats will have utterly no credibility in defending Biden. The media will not be able to explain it away. It is the level of crime necessary to repulse and revile basically all normies.

In total, the SARs flagged more than **$2 trillion in transactions** between 1999 and 2017 [the year President Trump interrupted and challenged their racket]. The documents included SARs filed by nearly **90 financial institutions.**" The top 10 most common abusers are well-represented by World Economic Forum members: Deutsche Bank, Bank of New York Mellon, Standard Chartered, JPMorgan Chase, Barclays, HSBC, Bank of China, Bank of America, Wells Fargo, and Citibank.

Hunter Biden's e-mails evidence his father's activity within this crucial criminal 4IR banking syndicate. As reported in the *Washington Examiner*,[589] Hunter Biden's "work with BHR (Shanghai) Equity Investment Fund Management Company isn't his only controversial connection[590] to Chinese business. While working with the multi-billion-dollar Chinese conglomerate CEFC, and this energy leader's official, Ye Jianming, on a natural gas deal in the United States worth millions, the Vice President's son agreed to represent Jianming's associate, Patrick Ho, who was eventually convicted by the Justice Department for bribing foreign governments. Jianming was reportedly detained in China.

"Hunter Biden said that by the end of October [2019], he'd be resigning from his position on the board of BHR (Shanghai) Equity Investment Fund Management Company."

BHR is a private equity firm established in 2013 with backing from Chinese state-owned finance companies. Their goal was to invest Chinese money globally. Their largest shareholder was the state-controlled Bank of China. This news came through Mesires, Hunter's lawyer, as first reported by the World Economic Forum member and Trump nemesis, **Michael Bloomberg**.[591]

Grading the Bank of China, and advising China's leadership on every aspect of economic development, was World Economic Forum's Coalition (WEC) director Schwab, and of course, the captured and corrupted intelligence agencies.

As mentioned, Schwab, with the world's most influential Big Banking and Big Pharma institutional investors, pioneered the concept of the "Great [Global] Reset."[592] The WEC's website heralds this and Schwab's other undertaking in "managing the direct consequences of the COVID-19 [contrived] crisis."[593]

More on WEC's Backing of China to Damage America

You should no longer be wondering who is financing America's destruction and global transition into transhumanism. For example, Black Lives Matter and ANTIFA were financed to commit criminal and political unrest by these same genocidalists. The pro-Chinese and anti-Trump stance of the World Bank and World Health Organization indicts the same racketeers, including George Soros's Fund Management, certified by Soros's presence in Schwab's World Economic Forum (WEF) coalition.

The world's wealthiest man, Jeff Bezos, donated $10 billion to WEF's **climate change agenda**.[594] Bezos created Amazon with CIA financing. This best explains why the *Washington Post*, also owned by Bezos, always outputs Deep State propaganda. Soon after the CIA's financing kicked in, Amazon was reported to be "the largest online shopping retailer in the world."

The WEF, as previously mentioned, is also financially and ideologically backed by Bill Gates.[595]

Perusing the WEF's "Partners" webpage creates a "long-lasting impact,"[596] Schwab's advertisements promise. Visitors viewed (on Oct. 17, 2020) articles that herald the 4IR 'paradigm shift' in global geopolitical, economic, and cross-cultural relationships all intertwined with data mining and mind-bending.

Featured articles here were titled: (1) "Why being an LGBT + ally can transform lives – yours included;" (2) "Shared learning platform – Innovation and health technology" heralding the world's most advanced biotech research and developments impacting health science and energy systems furthering "transhumanism," linking computers, telecommunications (e.g., 5G and Cloud technologies), robotics, human consciousness, and social behavior through the web; (3) "3 innovative ways that energy and resource extraction companies are going green."

So much for determining the source of the Democratic Party's 2020 platform, including their "Green New Deal."

Then, another WEF article heralds, "This is what the fight against COVID-19 can teach us about stopping climate change, according to Bill Gates."[597]

According to Schwab, who *China Daily* interviewed in 2014, "The World Economic Forum's latest Global Competitiveness Report show[ed] China as the highest-ranked 'BRICS economy.'"[598] The **BRICS** acronym was coined to "associate five major emerging national economies: Brazil, Russia, India, China, and South Africa."

Notice the United States is not among the 'favored nations,' as President Trump responsibly protested. This exclusion reflects the political and economic globalization bias and objectives enabling America's degeneration before the 4IR "Great Reset."

"Despite a slowdown from the breakneck pace of the past decade," Schwab reported enthusiastically about China, "we see encouraging signs that growth [in the Orient] is sustainable."

"China's leadership has established an enormously ambitious agenda that will require all the building blocks for an inclusive and accountable ecosystem that allows for entrepreneurship and innovation to flourish education and skills, cooperation between research and business, financing for SMEs, and removing barriers that limit competition."

Accordingly, there is no doubt Schwab's vision for the world following the Great Reset is a globalist undertaking. This New World Order vision competes directly against nationalistic leadership, conservative politics, and patriotic values.

This view of the world and its economic engine in the globalist coalition rejects America's dominance in the 4IR. That advancement features world interests, not national interests. World Bank financing is limited to nations that get aboard the 4IR train, adopting Big Tech, Big Energy, and Big Pharma's policies and products especially forced vaccinations.

Like the corporate-controlled media that neglects Klaus Schwab's father's service to the Nazi Party, the media likewise neglects the 4IR's transhumanist agenda and totalitarian assault on civilization.

The Third Reich's most advanced military neuroscience technologies involving sound and light bioweapons, drug research and pathogen developments, and Hitler's secret 'heavy water' nuclear weapons program matured into the current World Economic Forum members' monopolized industries.

Examining these facts provides probable cause for COVID-19 criminal indictments, beginning with the censorship surrounding the WEF's co-sponsorship with the Bill & Melinda Gates Foundation and Johns Hopkins University of the Event 201 coronavirus conference.

Influence peddling has consistently favored Schwab's 'Great Reset.' Now, with evidence discovered through Hunter Biden's laptop computer, this vision of world domination comes clearly into focus. The crime family's shady deals that involve the oil and gas company Burisma Holdings expand to Russia's true role in the globalists' geopolitical scheme. The Bank of China and Chinese Energy [Finance] Company (CEFC), too, fall into place in the global conquest now discerned thanks to Hunter Biden's forgetfulness from a meth pipe in his mouth.

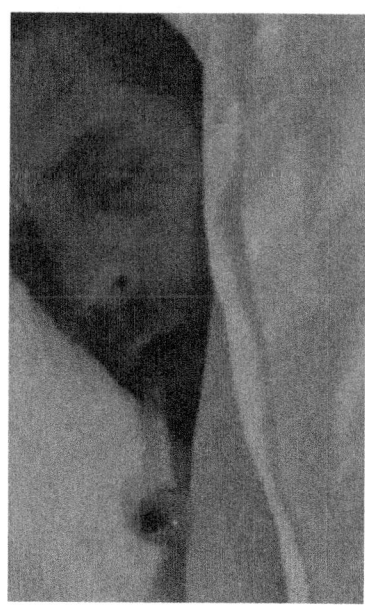

The Biden's Concealed Conflicting Criminal Interests

Hunter's laptop files provide substantial evidence, indicting the Biden Crime Family for sex, drugs, and political crimes.

As shown in the photo above, it is not uncommon for meth addicts to lose consciousness, repeatedly compelling instant sleep or bed rest. Drug 'burn-outs' become so dysfunctional and forgetful that losing things, such as one's lap computer, often occurs.

Consequently, this drug 'side effect' may be the best thing that has ever happened in American political history.

Biden's business cohorts in financial crimes include the CHINESE ENERGY [FINANCE] CORPORATION or "CEFC." according to the 2018 REPORT TO CONGRESS of the U.S.-CHINA ECONOMIC AND SECURITY REVIEW COMMISSION ONE HUNDRED FIFTEENTH CONGRESS SECOND SESSION NOVEMBER 2018.[599]

According to *Wikipedia*, CEFC is a company that used a "complex web of affiliated companies to facilitate fake deals, inflate trade figures, and obtain bank loans to fuel its aggressive expansion" internationally.[600]

As recalled in *Intelligence Quarterly*, "In May 2017, Hunter met privately with Ye Jianming (叶简明), chairman of Chinese energy company CEFC (中国华信能源), at a Miami hotel. The purpose of the meeting was for Hunter to use his contacts to help 'identify investment opportunities for CEFC.' Afterward, Hunter received a 2.8-carat diamond.[601]

Later, Chairman Ye was detained and put under investigation by federal officials. In March 2018, on suspicion of economic crimes, while CEFC was declared bankrupt in March 2020, Ye was alleged to have faked business deals and bribed foreign governments for oil rights. This pattern of white-collar crime was facilitated by Patrick Ho (何志平), a Chinese Communist Party member and the former Hong Kong Secretary for Home Affairs in Tung's administration.

On November 18, 2017, Patrick Ho was arrested at the John F. Kennedy International Airport on bribery and money laundering charges. He called Hunter Biden for assistance. Hunter later told *The New Yorker* that he didn't see Ye as a "shady character at all," and he characterized the outcome as "bad luck."

Democratic Party Politics, Policies, and COVID-19

The main Democratic Party objective was to blame Trump for COVID and the deaths of hundreds of thousands of Americans. An incredible number of voters subscribed to this red-herring blame game.

As shown above, Democratic Party officials arranged for Trump's 2016 Republican nomination. The leaked Democratic National Committee memorandum issued by Hillary Clinton and John Podesta was barely given media attention, even though it evidenced the seditious subversion of American politics and the CIA/NSA/Deep State's 'handling' of Hillary and the DNC.[602]

These incriminating facts extend to leading Democratic Party campaign contributors. I previously detailed media mogul Diller, who hired Chelsea Clinton to co-direct the IAC cartel with Seagrams crime family heir Edgar Bronfman, Jr.[603,283]

The Deep State Steals Elections

Did the Democratic Party, controlled by the Deep State agents I have already considered, steal the 2020 presidential election? The secreted evidence says, "Yes!"

The polls were wrong. The polling questions were deceitfully contrived. We now see a 'pattern-and-practice' emerging showing polls that are public persuasion, and propaganda ploys are as contrived as election results.

I am advancing here evidence that has been neglected showing a pattern-and-practice of election fraud, intertwined with science fraud. Recall that "science" journals are nearly all published by Reed-Elsevier. That publishing cartel is owned by a little-known company called "ChoicePoint Inc."[604] They control 'Medical MindWar,' the science and pseudoscience published in nearly all medical and scientific journals.

Here's why Biden's election was stolen:

In 2009, I contested gross conflicting interests in *The Lancet* that rejected my Op-Ed piece. I titled it "PHARMAGANDA: A Study in Conflicting Interests."[605] I made known that "the merger of Reed and Elsevier created a virtual monopoly over the entire medical and scientific publishing industry."

I wrote, "Elsevier alone publishes 250,000 articles annually in 2000 journals." I objected to ChoicePoint having purchased Reed-Elsevier in 2008 for $3.6 billion in cash. That merger effectively monopolized the medical, scientific intelligence industry.

I reviewed online advertisements promoting ChoicePoint. They describe the monopoly as a "prime intelligence service." I learned and revealed that the conglomerate "supplies governments and multinational corporations with demographic data, including confidential voter records." "The company is best known for its administration of corrupted records used to certify Past-President George W. Bush's contested 2000 election."

Yes, you read that right. Reflect on Biden's presumed "stolen 2020 election."

According to open sources, in the aftermath of the Bush-Gore contested election vote, ChoicePoint's subsidiary, DBT Online, was accused of conspiring with Florida Governor Jeb Bush, Florida Secretary of State Katherine Harris, and Florida Elections Unit Chief Clay Roberts in the voter fraud conspiracy involving the central voter file. ChoicePoint was also accused of knowingly using inaccurate data and also for racial discrimination.

I concluded that the medical science monopolists controlling ChoicePoint, Elsevier, and the worlds of scientific data-mining and media publishing cartel, had captured the electoral process and corrupted both parties equally.

That is why it was no surprise to me that in November 2021, contrary to the mainstream polls, Republican Glenn Younkin defeated Democrat Terry McAuliffe in the Virginia gubernatorial election. I knew, as soon as I read that McAuliffe had entered politics immediately after spending a quarter-century at the CIA-infested[606] and infamous private-equity firm called the "Carlyle Group," including becoming its CEO.[3]

In April 2003, the Carlyle Group was managed by a team of former US Government officials, including its president Frank Carlucci III, the former deputy director of the CIA before becoming Defense Secretary. Carlucci is accountable for overseeing the CIA's African AIDS virus deployment from his office in Portugal during the early 1970s.

Carlucci's deputy in Carlyle was James Baker III, US Secretary of State under George Bush senior. Recalling the fixed election of George W. Bush, the Deep State sent Baker III to Florida to oversee the vote recount in that election, to make sure Black votes were never counted, enabling Bush's victory over Al Gore.

Finally, it is public knowledge that the Carlyle Group managed the financial assets of the Saudi Binladen Corporation.

CHAPTER XI
COVID GENOCIDE, THE LAW & THE RISE OF THE FOURTH REICH

"Be not deceived: evil communications corrupt good manners...For God is not the author of confusion, but of peace..."
1 Corinthians 14 and 15:33

These are the strangest times. The vaccination 'Gestapo's propaganda and genocide stagger the mind. Parents are claimed to be "domestic terrorists" for simply voicing their interests in child protection from WOKE teachers and members of school boards. Others are smeared as evil "anti-vaxxers" and child abusers if they reject the jabs. Science journals spread "pseudo-science." Newsagents, agencies and so-called 'esteemed scientists' disseminate propaganda that extinguishes common sense and has led to millions of deaths.

The Elements of Vaccination Genocide Neglected by Government Officials and the Media

This chapter considers vaccination genocide in the context of the "ten stages of genocide"[607] as detailed by the honorable president of Genocide Watch,[608] Gregory H. Stanton.

Here I apply these ten stages as 'legal elements' to justify the claim of **vaccination genocide**. I claim the crime of genocide attributable to vaccines and vaccination advocates is worsening. I call this exploding vaccination genocide '**vaccinecide**' or alternatively '**injecticide**.'

The graphic below lists Stanton's "ten stages of genocide." I examine each of these as elements of vaccination genocide. They each apply to vaccines being used as "genocidal weapons of mass depopulation," as aforementioned and further detailed below:

TEN STAGES OF GENOCIDE

1. **Classification:** People are divided into "us and them."
2. **Symbolization:** People are forced to identify themselves
3. **Discrimination:** People begin to face systematic discrimination.
4. **Dehumanization:** People equated with animals, vermin, or diseases.
5. **Organization:** The government creates specific groups (police/military) to enforce the policies.
6. **Polarization:** The government broadcasts propaganda to turn the populace against the group.
7. **Preparation:** Official action to remove/relocate people.
8. **Persecution:** Beginning of murders, theft of property, trial massacres.
9. **Extermination:** Wholesale elimination of the group. It is 'extermination' and not murder because the people are not considered human.
10. **Denial:** The government denies that it has committed any crime.

1) We have obviously witnessed the "**divide to conquer**" strategy. Anti-vaxxers have been separated from pro-vaxxers by Big Pharma's bribed politicians,[512] the complicit media, and recently the President of the United States, and Deep State front man, Joseph Biden.

2) Anti-vaxxers have been **forced to identify themselves** for access to schools and workplaces. Soon, as California is modeling, drivers will need to evidence their vaccine record for a license to operate a motor vehicle or fly on a plane.

3) Anti-vaxxers have already faced **systematic discrimination**. Millions of 'vaccine hesitant' parents have been forced to home-school their children. Some hospitals and physicians are rejecting patients who refuse to take the jabs.

These are classic examples of "systematic discrimination." Physicians who have issued patients' vaccine exemptions for religious and medical reasons have had their licenses revoked. The rest of healthcare has been compromised by this discriminatory policy. Tens of thousands of healthcare and airline workers, and first responders, have been fired from their jobs for refusing COVID injections.

4) Un-vaccinated children are **being dehumanized** and said to act like vermin, disease carriers, threatening the rest of the 'herd.'
5) There are several classic examples of this.

The first example is sourced from Johns Hopkins University. Recall that this academic institution fed us COVID case and death statistics and co-sponsored the Coronavirus Predictive Programming "Event 201" planning conference.[92] Johns Hopkins was also heavily involved in hepatitis B vaccine research but neglected my peer-reviewed scientific thesis on the hep B vaccine origin of HIV/AIDS. Hopkins is also the school that mutates the AIDS virus to purportedly "treat" Blacks suffering from sickle-cell anemia.

To their credit, page 113 of Hopkins' "Care at Home for the Immunocompromised Patient"[609] (originally published in 2005 and revised in 2012) instructs non-defined "immunocompromised" people to "Avoid contact with children who were recently vaccinated."

That reasonable advisement respects the fact that "recently vaccinated" persons "shed" live genetically engineered viruses that could kill an immune-compromised person. The same is true today with COVID-19 injections that cause shedding and spreading of the S-protein (AIDS-laced) synthetic antigen.

Sadly, Hopkins' advisement[610] was changed in 2018[611] when the definition of "immunocompromised" was added. Officials arbitrarily and capriciously defined "immunocompromised" to mean "your immune system is weaker than normal."

But what is a "weaker than normal" immune system in this 'Age of Compounding Poisonings,' including toxic injections? Today, most people have been vaccinated numerous times and have developed multiple chemical sensitivities, allergies, and auto-immune diseases, largely caused by vaccination intoxications with foreign protein antigens. They are immune-compromised from exposures to chemicals and antigens in air, food, water, and/or vaccines manufactured in animals and chickens fed the same (or similar cross-reacting) foreign protein antigens.

According to one widely promoted study, more than half (54.6 percent) of the people surveyed in the U.S. "indicated that they had positive reactions to one or more allergens. Allergic rhinitis (hay fever) affects between 10 and 30 percent of all adults in the U.S. and as many as 40 percent of children."

More than fifteen percent of the population suffers from eczema, the most common vaccination side effect sourced from immunocompromising the antigen-antibody response. The immune system goes haywire in eczematous dermatitis patients. White blood cell bodyguards attack and destroy the victims' skin proteins, beginning with their subdermal connective tissue.

To their credit, Hopkins' cancer specialists noted this compromised immunity sources "probably from your disease or treatment." Vaccination is a 'treatment' that causes these and many more diseases. That means that all vaccinated persons who are sick and immune-compromised from the vaccines should not receive more vaccines or even socialize with healthy un-vaccinated people to reduce the risk of antigenic crossover.

Further evidencing this genocidal element, un-vaccinated children were dehumanized, smeared as disease carriers, and segregated for this fraudulent conversion of disease vectoring from those vaccinated to those not vaccinated. Hopkins ironically and hypocritically warns, "Tell friends and family who are sick not to visit [recently vaccinated children]."

Summarily, this entire element of genocide has been flipped, projected onto un-vaccinated people when, in fact, the actual dehumanized group is those vaccinated, since the vaccines alter DNA and genetic function, making people less human and more transhuman. Vaccinated people become "genetically-modified organisms" (i.e., "GMOs").

Accordingly, the dehumanization damaging those not vaccinated proves a classic loathsome fraud in "public health," tricking people into getting vaccinated, believing non-vaccinated infants, children, and adults are like vermin spreading diseases. This is a sales gimmick and subliminal social-engineering ploy.

6) The **government created and targeted 'vaccine hesitancy' groups to be 'neutralized.'** By passing mandatory vaccination laws, much like the Nazis passed eugenics laws, the American government created the enemy—groups of unvaccinated people.

7) Clearly, the **government broadcasts propaganda** to turn those favoring vaccines against those who remained "vaccine-hesitant." During the past few years, the CDC, WHO, and Deep State's media broadcast propaganda to turn the civilized people into mental zombies opposing any and all vaccine skeptics. Citizens are even attacked online by expertly trained critics, 'skeptics,' and ignorant scientists. This social dissonance was widely viewed on Capitol Hill,[512] in social media and mainstream news broadcasts. These actions are central to MindWar.

To find the facts and analyze the data, I reviewed the scientific literature and reputable open-source publications that conflict with the governing officials' misbehavior on Capitol Hill and in corrupted state governments. I found substantial evidence of bias from Big Pharma's influence.

In my article titled "Vaccination Hesitancy: A Red Herring,"[512] I compared the senators' hyperbole[612] with fact-based science to determine if 'vaccination hesitancy' is justified or not; whether blaming parents is reasonable, and if targeting vaccination hesitancy for extinction as a governmental policy is justifiable under the circumstances. I found that it was not.

I concluded complicity in genocide was present and unmistakable, despite diversionary arguments.

8) Clearly, the state and federal governments have removed or relocated **people** who have not been vaccinated. This has occurred at the Mexican border at claimed 'concentration camps' operating there and in schools across the nation.

Officials had acted criminally as complicit agents in genocide when they removed unvaccinated children from schools, smeared vaccine-hesitant parents, and removed or relocated non-vaccinated healthcare workers from their jobs. At the same time, by the end of 2021, the Biden administration enabled nearly 2 million unvaccinated, untested, illegal immigrants to enter the U.S. to sway elections in favor of Democrats and take jobs away from capable citizens.

9) The pro-vaccination forces have persecuted the unvaccinated and begun **stealing properties and conducting massacres**.

   Examples of property theft include parents having their homes taken and estates stolen by social service agents, prosecutors and judges persecuting, prosecuting, and robbing parents of their children and livelihoods. "Shaken baby syndrome" is one example of this, wherein parents are falsely blamed for deaths caused by vaccines, not shaking or suffocating infants. Many other parents have gone to jail and lost everything after being persecuted and prosecuted for vaccination hesitancy misrepresented as 'child abuse.'

   Another good example is the persecution of whistleblowers opposing mass murders and vaccine massacres, such as what CDC officials did to condemn Dr. William Thompson.[613]

   Nearing the end of his career, Thompson refused further complicity and became a whistleblower exposing CDC officials' scientific evidence tampering and data purging that occurred during the Atlanta Childhood Autism Study. The actual concealment of data and evidence tampering proved a government coverup of statistically significant increases in brain damage among African American boys caused by the MMR vaccine that contained neurotoxic mercury. The entire matter was white-washed and stonewalled by officials.

10) Generally, with vaccines, we are looking at **mass extermination** of "useless eaters"[614] –a designation Henry Kissinger gave to describe and dehumanize geriatric populations.

    As introduced earlier, Kissinger's company, Kissinger Associates, managed the world's leading vaccine maker, Merck & Co. Kissinger officiated Project Paperclip in the early days of the CIA that granted Nazi scientists' asylum if they agreed to work for Anglo-American controlled companies on secret government projects. Merck received the lions-share of the Nazi war chest at the end of WWII.

One example of Merck's subsequent acquisitions was Afluria.[615] Merck obtained exclusive rights to market and distributed this seasonal influenza vaccine from CSL Biotherapies. CSL Labs was largely financed by the Murdoch family, controllers of *FOX News, Wall Street Journal,* and the *New York Post.* James Murdoch is highly influential at Glaxo-Smith-Klein-Beecham, Pfizer's parent company, with James acting as their Non-Executive Director. At the time of this writing, *FOX News*'s hypocrisy is stunning. The 'right wing' voice repeatedly demeans Planned Parenthood yet refuses to report on the 'Deep State' Rockefeller-directed vaccinations-for-depopulation agenda openly promoted by the global elite.[411]

The most common systemic adverse events from Afluria were fibromyalgia pain and headaches in greater than 20% of recipients.[616] In adults 65 years of age and older, this side effect was seen in greater than 10% of recipients, prompting these elders to be more than **useless eaters**. Their chronic fibromyalgia kept them returning to drugstores and doctors' offices to buy more and more drugs. Additional drug side effects resulted in the "wholesale elimination of the group."

11) Finally, the vaccine genocide is evidenced by **governmental denials that it had committed any crime** with the passage of vaccine legislation mandating intoxications, and increased morbidity and mortality, nevertheless indemnifying the vaccine-makers from liability.

In so far as the secreted vaccine-cancer links, similar damages and denials have been seen in the white-washing by the U.S. General Accounting Office (GAO) investigating the origin of HIV/AIDS tied to the hepatitis B vaccine under the Merck & Co.'s SVCP Contract 71-2059,[617] as I explained and published previously.

Vaccine cancer links are unconscionably censored, as is honest VAERS data, to advance the Deep State's profitable depopulation genocide. No government agency or corporate-controlled media has permitted discussions on these crimes.

Intelligence sources that previously permitted vaccine controversies to be superficially considered (due to weak resistance and easily 'neutralized' critics) have now been suspended from public debate due to the worldwide explosion of opposition and the awakening of civilization to this genocide. As an alternative strategy to conceal the governments and media's complicity in the worsening genocide, these sources have especially suppressed science on the vaccine cancer links.[618]

I share this intelligence as the most censored consumer health and safety activist[619] in this field of vaccine-cancer links, the Editor-in-Chief of MedicialVeritas.org, and author of the best-selling scientific investigation into the man-made origin of the cancer virus called HIV.

Elements of the Crime of Genocide

To give you further clarity on this most serious "Crime of Genocide," I will quote liberally from the United Nations Convention on the Prevention and Punishment of the Crime of Genocide:[620-621]

"*[G]*enocide means any of the following acts committed with intent to destroy, in whole or in part, a national, ethnical, racial or religious group," such as vaccination-hesitant people, most of whom refuse the jabs for religious reasons. These acts include, but are not limited to:
   a) Killing members of the group;
   b) Causing serious bodily or mental harm to members of the group;
   c) Deliberately inflicting on the group conditions of life calculated to bring about its physical destruction in whole or in part;
   d) Imposing measures intended to prevent births within the group;
   e) Forcibly transferring children of the group to another group.

"The Genocide Convention establishes . . . that the crime of genocide may take place in the context of an armed conflict, international or non-international, but also in the context of a peaceful situation." In this case of 'vaccinecide,' although the military has helped finance DNA-altering vaccines, their international distribution "in the context of a peaceful situation" is anything but. The concomitant MindWar evidences military malice. According to the Convention, prosecutors are obligated to "prevent and to punish the crime of genocide."[621]

"The popular understanding of what constitutes genocide tends to be broader than the content of the norm under international law. Article II of the Genocide Convention contains a narrow definition of the crime of genocide, which includes two main elements: (1) *A mental element*: the 'intent to destroy, in whole or in part, a national, ethnical, racial or religious group, as such,'" and the ten *physical elements* detailed above.

The MindWar and genocide impacting vaccination-resistors especially demonize religious persons and groups, with 'intent to destroy, in whole or in part,' organized religions to convert civilization to the Luciferian 'Drug Cult,' as evidenced in a forthcoming chapter. Misplaced faith placed in the 'Gods of Science, Medicine, and Biotechnology' is the clear-and-present danger.

Here is the main challenge for prosecutors. The *"intent"* is the most difficult element to prove. To constitute genocide, there must be a proven intent on the part of perpetrators to physically destroy a national, ethnic, racial, or religious group. Cultural destruction does not suffice, nor does an intention to simply disperse a group. It is this special intent, or *dolus specialis*, that makes the crime of genocide so unique. In addition, case law has associated intent with the existence of a State or organizational plan or policy, even if the definition of genocide in international law does not include that element.

"Importantly, the victims of genocide are deliberately targeted - not randomly – because of their real or perceived membership in one of the four groups protected under the Convention (which excludes political groups, for example). This means that the target of destruction must be the group, as such, and not its members as individuals. Genocide can also be committed against only a part of the group, as long as that part is identifiable (including within a geographically limited area) and 'substantial.'"

The charge of vaccinecide targeting expressly the group of unvaccinated people, specifically identified based on 'public health' and 'National Security,' residing within the United States, fits these criteria. Today, we witness "the existence of a State or organizational plan or policy" to deprive the un-vaccinated group of Americans of their civil and Constitutional rights.

Damage Mounting from the Crime of Genocide

The ignorant masses have been devastated by the advancing genocide. Millions of naive people have bought into the vaccine safety, and efficacy lies and doesn't see the genocide unfolding in the 'bigger picture.'

According to the CDC's October 21, 2021, report, the government has "received 8,878 reports of death (0.0022%) among people who received a COVID-19 vaccine." (VAERS data.) But COVID vaccine deaths are wildly *under-reported* for the simple reason that doctors and nurses feel guilty when someone dies from their injection. Legal protection is provided by lying. Deaths can be attributed to 'underlying conditions' diverting from the damage vaccines deliver. To do otherwise would jeopardize the vaccine industry, as well as the medical establishment.

The medical and vaccine paradigm is flawed and fraudulent, as is the propagandists' mantra, "Vaccines are Safe and Effective." That is delusional, deadly, but 'generally accepted.'

As chronicled and evidenced earlier, several of America's most devious covert actors, all psychopaths, have committed mass shootings and killings to accomplish their globalization goals while enslaving people to drugs, more drugs, and even more drugs. Vaccines, more vaccines, and many more vaccines evidence a deadly capricious political agenda and addiction. These pharmaceuticals are fraudulently claimed to secure what God presumably failed to safeguard —human bodies in civilization's battle against germs. The genocidal operatives have extinguished the doctrine that had been generally accepted in science and medicine for a century, that "the germ is nothing, the terrain is everything."

On his deathbed, Louis Pasteur, the doctor that pioneered 'pasteurization,' confessed, "I was wrong. The germ is nothing. **The Terrain is everything.**" By terrain, he meant the internal pH of the body. That not only reflects the amount of energy in your body but the chemical and electromagnetic balance in the body between acids and bases (or alkalizers). This is most important in securing robust immunity against infectious diseases and cancers. The health of cells, tissues, organs, and immune defenses is damaged, not helped, by vaccinations that inject foreign proteins, genes, and chemicals into your body.

These threatening circumstances remind me of Joseph Goebbels — the propaganda minister for the Third Reich— who preached something like this: "If you tell a lie long enough, eventually it will be believed as truth; and the greater the lie, the more people will believe it." The phrase "vaccines are safe and effective" comes to mind because never a greater lie has ever been told.

Evidence of genocidal vaccinations is widespread and mounting. The pseudo-science hypocrisy is being revealed throughout the 'fake news' media (including *FOX News*).

On Nov. 8, 2019, *Fox News* hypocritically condemned competing propagandists on CNN, NBC, CBS, and ABC. I watched FOX's 'right wing' host debate a 'left-wing liberal' who supported Planned Parenthood that financed the RU486 'morning-after abortion pill. The 'liberal' depopulation spokesperson condemned anti-vaxxers as believers in 'pseudo-science.'[622]

Rather than slamming the liberal propagandist intelligently with real science, the *FOX News* host diverted to a safer topic as the networks always do. *FOX*, after all, must favor its owner and Merck vaccine company leading stockholder, Rupert Murdoch. Drug ad revenue also discourages 'biting the hand that feeds' the media and 'entertainment' industry.

A diversion is a form of **censorship**—substantive omissions evidence **fraud**, and in this case, also **reckless endangerment** of billions of people.

Toxic Chemicals and Supreme Court Decisions

The U.S. Congress passed the Chemical Weapons Convention Implementation Act in 1998. The Act forbids any person knowingly "to develop, produce, otherwise acquire, transfer directly or indirectly, receive, stockpile, retain, own, possess, or use, or threaten to use, any chemical weapon." 18 U.S.C. §229(a)(1).

It defines "chemical weapon" in relevant part as "[a] toxic chemical and its precursors, except were intended for a purpose not prohibited under this chapter as long as the type and quantity is consistent with such a purpose." §229F(1)(A).

"Toxic chemical," in turn, is defined in general as "any chemical which through its chemical action on life processes can cause death, temporary incapacitation or permanent harm to humans or animals.

"The term includes all such chemicals, regardless of their origin or of their method of production, and regardless of whether they are produced in facilities, in munitions or elsewhere." §229F(8)(A).

The drug/vaccine industry has been able to get around charges of reckless endangerment and killing millions of people using toxic chemicals, especially fragile infants, by claiming its exemption under this chapter of law.

Congress exempted the use of toxic chemicals for "[a]ny peaceful purpose related to an industrial, agricultural, research, medical, or pharmaceutical activity or other activity," and other specific purposes. §229F (7).

Alternatively, according to the U.S. Supreme Court's discussion in ***Bond v. US***,[623] 134 S. Ct. 2077 (2014), "A person who violates section 229 may be subject to severe punishment: imprisonment 'for any term of years,' or if a victim's death results, the death penalty or imprisonment 'for life'" may be ruled. §229A(a).

This law, and the circumstances arising here, wherein the vaccine Gestapo "threatens to use" "toxic chemicals" for "medical or pharmaceutical activity, or other activity," or manufactures the COVID-19 synthetic bioweapon's 'gain-of-function' for increased infectivity and pathogenicity, begs the question of actual *intent* ('*mens rea*' in law).

A legal argument and question arise here. The industry and those who aid and abet the deadly use of toxic chemicals and genetic materials in vaccines rely exclusively on their 'general acceptance' in medicine — general acceptance of safety and efficacy. The general acceptance of vaccines is based on the "peaceful purpose" of presumably preventing diseases.

But that presumption is challenged by certain facts:
1) the world's leading vaccine promoter and distributor, Bill Gates, openly claims vaccines will **reduce populations** by 15%. Gates's arbitrary and capricious claim lacks general acceptance and solid science. The speculation is a fringe theory at best. Yet, it is being administered mainly through Gates's influence for corrupt conflicting interests;
2) The media that promotes the mantra that "vaccines are safe and effective" is clearly untrustworthy and biased by similar conflicting interests; and
3) the top experts in immunology and vaccinology are untrustworthy, as evidenced by conflicting medical opinions aired by the MindWar networks.

Challenging the 'General Acceptance' of Vaccines

The U.S. Supreme Court in *Daubert v. Merrell Dow Pharmaceuticals, Inc.*,[624] 509 U.S. 579 (1993) concluded that the general agreement in medical science is sufficiently untrustworthy to permit opposing science and arguments to be entered into courts of law.

The "Daubert standard" emerged from Daubert's case against the Dow chemical cartel. Daubert challenged the trustworthiness of so-called 'experts' in science. Society's 'general acceptance' of science, or 'general agreement' in medicine, is often flawed and untrustworthy, Daubert proclaimed, and the Supreme Court agreed. The "Daubert standard" was created from that decision.

Law and science now presume that contrary evidence must be admissible in courts to oppose the general acceptance standard as needed and constructive to truth, justice, public health, and safety.

Since then, the phrase 'general acceptance' has evolved as a legal "term of art" that holds **special meaning.**

According to the Daubert standard, science has often been pseudo-science published by yellow-presses and charlatans with conflicting interests. Their propaganda amounts to consumer fraud and unfair and deceptive trade.

Therefore, vaccines' 'general acceptance' cannot be trusted. The defense that it is used for a "peaceful purpose" is ridiculous. The action presumes war against the microbial world.

The "peaceful purpose" defense is also undermined by the fact that all vaccines, including the Pfizer and Moderna mRNA vaccines, inject immune-system poisons that cause immune cells to defend against the attack. The cells are aggravated, not pacified. That is what 'inflammation' and immune reactions entail.

Your white blood cell bodyguards—WBCs—go haywire. They often encounter the aforementioned 'antigenic complexes' rather than normal cells and tissues. This confuses your WBCs. They start attacking your own body. That is called 'auto-immune disease.' Allergies are simply one of nearly 100 new auto-immune diseases arising from vaccines produced in growth media such as chickens' eggs, cows, or monkeys that ate grass and/or foreign proteins. Then, during the manufacturing process, they conveyed these 'antigens' to the vaccines.

These animals' immune reactants are thereby transmitted to human bloodstreams and immune cells that way. That is how the human cells are induced to **declare war inside your body**. When grass pollen fills the air, your intoxicated body reacts with stuffy sinuses and teary eyes every springtime. That's a 'normal' response to being poisoned and suffering the long-term consequences of 'fake science,' greedy salespeople, and criminal psychopaths like Paul Offit and Anthony Fauci.

All of this falls under 'biological warfare.' The sickening and enslaving practice of vaccination is justified by Mind War, and today the War on COVID-19. Like the "War on AIDS," the "War on Cancer," and even the "War on Drugs," this war is not peace-making. The military finances this non-peaceful purpose for the express purpose of biological warfare and bio-warfare defense. COVID-19 is, after all, claimed by officials to threaten National Security and is certainly a lab-engineered bioweapon.

Add the fact that COVID-19 sourced from the military bioweapons facility in Wuhan. Finally, consider the synthetic S-protein antigen 'gain-of-function.' That was manufactured to increase the virus's transmissibility and increase disease and damage it causes. That is surely not "peaceful" by any stretch of the imagination.

In this instance, the 'Court of Public Opinion' weighs heavily in favor of anti-vaccinationists. This troubles the Deep State's business and transhumanist agenda. Vaccine hesitancy is medically and legally troubling for the oligarchy. The World Health Organization —the World Bank's 'captured' propaganda mill, along with the U.S. Centers for Disease Control (CDC)— has condemned 'Vaccination Hesitancy' among its top risks to public health. Now you know why.

Nazi-like Propaganda Politicizing Science

Enter Kai Kupferschmidt for the "esteemed" journal *Science*.[625] Because when all seems lost, officials divert to a smear campaign to 'neutralize' messengers rather than refute their messages. Here is a good example.[626]

Kuperferschimidt wrote the article titled "Top Israeli immunologist accused of promoting antivaccine views"[626] (Nov. 6, 2019). That propaganda was published by the American Association for Advancement of Science (AAAS) in their online journal, *Science*. AAAS is complicit with Big Pharma in the global genocide (aka, vaccinecide).

The article gives readers the impression that Kuperferschimidt is the vaccine Gestapo's agent, assigned to 'neutralize' "Yehuda Shoenfeld" —a world-renowned immunologist. Shoenfeld had published works offensive to Big Pharma.

Relatedly, in 2013, Kupferschmidt was awarded the Journalism Prize granted by the German AIDS Foundation.

Quoting Kupferschmidt, who condemned the Jewish expert's immunology science opposing vaccinecide, "Formerly at Tel Aviv University in Israel, [Shoenfeld] now runs a center for autoimmune diseases at Sheba Medical Center, Israel's largest hospital. He is editor-in-chief of both journals of the Israel Medical Association (IMA), serves on the editorial board of dozens of other journals, and was elected a member of the Israel Academy of Sciences and Humanities in June. "Yet a group of Israeli doctors says his ideas are a danger to public health."

The German drove a divisive stake into the heart of the Jewish medical community.

"Shoenfeld has long espoused theories popular among antivaccine advocates and spoken at their meetings," Kupferschmidt continued. The Jewish doctor caused "tensions with the Israeli medical community.

"In September, the issue came to a boil when Shoenfeld decided to publish a positive review of an anonymous antivaccine book in *Harefuah*,[627] IMA's Hebrew-language journal. The two reviewers, who did not have a medical background, wrote that the book "raises a strong suspicion that key aspects of vaccine safety have not been properly tested."

To pro-vaccine forces, that's like writing that the Nazi concentration camp gas chambers claimed good for hygiene, 'public health' and "disinfection,' were not "properly tested."

Kupferschmidt's propaganda was transparent and consistent with the CDC's data censorship during the infamous Atlanta Autism Study. Therein, the injury data for Black boys given the MMR vaccine was suppressed.

Making the scholarly Jewish immunologist the bitter enemy of the medical community was well-coordinated with the 2019 attack against the orthodox Jewish community that opposes the unfolding vaccine holocaust for religious, moral, and ethical reasons.[628]

The 'Vaccine Gestapo' Hides its Shame

"The response was swift," the German propagandist continued. "Shmuel Rishpon, chair of Israel's Advisory Committee on Infectious Diseases and Immunization, resigned as editor of *Harefuah*. Israel's Association of Public Health Physicians called for an investigation and Shoenfeld's resignation as editor-in-chief. The association is considering calling on authors and reviewers to boycott *Harefuah*, its chairman, Hagai Levine, told *Science*."

German persecutor Kupferschmidt continued. "[The Jewish scholar] Shoenfeld defend[ed] his decision [to permit two non-medical referees to allow publishing a book review in *IMA*, Shoenfeld's science journal, that pro-vaccine zombies did not like] and says the medical community is trying to silence vaccine critics…

"'If you write something about vaccines which doesn't say that everything is OK ... all the vultures are jumping on the one who writes it,' [the Jewish doctor] said. *IMA* did not respond to *Science*'s questions."

I'm sure Kupferschmidt wouldn't respond to my questions either, beginning with, "How much pro-PhRMA bias has influenced **Science**'s smearing of Shoenfeld?"

"Shoenfeld's association with the antivaccine movement goes back many years," Kupferschmidt added for defamatory impact. "In a 2011 paper,[629] he proposed that adjuvants, compounds such as aluminum added to vaccines to boost an immune reaction, can lead to chronic activation of the immune system that he called autoimmune/inflammatory syndrome induced by adjuvants (ASIA). Shoenfeld has published many papers on the syndrome, which vaccine critics cite as a danger of vaccination."

It is noteworthy that aluminum, the very ingredient Kuperferschmidt admits is used to "boost an immune reaction," was the precise adjuvant used to test Gardasil, the HPV vaccine made by Merck. Their own package insert showed the mixing-up of two control groups, one receiving saline and the other amorphous aluminum hydroxyphosphate sulfate (AAHS) adjuvant.

The AAHS "can have a profound influence on the magnitude and quality of the immune response to the HPV vaccine," stated Caulfield MJ, Shi L, et al., in their article, "Effect of alternative aluminum adjuvants on the absorption and immunogenicity of HPV16 L1 VLPs in mice." (Hum Vaccin.[630] 2007 Jul-Aug;$3^{631(4)}$:139-45. Epub 2007 Apr 5.

Combining the two control groups, purposely confused (i.e., confounded) the true assessment of risk. But Kupferschmidt, a so-called 'award-winning' AIDS journalist, undermined his own smearing of Shoenfeld by neglecting aluminum's abuse in corporate-biased science. His position that vaccines are safe and Shoenfeld's work is bogus is completely discredited by real science.

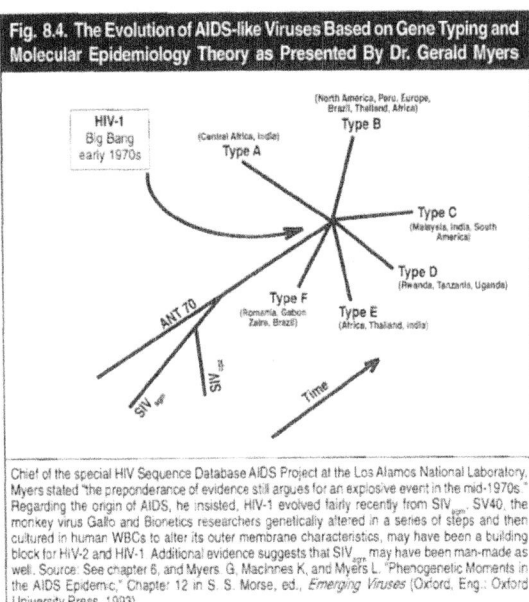

Fig. 8.4. The Evolution of AIDS-like Viruses Based on Gene Typing and Molecular Epidemiology Theory as Presented By Dr. Gerald Myers

Chief of the special HIV Sequence Database AIDS Project at the Los Alamos National Laboratory, Myers stated "the preponderance of evidence still argues for an explosive event in the mid-1970s." Regarding the origin of AIDS, he insisted, HIV-1 evolved fairly recently from SIV_{agm}. SV40, the monkey virus Gallo and Bionetics researchers genetically altered in a series of steps and then cultured in human WBCs to alter its outer membrane characteristics, may have been a building block for HIV-2 and HIV-1. Additional evidence suggests that SIV_{agm} may have been man-made as well. Source: See chapter 6, and Myers. G, MacInnes K, and Myers L. "Phenogenetic Moments in the AIDS Epidemic," Chapter 12 in S. S. Morse, ed., *Emerging Viruses* (Oxford, Eng.: Oxford University Press, 1993).

These transactions were much like the smearing and censorship I received from the World Health Organization's AIDS Science Working Group when I advanced the U.S. Govt. contracts under which numerous AIDS-like and Ebola-like viruses were bio-engineered by Litton Bionetics during the late 1960s. The immune-system ravaging inventions (i.e., recombinations) were tied to the advanced bioweapons project involving Litton and Merck's contract. Like Rockefeller's Dulles Law Firm and Standard Oil Company, Merck worked both sides in WWII.

"Adjuvants' side effects are a 'valid issue,'" says immunologist Ruth Arnon of the Weizmann Institute of Science in Rehovot. Although Arnon has not read the review, she says she considers Shoenfeld "one of the most prominent world experts on autoimmunity" and welcomed his election into the Israeli academy, of which she is a past president.

The current president, Nili Cohen, a legal scholar, says Shoenfeld was "supported by the best scholars relevant to the field."

"Science is not necessarily based on consensus," she wrote in an email [akin to the 'Daubert standard' of questioning the 'general agreement.'] "Progress in science has been occasionally fueled by controversies."

But never in the realm of sacrificing vaccines —the 'sacred cow of public health' and obvious genocidal weapon of mass depopulation, according to Bill Gates' lectures.[411] See no evil, hear no evil, and speak no evil is the Nazi fascist doctrine Kuperferschmidt preferred.

Next, consider the demonically-possessed Paul Offit[332] of the Children's Hospital of Philadelphia and his support for vaccinations and the 'Gestapo's reputation. Offit provided Kupferschmidt's refutation of the autoimmune disease reactions killing millions of infants, children, and adults from adjuvants and foreign proteins.

"But ASIA is ill-defined," pediatrician Offit said. "Symptoms are said to appear up to 2 decades after exposure to an adjuvant and include anything from dry mouth to memory loss."

Wow!

For Kupferschmidt and *Science* to overlook this condemning statement admitting that vaccine risks may compound for "decades after exposure" is chilling evidence of the **chronic genocide, officials' recklessness,** and **negligent misrepresentation** that has permitted cancers and auto-immune diseases to skyrocket during the 'vaccination era.'

Offit, the Josef Mengele of pediatric consciousness, continued subverting his first statement, common sense, and honesty in public health policy. "Nor do epidemiological studies support the idea... A 2012 study[633] analyzed data from more than 18,000 people who were given a high amount of aluminum under the skin as part of immunotherapy against allergies; as researchers pointed out in a 2017 paper,[634] they had a lower rate of autoimmune disease than controls.

"Shoenfeld continued to push this frankly ill-conceived and arguably disproven notion despite the evidence,'" Offit criticized.

Researching Offit's citations, however, exposed Offit's scam. The lead author of that 2012 paper, A. Linneberg, honestly disclosed that he received "honoraria for lectures from ALK-Abelló, GlaxoSmithKline, and Siemens Medical Diagnostics." That means the study was ill-fated to sell what vaccine makers financed —false safety assurances for poisonous injections.

The 2017 study cited by Offit was published by the additionally untrustworthy medical cartel, Elsevier and the RELX Group, with institutional investors heavily profiting from diseases caused by GlaxoSmithKline's vaccines.

The aluminum injection in that study was subcutaneous, not intramuscular. Toxic aluminum was not injected systemically as it would be during a standard vaccination. The allergy test was recklessly misrepresented as "immunotherapy" akin to vaccination, which it is not.

More importantly, the propagandists, Kupferschmidt and Offit, omitted and neglected the dozens of other dangerous and damaging vaccine adjuvants other than aluminum shown in the above table from a 2002 science study.[635] Since then, many more adjuvants have been added, causing more side effects than Alzheimer's disease and memory loss associated with chronic aluminum poisoning.

Notice also how the world's most prolific medical con artist, Paul Offit, a patent holder on six vaccine patents,[636] promotes adjuvants most profitably. Here Offit suddenly appears from the United States to back the German Gestapo for *Science* and protect the interests of the globalist cartel.

Mineral salts Aluminum phosphate* Calcium phosphate*	Aluminum hydroxide*
Immunostimulatory adjuvants Saponins e.g., QS21 MDP derivatives Bacterial DNA (CpG oligos) LPS MPL and synthetic derivatives Lipopeptides	Cytokines e.g., IL-2, IL-12, GM-CSF
Lipid particles Liposomes Virosomes* Iscoms Cochleates	Emulsions e.g., Freund's, SAF, MF59*
Particulate adjuvants Poloxamer particles Virus-like particles	PLG microparticles
Mucosal adjuvants Cholera toxin (CT) Mutant toxins e.g., LTK63 and LTR72 Microparticles Polymerized liposomes Chitosan	Heat labile enterotoxin (LT)

Note: With the exception of cochleates and polymerized liposomes, all of these adjuvants have been evaluated in clinical trials. However, only those marked* are currently included as adjuvants in approved vaccine products.

Further disparaging Shoenfeld, their commercial enemy, Kupferschmidt, and Offit continued, "A 2016 study by Shoenfeld claimed to show abnormal behavior, caused by ASIA, in mice given the human papillomavirus vaccine or an aluminum adjuvant. The study was published in *Vaccine* but later withdrawn. Review by the Editor-in-Chief and evaluation by outside experts confirmed that the methodology is seriously flawed, and the claims that the article makes are unjustified," a note on Vaccine's website states.[637]

The paper was republished a few months later under almost the exact title in *Immunologic Research*, where Shoenfeld is on the editorial board.[638]

OBJECTION! The journal *Vaccine* is owned by the same company that owns *Science*, which is grossly biased by its institutional investors' holding massive amounts of stock in drug companies making vaccines. A few of these are listed above. The investors in the parent company of Reed Elsevier, RELX Co., include BlackRock Inc., CitiGroup Inc., etc.

Can you imagine how many **trillions of dollars** these stockholders would lose if vaccine adjuvants were suddenly confirmed deadly and liability for their injuries was advanced internationally? What impact would that have on the global economy?

Chilling!

This sorry state of affairs makes the German propagandists Kupferschmidt and Offit MindWar mercenaries for the Deep State — despicable creatures who make their livings aiding-and-abetting genocide. That's why both of these menaces to society won their bogus awards: publicity stunts and photo ops for the drug cartel.

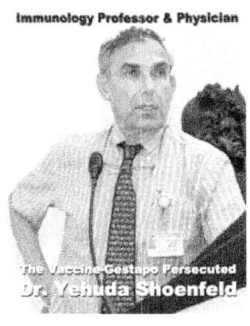

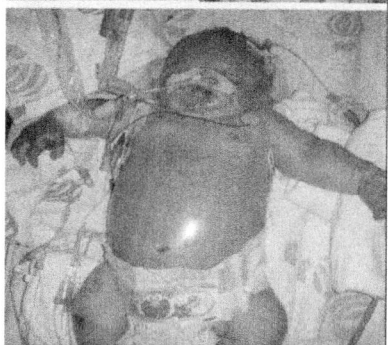

CHAPTER XII

TRANSHUMANISM AND THE GREAT GLOBAL RESET

Transhumanism and the Great Global Reset depend heavily on advancing nano-bioelectronic new vaccine technology. COVID-19 is a contrivance to accomplish these three objectives.

Nano-bioelectronic devices in vaccine hydrogels modulate DNA and genetic expression. They also are capable of subverting human consciousness, human behavior, physiology, and metabolism via neurological impacts in the "brain-Cloud connection."

"Nanobots are a real concern about wiping out humanity because they can be weapons of mass destruction," said Louis Del Monte, a Minnesota-based physicist and futurist interviewed by *CNBC* —Bill Gates's propaganda mill. The 'predictive programming' is unsettling, especially considering the MIT and Harvard advances in the Langer and Lieber labs, respectively, being applied in vaccinology.[639]

I am talking about nanotechnologies being deployed now in many fields. The devices are so small they are measured on a nanoscale of between 1 and 100 nanometers. One nanometer is many times smaller than your cell's DNA.

Journalist Jeff Daniels wrote in 2017 that "mini-nukes" and "mosquito-like weapons" were being mass-produced by U.S., Russian and Chinese governments to gain military superiority.[639]

The Defense Advanced Research Projects Agency, or DARPA, "has a program called the Fast Lightweight Autonomy[640] program to enable autonomous drones to enter a building and avoid hitting walls or objects. DARPA announced a breakthrough[641] last year after tests in a hangar in Massachusetts, Daniels reported.

"Previously, the Army Research Laboratory announced it created an advanced drone the size of a fly complete with a set of "tiny robotic legs"[642] —a major achievement since it presumably might be capable of entering a building undetected to perform surveillance, or used for more nefarious actions.

"Frightening details about military nanotechnologies were outlined in a 2010 report from the Pentagon's Defense Threat Reduction Agency," Daniels continued, "including how 'transgenic insects could be developed to produce and deliver protein-based biological warfare agents,[643] and be used offensively against targets in a foreign country.'"

"It also forecasts 'micro explosives' along with 'nanobots serving as [bioweapons] delivery systems or as micro-weapons themselves, and inhalable micro-particles to cripple personnel.'"

"In the case of nanoscale robots, Del Monte said they can be the size of a mosquito or smaller and programmed to use toxins to kill or immobilize people; what's more, these autonomous bots ultimately could become self-replicating." That means they are virtually alive. And once they enter your body, they can overtake every function from physiology and metabolism to brain neurology and consciousness.

Even denialists and censors finally admit to these capabilities enabling the transhumanist movement to grow. The industries involved merge Big Tech products and services with commercial and geopolitical transhumanist objectives. Cyborg civilization is being commercially and culturally induced to grow alongside other globalist impositions, including the "mandatory" vaccinations that deliver these "novel" bioweapons.

What's really at stake is the global elite's plot to impose their "Final Solution" to control, reduce, even replace world populations with transhumans, humanoids, or pseudo-'aliens'. Previously, that sounded completely far-fetched, beyond a "conspiracy theory." But now, people everywhere are awakening to this unfolding, as evidenced by the advancing biotechnologies and 'data-mining' capabilities in the untrustworthy vaccines.

The 'Great Reset' discussed in the financial world is akin to the "Final Solution" advanced by Bill Gates and the original Nazis–eugenicists representing the global elite.

Why is this happening so quickly? For one, the Lords of Misrule is now threatened by burgeoning populations awakening to their plot. Anti-vaxxers, for instance, are awakening from their naïve and trusting 'hypnotronic' trance states.

644

This mass awakening and resistance threaten the international crime syndicate. They heavily invested in genetically engineered bioelectronic COVID-19 vaccinations and heralded these in their media to condition society's acceptance of these "novel" technologies. They have advertised these as remedies against illnesses that vaccines have largely generated. This is a subtle form of "crisis capitalism." Create the problem, then hype and sell the solution. This criminal enterprise is central to the COVID Coup and pushes for mandatory injections at the time of this writing.

The media censorship surrounding these most important facts is *pathognomonic* of the MindWar ongoing in this 'Technotronic Era' administered to secure global fascism. This is the essence of the 'Fourth Industrial Revolution' or "4IR" featuring AI and AGI.

The globalists now disclosed secrets and special interests evidence their scheme and condemn the Big Banking, Big Pharma, and Big Tech syndicate advancing this Brave New World.

Make no mistake; this is the 'Rise of the Fourth Reich.' Its 'predecessors-in-interest' include humanity's greatest nemesis--the IG Farben/Bayar/IBM/Standard Oil cartel that expanded and consolidated its power. Today this syndicate pollutes politics, science, economics, law enforcement, and threatens civilization with extinction via transhumanism.

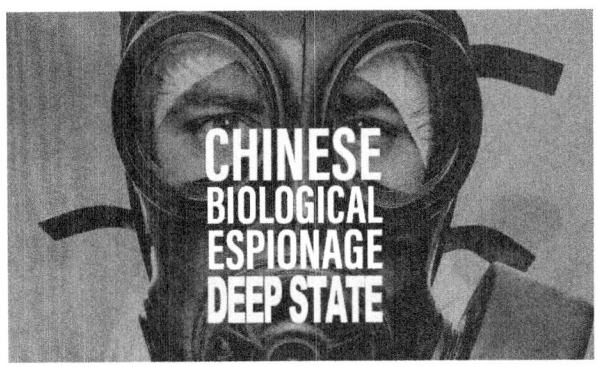

The World Economic Forum 'Coalition' and Civilization's 'Great Reset'

As the Bidens applied their influence in the Ukraine and China, on July 23, 2015, The Pirbright Institute in London, closely tied to the Gates Foundation, German Boehringer Co, the Merck Drug Co., with a "major stakeholder"[645] being the James Farrar-directed Wellcome Trust advising Fauci, applied for a patent on the leading suspect lab coronavirus.[646]

The Pirbright Institute heralds its role in preventing and controlling viral diseases on behalf of the global elite. As detailed in the screenshot below, the joint venture between Pirbright, Merck, Sanofi, and the Wellcome stakeholders, advanced bio-data collection that was done the prior year as members of the **World Economic Forum**. The Pirbright patent on the suspected COVID-19 virus progenitor was granted on January 28, 2016–a year before President Trump took office.

This is how Fauci knew in advance, by coordinating with Farrar that, as President Trump took office, his administration would face a terrible biological crisis.[88] Pirbright's coronavirus patent was granted the year before January 12, 2017, when Fauci delivered his historical and prophetic address to his colleagues at Georgetown University.[647]

Further evidencing the COVID Coup and Fauci's complicity in the World Economic Forum-administered racket, the Georgetown press release stated, "In addition to the upcoming inauguration of President Trump, the event's timing coincided with leadership changes at major global health organizations including the UN, World Bank, WHO, and the Global Fund to Fight AIDS, Tuberculosis, and Malaria, making this an important moment for pandemic preparedness."

Each of those "major global health organizations" is precisely those most heavily lobbied and financed by the Bill & Melinda Gates Foundation and Clinton Foundations, with Democratic Party hopeful Chelsea Clinton promoting them all from her faculty post at New York University.[603]

At that time, Western Intelligence was overseeing infectious diseases worldwide, and their liberal media were joining the DNC's "coalition" to damage the Republican Party and its "Pied Piper" candidate Donald Trump.

On January 14, 2016, two weeks before the Pirbright coronavirus was patented, Klaus Schwab, Director of the World Economic Forum that included Merck, the Bill & Melinda Gates Foundation, Blackrock, Blackstone, Bloomberg, and other heavy-weights in the banking and pharmaceutical industries backing Democrats, liberal policies, and American Marxism, introduced his "Fourth Industrial Revolution" (4IR) thesis to these global elite directing world finance, science, medicine and 'public health.'

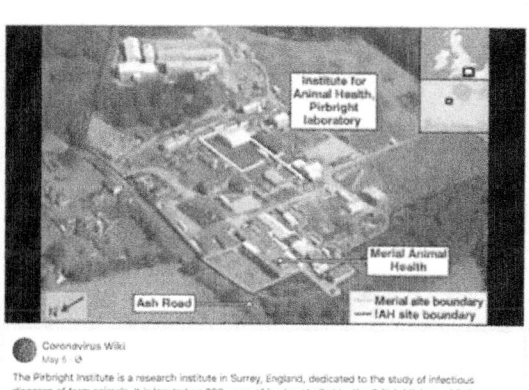

As reviewed by Western intelligence researchers,[648] Guy Garrett and Beverly Hayes in England and the US, this 4IR term "originated in 2016 when described by Klaus Schwab (Founder and Executive Chairman of the World Economic Forum), as a 'technological revolution that will fundamentally alter the way we live, work, and relate to one another.'"

Schwab Describes "4IR" and the Rise of the Fourth Reich

"Will the Future Be Human?" asked Yuval Noah Harari[649] rhetorically.

This Professor in the Department of History, Hebrew University of Jerusalem, lectured on January 25, 2018, at the annual meeting of the World Economic Forum. His chilling message was clear. The "future of our species and really the future of life" is advancing from the 4IR and data systems administration controlled by criminal psychopaths, Harari neglected to mention.

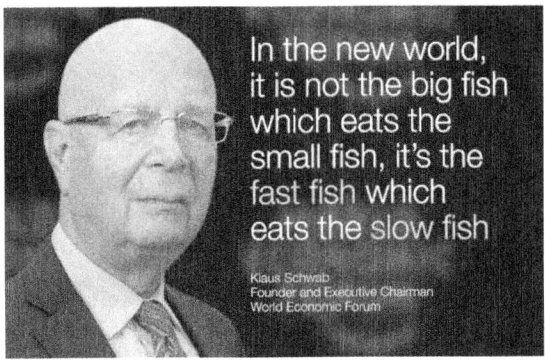

Schwab described 4IR as "a digital revolution with innovative uses of a combination of technologies that build upon the premise of the third revolution (i.e., electronics and information technology to automate production). As a result, emerging technologies have brought forth advancements in artificial intelligence, robotics, the Internet of Things, autonomous vehicles, 3D printing, nanotechnology, biotechnology, materials science, energy storage, and quantum computing. This rapid evolution will undoubtedly affect industries worldwide... The pharmaceutical industry is also experiencing the impacts of I4.0. Digital and mobile technologies have brought on significant advancements in data acquisition and accessibility related to health care and patient data... As with other industries, data will become a critical asset to their business, and the effective utilization of this data can play a critical role in driving growth in the business and bringing novel therapies to the patients who need them."

"Earth will be dominated by **entities** that are more different from us than we are different from Neanderthals or chimpanzees," Professor Harari lectured. "We will learn how to engineer bodies, and brains, and minds. These will be the main products of the economy…

"How exactly will the future masters of the planet look like?" Harari rhetorically asked. "This will be decided by the people who own the data [such as Silicon Valley moguls]. Those who control the data control not just humanity, but the future of life itself. Because today, **data is the most important asset in the world**."

In 2004, The World Bank gave special recognition to Klaus Schwab following the publication of *Mind, Heart and Soul in the Fight Against Poverty*, stating, "This framework owes a debt to Klaus Schwab, executive chair and founder of the World Economic Forum, who highlighted the need for head, heart, and soul as a framework for partnerships at the 2004 Davos annual meeting."[650]

Accordingly, Big Brother is going after your "head, heart, and soul." Your "head" will be captured by the nano-bioelectronic vaccine-transmitted technologies that merge with your neurology and link your brain to the Cloud. Your "heart" will be attacked wirelessly by sound frequencies that disrupt your normal heart rhythm and compete against the 528Hz frequency resonance strengthening your heart. And your soul, that is fundamentally spiritual energy, will be likewise stifled or demonically disposed to accommodate evil.

Background on Schwab with Ties to the Third Reich

Who is Klaus Martin Schwab, and how did an obscure German professor of 'business policy' become the leader of the '4th Industrial Revolution and the New World Order?

Schwab was born on March 30, 1938, in Ravensburg, Germany. His father was the managing director of a subsidiary of the Zurich-based engineering firm, Escher Wyss. Besides the fact that his father could move freely back and forth across the Nazi border, nothing else has been publicly disclosed about Klaus's father's work for the Third Reich. This intelligence has been heavily censored.

"It would appear," wrote one anonymous researcher, "that the name of Escher Wyss's subsidiary in Ravensburg (where Klaus' father was managing director) was Andritz.

In 1941, Andritz became part of the Duisburg Demag Group (DEMAG). From September 1942 to April 1945, DEMAG used slave labor from the Sachsenhausen concentration camp to build Panzer armored vehicles for the Nazi War Machine.

Alternatively, the elder Schwab contributed to the Nazi nuclear program. "A/G Esher Wyss (the parent company of the Nazi subsidiary which employed Klaus' father) had engineered the 14,500 HP turbine for the Norsk -Hydro Works heavy water plant in Rjukan, Norway. The Norsk-Hydro facility powered by Escher Wyss was the only industrial plant under German control capable of producing heavy water (deuterium oxide, D2O) essential for making plutonium for the Nazi atomic bomb."

We would like to know the full truth about this man who now advances the world's Great Reset.

Data Mining, GMOs, Nano-gels and 4IR

A recent article in *Genetic Engineering and Biotechnology News* details a couple of practical applications of advanced data mining in public health and COVID-19 response. The article heralded MIT's research and development of the latest vaccination methods and materials.[651]

Imagine a skin patch that is peppered with "microneedles." The needles inject "fluorescent microparticles called quantum dots (QD)" into your skin. These QDs deliver 'immunizations' and "create new possibilities for data storage" and "biosensing," the author cheers.

The QDs "invisibly encode vaccination history directly in the skin," thereby broadcasting vaccination records using "near-infrared (NIR) light that can be detected by a specially equipped smartphone."

The electromagnetic wave telecommunications withstood five years of "photobleaching" that simulated long-term exposure to sunlight.

"It's possible… this 'invisible' approach could create new possibilities for… vaccine applications that could improve how medical care is provided, particularly in the developing world," reported senior researchers Robert Langer and David Koch at MIT. (Langer, you may recall, is the lead entrepreneur and chief scientist at **Moderna**.)

Alternatively, in untrustworthy hands, the same biotechnology could be used to induce cancers, myriad diseases, and disastrous behavioral impacts consistent with the '1984' 'Brave New World' agenda.

Don't forget Bill Gates, the pedophile psychopath Jeffrey Epstein, and DARPA financially fueled this totalitarian enterprise.[652]

It is clear that these bioengineers engaged with Gates, Microsoft, and MIT's infamous Media Lab, collaborating with Langer's lab, advanced "neurobiology, biologically-inspired fabrication, socially-engaging robots, emotive computing, bionics, and hyper instruments," as detailed in *Wikipedia* and in the "ID2020" propaganda.[653]

If you have any doubt about what side of the political spectrum MIT lies in, consider the entity's sponsorship of Noam Chomsky.[654] "The world-renowned public intellectual" told you how to vote in the 2020 presidential election. He made "a plausible case for a chilling assessment of arguably the world's most powerful man," Donald Trump.[655]

"Voting for U.S. President Donald Trump is worse than voting for Hitler," Chomsky affirmed to interviewer Linda Solomon Wood during a *Canada's National Observer*-sponsored webinar in April.⁶⁵⁶ "Hitler was maybe the worst criminal in human history." He wanted to murder millions of Jews, Slavs, homosexuals, and others. "But what does Trump want to do? He wants to destroy the prospects for organized human life."

Chomsky neglects that "organized human life" after the "Global Reset" would euthanize him for dementia. The executioner will be a transhuman.

Hawaii's Importance in the Globalist Underworld

During the Senate confirmation hearings evaluating Supreme Court Justice nominee Amy Coney Barrett, you may have noticed the outrageous opposition and interrogation waged by Hawaii Senator Mazie Hirono. Insulting the integrity of the Senate and raising the specter of suspicion, Hirono asked, "Since you became a legal adult, have you ever made unwanted requests for sexual favors or committed any verbal or physical harassment or assault of a sexual nature?"⁶⁵⁷

"No, Senator Hirono," Barrett replied.

Hirono continued the interrogation by asking if Barrett had ever been legally involved as a party in a case concerning sexual assault. "Have you ever faced discipline or entered into a settlement related to this kind of conduct?" Hirono asked.

"No, senator," Barrett replied again.

I know Senator Hirono personally, having corresponded with her and her staff repeatedly concerning our opposition to organized crime in Hawaii, impacting U.S. mainland security.

Ms. Hirono was endorsed by the late U.S. Senator Daniel Inouye (D), the most famous and beloved Hawaii Senator whose leadership helped expose the CIA's MKULTRA mind-control program and "Deep Throat" through the Senate Watergate Committee. Inouye chaired the Senate Select Committee on Secret Military Assistance to Iran and the Nicaraguan Opposition from 1987 until 1989. Like the Bidens, Inouye was accused of sexual misconduct.[658]

Inouye referred to the 'Deep State' by a different name. He called it the "Secret Government." Others called it the "Illuminati," the "Oligarchy," the "Octopus," the "Shadow Governors," "Big Brother," or simply "The Mob." President Trump called it "the Swamp."

Sen. Inouye warned:

"[There exists] a shadowy Government with its own Air Force, its own Navy, its own fundraising mechanism, and the ability to pursue its own ideas of the national interest, free from all checks and balances, and free from the law itself."[659]

Sadly, Senator Hirono did nothing to help me. I relay this intelligence in the interest of public health and safety. Hawaii's Democratic Party-dominated legislature, judiciary, and law enforcement are totally 'out of control' under the control of the 'Deep State' or Asian-American banking syndicate in the World Economic Forum consortium.

The strategic location of Hawaii between China and the mainland U.S. makes it of critical importance to military and globalist commerce. (As detailed below, the first China-U.S. Sub-national Legislatures Cooperation Forum was held on June 30, 2016, in Hawaii.)

Hawaii functions most importantly for Western and Eastern intelligence agencies. The state was severely corrupted at its inception. I made this clear in several popular articles. If you think New Jersey is a hotbed of corruption, Hawaii is shamefully MUCH worse. If you think the Deep State genocide is fictitious, look at what they have done to native Hawaiians.[660]

Material to Hawaii's genocidal corruption, Barack Obama's grandmother agented for both The Bank of Hawaii and the Western intelligence community. Edward Snowden's National Security Agency leaks were sourced from Booze-Allen Hamilton in Honolulu. The Hawaii Nuclear Missile Alert fiasco heavily implicated the Hawaii Electric Company (HELCO).[661] HELCO partnered in the "Fusion

Center" with America's top intelligence agencies working with the civil defense group.

According to a Congressional hearing, Hawaii traffics much of America's methamphetamine supply.[662] Hawaii produces all domestically manufactured dimethyltryptamine —"DMT,"[663] that acts like the new hallucinogenic "designer LSD." It conditions the brain for the new transhuman 'reality.'

Hawaii is the gateway for opium traffickers moving heroin, fentanyl, and pain killers through the Pacific Rim nations and islands. A substantial amount of cocaine trafficked from Mexico to North America by the Salinas Cartel comes through the ports of Honolulu and Hilo.[664]

Hilo's mayor, the former prosecutor, Mitch Roth (D),[665] has turned blind eyes to his criminal cartel associates for the past dozens of years; since I first met with him to prosecute DMT drug kingpin and professional forger, Paul Sulla in 2009.

A live news report in Hawaii[666] on the lab virus origin of COVID-19 was instantly censored. Similarly, all news of DMT's manufacture and trafficking from Hawaii to the mainland by a drug kingpin Paul J. Sulla, Jr.[667] has been censored.[668]

Much like the COVID Coup tracks through the political underworld, domestic DMT is sourced from the University of Hawaii's botanical garden.[669] The leading beneficiary, the Heffter Institute,[670] now collaborates with federal drug agencies, the Multidisciplinary Association for Psychedelic Studies (aka "MAPS"), Yale, NYU, Johns Hopkins, and Harvard. Their nefarious perverted movement intertwines brain 'consciousness' studies with military neuroscience interventions for psychological/emotional problems such as PTSD and the development of AGI —Artificial General Intelligence.

TRANSHUMANISM AND THE GREAT GLOBAL RESET

Psychedelic Times showed how the "neural plasticity" of human brains induced by the DMT in ayahuasca would facilitate the merging of humans with artificial intelligence (AI). This also ties into Gates's purpose for financing with Epstein MIT's and Harvard's neuroscience labs. The nano-bioelectronics are deployed in the hydrogels encasing the recombinant RNA and DNA vaccines. Once a bio-electrifying "vaccine" is injected into a human, the nano-bot-like bioelectronic and genome becomes genetically modified; thereby subject to external ownership, wireless interference, and brain-Cloud uploads and downloads.

60 Minutes broadcast a related hour promoting genetic engineering and free medical education at NYU, with the central segment featuring psychedelic research at Johns Hopkins.[671]

The Bidens' Agency in Advancing China, and the 'Great Reset,' is Not Promoting Peace.

The Biden's involvements in China and Ukraine were important for the globalizing syndicate.

Much like the censorship evidencing the media's complicity in the "Great Reset," the first China-U.S. forum was widely censored by the American news media.

Hunter Biden's transactions within the **Chinese conglomerate CEFC** received little attention from the mainstream presses. Photos from that forum "show a low-key, casual affair, not in keeping with a major political event. "That is to say, the event tends to have only been publicized in Chinese state media or on Chinese government websites, rather than by the US government or in US media," reported *Intelligence Quarterly* in an article subsequently censored.[672]

In 2015, CEFC opened its new European headquarters in Prague, and Prime Minister Zeman hired Ye Jianming, the company's chairman, as an adviser.

China sponsors a think tank called CEFC that Mr. Ye said in a September 2016 interview that CEFC China "closely follows [China's] national strategies" and maps out its corporate strategy based on China's strategic priorities. Project 2049 Institute, a think tank in Washington, DC, assessed a CEFC think tank as a "political warfare platform affiliated with [the former General Political Department of the People's Liberation Army] and the CCP propaganda and ideology system."

According to Czech sinologist Martin Hála, "CEFC not only follows the PRC's state policies closely... but aligns itself with the most conservative elements in the CCP and [People's Liberation Army]."

The *China Digital Times*, on February 8, 2018, published Martin Hála's "CEFC: Economic Diplomacy with Chinese Characteristics: A Mysterious Company Paves the New Silk Road in Eastern Europe and Beyond. Similarly, Scott Cendrowski published in *Fortune*, September 28, 2016, "The Unusual Journey of China's Newest Oil Baron." Both articles considered the dangers of this crime syndicate's financial transactions.

For example, Cendrowski wrote, "Hungarian Prime Minister Viktor Orbán has championed building an 'illiberal state on national foundations," which is counter to the EU's model of liberal democracy. He endorsed Beijing's rejection of universal values.

A similar condemnation of the Biden-Chinese business alliance was published by Thorsten Benner et al., in "Authoritarian Advance: Responding to China's Growing Political Influence in Europe," Global Public Policy Institute and Mercator Institute for China Studies, in February 2018.

Extending this nefarious influence into media censorship and propaganda, according to a study by the Czech think tank Association for International Affairs (AMO), "Chinese ownership [of media institutions] equals zero negative comments on the country," raising concerns about future acquisitions of media companies by entities connected to Beijing.

Meanwhile, in America, censored and persecuted investigative journalist David Daleiden published a short video exposing Joe Biden's running mate, Kamala Harris, as a 'dictator' who targets whistleblowers and covers up Planned Parenthood crimes of selling aborted 'fetuses body parts.'[673] That production was similarly censored.

Additional CEFC investments have been secreted. These include the "Select Advanced PLA Air Force Systems…" and "the H-20 long-range stealth bomber, [that] provides 'Strike'; Nuclear Deterrence; [and] A2/AD [by 2025]."

673

According to the DOD, this H-20 war machine integrates "fifth-generation technologies capable of carrying nuclear weapons. Replacing the H-6, the H-20 will have an increased range of at least 5,000 miles (mi), boosting China's ability to operate farther from its shores, putting Hawaii certainly at risk, and likely the West Coast of North America as well.

CHAPTER XIII
THE TOP TEN MOST DAMAGING CORONAVIRUS LIES

Everything that is happening pursuant to the coronavirus, and international response to it, can be viewed as the 'fruit' of a 'magical deception.' The purpose of this *Twilight Zone* or "New Normal" is to advance the "INOCULATION OF DISEASES" as warned in the **Book of Revelation** 18:23-24 and the *Protocols of the Elders of Sion* 10:19.

MindWar for the Soul of Humanity

"And the light of a candle shall shine no more at all in thee; and the voice of the bridegroom and of the bride shall be heard no more at all in thee: for thy merchants were the great men of the earth; for by thy sorceries were all nations deceived. And in her was found the blood of prophets, and of saints, and of all that were slain upon the earth." Revelation 18:23-24

Let me first review Chapters III and IV, in the context of Chapter VI, wherein I considered MindWar.

This COVID 'MindWar' is for the Soul of Humanity. Civilization has been hypnotized to accept bioengineering the transhuman 'Master Race.' The multi-national corporate-controlled media dare not mention this. Instead, they broadcast voices of experts and government officials who lie to us. "For by thy sorceries were all nations deceived." Rev 18: 24

First and foremost, officials lied to us about the origins of AIDS, Ebola, and COVID-19 from purportedly eating monkeys, bats, or pangolin meat at the Wuhan market.

Then they lied about the pathogenesis of these diseases, that is, how the viruses cause disease, omitting the most important 'gain-of-function' 'spike protein' attachment mechanism. Then they lied about skyrocketing deaths falsely attributed to air travel, never lab virus bioengineering and deployment by the CIA or military.

More honestly, the deaths came mostly from medical malpractices, misdiagnoses, negligent treatments, multiple intoxicating assaults from drugs and vaccine ingredients, pre-existing disorders, and damaging lifestyle risks causing immune dysfunctions and terminal complications.

Then they lied about 'proper medical treatment' of victims placed wrongly on deadly drugs and lung destroying ventilators that did more harm than good. With AIDS, it was pneumocystis pneumonia; with COVID, it was acute respiratory distress syndrome or "ARDS." The disease reflected immune dysfunction and oxygen deprivation from histamine release, alveolar obstruction from mucous, and in many cases, hemoglobin deficiency and neurological damage probably caused by drug side-effects. COVID-19 is no 'atypical pneumonia.' This is suffocation by oxygen deprivation from biowarfare against *We The People*.

Medical malpractice, negligence, and scientific fraud by omissions and misrepresentations are civil torts. In this case, they compound **'biocrime'** because people have died by the millions of what amounts to bioterrorism with all of its illegal and genocidal elements satisfied. Bad intelligence and false instructions were given to everyone frightened and submitting. This includes the first responders and healthcare workers who served as 'evil obedient' pawns in the globalist game.

> *"This is [also] 'MindWar' for the Soul of Humanity. . . . Civilization has been hypnotized to accept bioengineering the 'Master Race.'."*

Nonetheless, the MindWar, and chosen psychological operations (i.e., PSYOPS) and "pandemic response" we witnessed perfectly devastated civilization and the global economy in favor of the descendants of those who wrote the previously mentioned **Protocols of the Elders of Sion**. The result fits too perfectly with Bible prophecy also to be dismissed or ignored.

We suffer from what the **Book of Revelation** refers to as "sorcery." Advanced by "magic spells," the magicians deceived all the wealthiest merchants on earth. They stole and poisoned people's blood globally. They, thereby, extinguished the soul or spirit of divine communion in God's people — "The Bride." That endowment and sustaining force, generating health and healing, was terminated in those deceived.

The facts mentioned above surrounding the suppressed AIDS virus envelop gene sequences in the coronavirus' spike protein, plus the bioelectronic technologies presumably administered through the vaccines for the 'brain-Cloud connection' has been published and confirmed by reputable scientists and intelligence analysts. These facts implore this presumption: That COVID-19 is a highly unstable mutagen of lab virus origin loosed for bioterrorism, vaccination impositions, and transhumanism. And despite this knowledge, the War on We The People is unlikely to end soon.

This prediction of further impositions is based on the global genocide that is to consummate the Great Global Reset heralding global fascism. Babylon has been *captured* and is now failing and falling under the guise of 'public health.'

No other reasonable presumption can be made, nor conclusion drawn, given the evidence in hand. And this 'differential diagnosis' features members of Congress. They have been bribed or extorted by the Deep State, Big Pharma, Big Tech, and globalist conspiracy to act-out-their willfully blind roles spinning the corporate-controlled media like sorcerers.

Many of us have already discovered, reported, and actively opposed this genocidal sorcery.

Given this introduction, the following "Top Ten Most Damaging Coronavirus Lies" are considered:

CLAIM 1: CoV-19 is not a 'lab virus.'

With COVID, scientific evidence tampering and fraud were first evidenced by the suppression of the science published and later "withdrawn" (i.e., censored/concealed) by Pradhan et al. Chapter V discusses the vitally important and suppressed "Indian paper."

In this case, where lives were and are at stake, censorship is a felony. It is also **treasonous** under the circumstances of international bioweapons treaty violations.

As mentioned, Pradhan et al.'s[13] group of nine scientists were affiliated with IBM and multiple academic institutions in India and Asia. They noted the "novel" attachment of HIV-1 genes in the new pathogen's S-protein apparatus enabling the 2019 coronavirus plague. Presumably, this virus culture had been incubated first at the University of North Carolina and later in Wuhan, both largely **financed through the NIH**.[674]

Subsequently, Pradhan et al.'s honorable and heroic revelations were corroborated by a team of Chinese scientists, Wang et al. These investigators published on April 7, 2020, in the *Journal of Cellular and Molecular Immunology*,[675] that "SARS-CoV-2 infects T lymphocytes through its spike protein-mediated membrane fusion" precisely as Pradhan's group detailed. Matt Steib in *New York Magazine* heralded this bad news.[676] In Steib's words, "The study found that COVID-19's damage to the T lymphocytes resembled that caused by HIV[-1/AIDS]."

This corroboration verified the alarm sounded by this author/whistleblower a quarter-century ago with the publication of Emerging Viruses: AIDS & Ebola—Nature, Accident or Intentional? I lectured and published prolifically about the risk to humanity of the lab viruses issued from the NIH, NCI, U.S. military, drug companies, and vaccine makers. Government officials and legislative bodies have turned blind eyes and deaf ears to my warnings. Now we are all paying the price.

"Dual purpose" lab viruses bioengineered to jump species and destroy human immunity, justified as 'cancer research' or 'vaccination immunology,' must be **outlawed** as violations of international bioweapons treaties.[677]

Those who produced and released the mutating coronavirus have concealed the 'mutagenic risk' of recombining the highly unstable man-made virus with other far more deadly viruses. These matters could now trigger a biological apocalypse, the Armageddon.

Many people dismissed the lab origin theory prior to the NIH's October 20, 2021 admission that Fauci et al. funded Baric's and Daszak's gain-of-function coronavirus weaponization. The "natural evolution" of this plague became the first and most damaging false claim.

This lie is most damaging because it buries the COVID Coup, threatens further 'outbreaks,' hides remedial measures, and delays criminal justice. The 'bioterrorism' criminal psychopaths administering the 'biocrime, and the transhumanist underpinnings evidencing the nano-bioelectronic vaccine technologies developed at Harvard and MIT risk civilization's extinction and transformation into an AI colony.

The fraudulent concealment of the four AIDS virus gene sequences in the n-CoV/SARS bioweapon proves bats, or exposure to contaminated meat at the Wuhan market, **did not** source the outbreak.

If they lied about the origin, what makes you think they wouldn't lie about the 'End Game' and 'Final Solution?'

The lab origin admission is a 'smoking gun' proving the biocrime. The fraudulent concealment of this discovery was akin to scientific evidence tampering that is a felony under these deadly circumstances.

The censorship proves an organized criminal conspiracy to commit genocide.

The deadly outcomes of this imposition and biocrime are compounded by officials' extortions, social isolation called 'distancing,' mass quarantining generating **mass psychogenic distress**, mental illness, and skyrocketing rates of suicide.

Stress-related physical ailments were generated by this lie and 'madness,' along with cultural degeneration. We witnessed increased riots, arson, social discord, and exploding murder rates, all resulting from the shutdown.

If they lied about lab origin and dismissed the massive damage lockdowns, social distancing, masking, and vaccinations would cause, what makes you think these people don't deserve criminal prosecution to terminate their globalist agenda?

As you can see, the Greatest and Most Damaging Coronavirus Lie is that the bioweapon used to justify all of this misery evolved as a mystery or infected humanity "naturally."

> *"This imposition is likely to continue to justify unconscionable acts of global genocide to consummate global fascism. In the name of 'pubic health' and 'society's safety,' Babylon has been captured, is failing and falling."*

CLAIM 2: Mass antibody testing is required.

It is said that the best coronavirus defense and cure is complete 'herd immunity' acquired by mass infections or vaccinations.

It is also said that testing is most useful in predicting and preventing future COVID outbreaks.

These are intertwined 'red herring' arguments. They neglect (or fraudulently conceal) the Deep State's primary interest in the bioweapon and 'coronavirus testing' —social conditioning for DNA data-mining, genetic engineering, and genetic profiling required to influence, or completely control, individuals electro-genetically and psychotronically using the latest military advancements in 'neuroscience.'

THE TOP TEN MOST DAMAGING CORONAVIRUS LIES

This is called the "6th domain of warfare" and also 'MindWar.'

A recent review of this little-known operation administered by 'shadow governors' to transform civilization for transhumanism and a monopoly over world culture, commerce, religions, and war has been published by *State of the Nation*[678] and expanded in *Ayahuasca Death Cult: The Psycho-Spiritual Delusion.*

To get people tested, you need to persuade them. That requires MindWar and "thought control". The leading propagandists here include Johns Hopkins University, the Bill and Melinda Gates Foundation, the corporate-controlled media, and the Pirbright Institute in Surrey, England, "dedicated to the study of infectious diseases of farm animals." This enterprise is controlled by the British Ministry of Defense (MOD)..." The "farm animals" they treat include you and me.

The Pirbright Institute has close affiliations with vaccine and drug makers, including British Merial (originally a joint venture between the drug companies Merck and Sanofi-Aventis), Germany's Boehringer-Ingelheim, Britain's Wellcome Trust, and the Bill and Melinda Gates Foundation. The Wellcome Trust and the Gates Foundation heavily fund The Pirbright Institute...

"On the same property as The Pirbright Institute is Cobbett Hill Earth Station[679] —a satellite teleport that boasts 'more than 25 antennas with active operations on 13 satellites' using C and KU-band antennas that are powerful enough for directed energy weaponry." That frequency technology used for crowd control can also be used for population control.

What if the nasal swab COVID tests could be used to implant nano-bioelectronic devices adjacent your brain; perhaps so small they might migrate into your brain to enable a transhuman wireless energy connection to the Cloud? This technology exists, and what makes you think the forces of evil are not deploying this capability right now?

According to the FDA,[680] Abbott Labs, the **maker of the heavily touted coronavirus testing technology**, maintains a commercial alliance, virtual partnership, with Boehringer-Ingelheim.

The Wellcome Trust controls London's Biocentre, the UK's largest non-governmental source of funds for biomedical research. It heavily funds research into alleged 'mysterious' neurodegenerative diseases linked by censored science to thimerosal mercury in vaccines. The study of neurodegenerative diseases by drug industrialists, as conducted at Johns Hopkins University, advancing genetic engineering and studies impacting brain neurology and 'expanded brain circuitry,' conceals the Deep State's advancement of MindWar research and AI developments.

Hitler watching festivities at Nuremberg - 1934.

Legislators are bribed to aid-and-abet this cartel and its agendas.

I previously reported,[518] "Abbott's press officials over the years distanced the company from its cartel agreements with the Rockefeller-IBM-partnered IG Farben conglomerate that administered Auschwitz and financed the death camps of WWII.

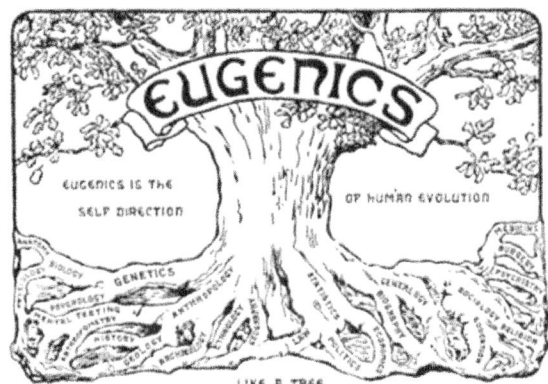

"Decartelization following Nuremberg Trials gave rise to Bayar AG, Hoechst, and BASF. Abbott acquired the latter in late 2000 for $6.9B (US) in cash.[681]

Since that time, Abbott has been a leading funder of the American Legislative Exchange Council (ALEC)[119] legislative 'bill mill' pushing mandatory vaccines through campaign financing of candidates in both parties. The company that manufactures the AIDS virus spike protein attachment 'protease inhibitor' Norvir has a history of bribing medical doctors and public officials, resulting in its 2006 rebuke by the Association of the British Pharmaceutical Industry (ABPI)." [120]

Meanwhile, the Gates Foundation has been royally exposed for continuing genocidal 'eugenic' operations begun for "Master Race" white supremacy and ruling elite protectionism.[682] Bill Gates personifies 'Babylon the great, the mother of prostitutes and the abominations of the earth." Together with allies such as Peter Theil, and alleged competitor Elon Musk, Gates backs population control technologies such as 5G, 'Promis' software, AI, the Cloud, transhumanism, psychotronic warfare, and GMO infected mosquitoes that can transmit their viral bioweapons. This public and private enterprise intertwines with 'census-taking,' genealogy tracking, and coronavirus testing.

Forced antibody testing invites unprecedented violations of civil liberties and human rights. Few people are talking about this electro-genetically controlled **Brave New World** or its "conspiracy theory."

From this vantage point, forced genetic or antibody testing extortionately leverages the coronavirus pandemic consistent with a Deep State/Big Pharma/Bill Gates and Peter Theil ploy.

CLAIM 3: China is to blame for CoV19.

This is how "magic" is administered. Everyone is directed to look at the hand that isn't administering the illusion. Here, everyone is looking at Chinese v. U.S. military bioweapons labs. That tells us it was more likely a secret Deep State pharmaceutical lab that produced the virus, released militarily to incite war between the U.S. and China. That is a 'conspiracy theory' that is unproven but reasonable under the circumstances.

Most suspect, as mentioned above, is The Pirbright Institute that patented on Nov. 20, 2018, U.S. Pat. No. 10,130,701.[683] This attenuated coronavirus patent may or may not be the source of the circulating bioweapon, but its claim for use as a vaccine certainly raises reasonable suspicions under criminal circumstances.

This allegation is also justified by China overtaking Big Pharma by law on Dec. 1, 2019. This was days before the first outbreak in Wuhan. In effect, the People's Liberation Army (PLA) committed a hostile takeover of China's pharmaceutical industry' operations. No one is talking about this justification for retaliation against China by Big Pharma other than this author. Why?

The Magical Disappearing Act

the White House probe.

Read between the lines of the current TV propaganda. It's all about US over-reliance on China for consumer products, drugs, and even some vaccines, we were told.

Similarly, recall President Trumps TV ads. 'Big Pharma is charging too much for drugs.'

Was China, not Big Pharma, actually sourcing the drugs? No. That is the illusion spun by the media and politicians to justify military budgets and saber-rattling against China.

Don't forget the "DEEP STATE IS BIG PHARMA."

Did China create HIV-1?

No.

Did China create the hepatitis B vaccines transmitting AIDS?

No. Big Pharma (Merck/NCI/SVCP) did.[66]

Did China conceal the AIDS-virus genes spliced into the S- protein antigen attachment apparatus of the CoV-19?

Yes.

Why?

Because China depends on Big Pharma and its global investment bankers and stockholders for money. So Chinese officials, like Western officials, have kept quiet about Deep State military bioweapons programs and depopulation investments financing black ops.

Indeed, China cannot be trusted, but anti-Chinese arguments blaming China on the Wuhan outbreak are unconvincing.

The propagandists and legislators now urge China to "close down your meat markets." That is a red-herring diversion and obfuscation.

Did a bat, or dog meat, splice HIV-1 gene segments into the deadly CoV-19? No.

But those contriving of this **diversion** are complicit. Those recklessly neglecting this science have been aiding-and-abetting the global biocrime.

The heavy favorite among CoV-19 bioterror suspects is the Deep State/Big Pharma, and increasingly the 'Octopus' controlling The Pirbright Institute, not China nor even the American government under Trump's administration.

CLAIM 4: There are no other treatments for CoV-19.

This lie compounds the indictment of the COVID Cartel suspects because what is obvious to anyone knowledgeable about health and science is the central importance of natural immunity and natural antioxidants.

Antioxidants are electron-donating molecules powerfully effective against all positively charged pathogens, albeit fraudulently concealed in the international coronavirus response. The coronavirus pathogen is "positively charged" on its surface because of the synthetic 'gain-of-function' S-protein antigen. That positive charge is duplicated on the vaccines' coating. Antioxidants are the best, safest, and most natural way to neutralize this positive, electromagnetically attractive attachment mechanism.[13]

The active ingredient in 'natural cures' is the 'hydroxy radical.' This chemistry delivers naturally effective negatively-charged alkalizing antioxidants. Vitamins C and D, zinc, chlorophyll, even certain natural frequencies such as 528Hz resonating in the broad spectrum anti-microbial OxySilver, provide powerful antioxidant immune-boosting properties.

Another example is hydroxychloroquine. Its functional group also operates as an antioxidant, but unfortunately, this is a drug that can cause damaging side effects.

All of these remedies oppose the 'positively charged' CoV-19 spike protein attachment apparatus. In the presence of these antioxidants, the virus cannot attach to negatively charged human cells to cause the disease.

Accordingly, it is a total lie that "there are no effective treatments for coronavirus disease other than developing drugs and vaccines."

CLAIM 5: Johns Hopkins accurately tracks and reports.

This claim falsely presumes that coronavirus testing methods administered worldwide are accurate. This also erroneously presumes Johns Hopkins Univ. is honest, unbiased, and has no conflicting interests. Just the opposite is true.

Also, this claim falsely represents that people who died of supposedly "coronavirus disease" did not die of multiple other known causes of death, including preexisting medical conditions, immune suppression caused by lifestyle risks and previous vaccinations, nosocomial infections acquired in healthcare facilities, and iatrogenic causes of death from medical malpractices. This latter risk includes the improper diagnosis preceding the shredding of the lungs from unwarranted use of ventilators to treat the underlying unrecognized oxygen deprivation syndrome. Science tells us that the main three causes of deaths from COVID include: (a) histamine-induced 'vasoconstricting reducing blood flow adjacent lung air-sacks called "alveoli;" (b) sticky mucous build-up reducing oxygen uptake by the alveoli; and (c) damage to the alveoli that takes time to heal.

COVID case numbers are also false and misleading because three different types of tests are administered to confirm a "case,"[684] and each one often produces false positives.

1) The 'slow' antibody test is an "enzyme-linked immunosorbent assay (ELISA) that is not a test for [ongoing] infections. Alternatively,

2) the most popular testing technology is produced by Abbott Labs. It is called the Abbott RealTime SARS-CoV-2 assay.[685] This is a "real-time reverse transcription-polymerase chain reaction (rRT-PCR) test administered on the Abbott m2000 System.

Here, a SARS-CoV-2 primer and probe set is used to detect RNA from SARS-CoV-2 in nasopharyngeal and oropharyngeal swabs from patients with signs and symptoms of an infection; and

3) The third type of test is a viral DNA test. The fastest one to date is called ID NOW. It, too, is manufactured by the infamous Abbott Labs.

As briefly mentioned above, Abbott Labs bought out the Nazi-successor-in-interest BASF and virtually partnered with the infamous Boehringer-IG Farben syndicate, also responsible for illegalities in the marketing of Humira and Depakote.

Abbott's ELISA antibody test preceded its quick test and supposedly identified coronavirus genes using the "reverse transcriptase enzyme" popularized in 1970 by the claimed "simultaneous discovery" of this "novel" enzyme in RNA tumor viruses identified by **David Baltimore** and **Howard Temin**.

Baltimore's commercial isolation of the "AIDS-enzyme" at MIT occurred at the same time the AIDS virus began to be injected into people through hepatitis B vaccines. Health officials have censored this intelligence.

David Baltimore's commercializing-isolation of the AIDS-virus enzyme reverse transcriptase enabled the study of never-before-seen leukemia/lymphoma/sarcoma cancer complex developed during the Special Virus Cancer Program.12 According to a leading review article, this enzyme isolation "revolutionized molecular biology and laid the foundations for retrovirology and cancer biology."[686]

David Crow has critically examined the serious flaws in coronavirus testing methodology and accuracy in his April 10, 2020, publication of "Flaws in Coronavirus Pandemic Theory."[687]

Crow evidenced the arbitrary nature of the reverse transcriptase (RT) PCR test for viral DNA, Crow analyzed several studies. In one example, "[a] group of doctors in Marseille, France, working in a very experienced lab that regularly does testing for respiratory viruses, reported testing 4,084 samples for COVID-19 using several systems approved for use in Europe, without a single positive. This included 337 people returning from China tested twice and 32 people referred because of suspected infection.

Crow concluded, "It is statistically improbable that this lab was just lucky not to get any COVID-19 cases. It is more likely that they used more stringent criteria, illustrating that the performance of not just test kits, but labs, with this new test, is completely [arbitrary and] unknown. Yet, a positive test remains unquestioned in every 'case' [tallied by Johns Hopkins University]."

Johns Hopkins, as evidenced in previous chapters, operates at "the heart of the beast" —the 'Plague Syndicate'.[86]

Consequently, imagine a drug and vaccine syndicate overriding the worlds of science, medicine, and the media, which persuade governments and world commerce to shut down, and justify the hostile takeover of civilization based on totally questionable reporting of test results. These test results, promoted by the troublemakers, erroneously presume to identify the presence of coronavirus. But the tests, even if they were accurate and reliable, don't actually test for the presence of the virus or its disease.

That summarizes the absurdity and criminal complicity of what is ongoing in the name of "coronavirus testing."

CLAIM 6: Dr. Anthony Fauci is most trustworthy.

In this video,[688] you can view Dr. Fauci testifying before Congress like a lying idiot, claiming ignorance about the brain damage many measles vaccine-injured victims suffer —pathology openly acknowledged by the vaccine maker in every package insert.

In another video,[157] you can view *FOX News*' Tucker Carlson 'spin' Anthony Fauci's wrongdoing into praise for the CIA's leading infectious disease propagandist. According to the *Washington Post*, as corroborated by the CIA,[689] U.S. National Security demands oversight of Fauci and his agency by the CIA and Department of Homeland Security.

Fauci's conflicting interests are increasingly appearing. Evidence for this criminal indictment and syndicate is found in Chapters III and IV and many open-source publications, including Fauci's own NIAID heralding their funding of studies in 2018 in collaboration with the Chinese Government's dual-purpose pharmaceutical/bioweapons lab — the Wuhan Institute of Virology.[690]

As already mentioned, in late October 2021, the NIH finally admitted funding 'gain-of-function' coronavirus research in Wuhan, contrary to Fauci's repeated 'prevarication.'

Summarily, Anthony Fauci is a flim-flam man for the Deep State's 'Death Sciences' syndicate. He operates at the forefront of fraudulent concealments, scientific evidence tampering, and issuing propaganda for the global racket exploiting emerging biotechnology, resulting diseases, and genocidal pandemics.

Unconscionable evidence of 'bad faith' is compounded here, listing a dozen Big Pharma patents held by Anthony Fauci et al. This evidence best explains this front man's financial motives and conflicting interests tainting his 'coronavirus response' leadership.

Aside from Fauci's fraudulent concealment of Pradhan et al.'s determinations of the AIDS virus gene sequences spliced into the n-CoV/SARS/HIV-1 bioweapon, additional Fauci patents show the NIAID "AIDS Czar" capitalized on every aspect of the AIDS industry's commercialization. This includes diagnostic and therapeutic products and services, from antibody testing and vaccines additives (such as "intermittent interleukin 2) to anti-retroviral drug therapies.

The evidence in these references and links provide substantial 'probable cause' to indict Fauci and his partners for treason and genocide.

COVID COUP: "The Rise of the Fourth Reich"

Awaken to the 'Elements of Vaccination Genocide.'

CLAIM 7: 'Social Distancing' will defeat the coronavirus.

On *Fox News*, on April 13, 2020, Tucker Carlson declared, "We were wrong." Officials and the media had over-estimated the numbers of coronavirus cases and deaths that would "overwhelm healthcare." The initial objective of the governmental response was to prevent this.

Johns Hopkins University mainly administered the decisional data. This criminally complicit enterprise co-sponsored the "Coronavirus Predictive Programming Conference"[86] that Carlson and the rest of the news media recklessly neglected. Carlson surprisingly admitted that the specious "mathematical models," predictive data, and traitorous negligence, were abused to commit and justify a "bait-and-switch" consumer fraud.

The liars bated citizens with the infamous "Flatten the Curve" argument. This reasonably justified shutting down American commerce and social contacts. They recklessly and erroneously predicted these actions would slow the spread of the virus and prevent shortages of supplies overwhelming healthcare.

Then, while the devil-doers fanned irrational fears to defraud the public, officials and the complicit media suddenly switched the message. "Crush the curve" by defeating the coronavirus altogether by all means, especially by "testing" and developing vaccines as 'predictively programmed.'

> *"Summarily, Anthony Fauci is a veteran Deep State 'Death Sciences' agent operating at the forefront of fraudulent concealments, scientific evidence tampering, and issuing propaganda for the global racket exploiting emerging biotechnology, resulting diseases, and genocidal pandemics."*

We must presume that Anthony Fauci et al. completely conned Trump.

In arguing for social distancing and shutdowns, Fauci extorted the administration with the threat of mass morbidity and mortality. He even advocated terminating handshaking forever, with no possibility of returning to traditional 'normalcy.' He endorsed vaccination certification for travel and commerce.

Such "leadership" exclusively favored the Deep State's 'Pharma Cult.'

Personal Protective Behaviors Engineered by the Devil-doers

Irrational and irresponsible, you may have noticed the conflicting opinions and confusion generated by 'experts' concerning using personal protective equipment (PPE), especially masks. Do masks prevent coronavirus transmissions and the eventual spread of the bioweapon across society?

The latter is unrealistic.

Also, handwashing has its drawbacks, especially against airborne and allegedly 'meat' borne viral transmissions, especially when animals and pets don't wash their hands and are reportedly susceptible to this "novel" disease.

Likewise, a predicted shortage of ventilators, supposedly central to competent medical care, was false on both counts. The ventilator shortage never really happened because the disease tracking 'models' were false. And the use of ventilators to 'save lives' compounded damage from coronavirus disease and deaths, as explained below by New York physician Dr. Kyle-Sidell.

The Questionable Utility of Masks

Returning to the supposed benefit of wearing masks to prevent the spread of coronavirus disease, officials have been 'all over the board" on this issue.[692]

First of all, masks do nothing to increase your main defense against the disease, that is, natural immunity. As previously mentioned, "The germ is nothing, the terrain is everything," confessed Louis Pasteur from his death bed.

So if masks work, they deprive healthy people of natural airborne exposures, best enabling natural immunity and herd immunity against coronavirus disease.

Besides this, masks make breathing more difficult. Masks, thereby, reduce oxygen intake and increase risks, as I detailed earlier.

Lowering oxygen intake depletes natural antioxidants in the blood and reduces disease resistance, especially against coronavirus disease.[693]

Masks are also uncomfortable and 'hot.' They increase sweating in warm places. Perspiration is a 'detoxification reaction.' This body fluid may release, and even spread, more viruses and S-protein antigens. Suppose your arm sweats from the heat and wearing a mask, and you 'elbow shake' with someone? All of this is absurd.

Humans jumping through hoops to prevent death is exactly what the COVID Cartel wants to divert attention away from their genocidal actions for their own criminal protection.

Besides 'shedding' viruses through sweat, 'dirty' used masks can spread disease. Officials require proper 'biohazard' containment and disposal of potentially infected materials. Are people disposing of soiled masks properly, according to OSHA standards? No. Yet, OSHA is threatening to fine businesses for not imposing mandatory vaccinations. This extreme hypocrisy evidences criminality.

Masks worn publicly are also visible reminders of distressing messaging and social conditioning. To see people wearing masks breeds worry, not safety or security. Are these faceless people you see wearing masks 'Zombies'? Are they infected? Or are they simply forced by officials and induced by worry to wear this 'PPE'? All of this has a damaging impact on mental health, society, and quality of life?

"Social distancing" devastating the economy is delaying and inviting much more tragedy."

Similar drawbacks accompany "social distancing" impositions, including "elbow shaking." The utility (or benefit) of these actions is complete conjecture. Where is the science on these behaviors? Non-existent.

Consider this question: What if one year from now science proves that social distancing and mask-wearing delayed 'herd immunity,' which fueled a resurgence of the coronavirus pandemic? Then what?

Even worse, what if the delayed herd immunity gave time for a further mutation of the coronavirus — recombination with a far more deadly virus— that results in far greater morbidity and mortality? In that case, social distancing would be contributing to a 'biological apocalypse.'

Meanwhile, while neglecting these most important questions and concerns, the media focuses on less weighty issues. In this context, the hotly contested questions, to mask or not, or stay "6 ft. away" from people in public, serve as 'red herrings.'

These secondary issues brilliantly divert people from most protective and preventive self-care strategies. Likewise, these diversions occupy attention spans, keeping people from awakening to the fraudulently concealed science evidencing the lab creation of this bio-terrorizing imposition and criminal liability.

These 'red herrings' prevent or delay the investigation, prosecution, and conviction of the devil-doers that have criminally imposed such restrictions on humanity in violation of bioweapons treaties. The suspects produced and released COVID-19 and then imposed unreasonable, irresponsible, and uncivilized restrictions on society and human behavior.

Consequently, the 'social conditioning' for 'social distancing' administered by such media messaging and public engineering to isolate people clearly serves Deep State objective #1 —to condition 'dehumanization' of civilization for the emergence of **transhumanism** and its commercial and global governmental goals.

As I reported earlier and elsewhere, this scheme, damaging civilization, benefits pharmaceutical industrialists advancing artificial intelligence (AI). These facts heavily incriminate Johns Hopkins, Bill Gates, Klaus Schwab, George Soros, Peter Theil front men like Anthony Fauci, and other Deep State agents who have served military MindWar operations financed through black ops such as those at MIT's Media Labs.[283]

Given the facts and criminal orchestration of the COVID Coup, it is no accident that Big Tech advanced the 'Metaverse' as humanity was struggling to cope with this new traumatic and distressing 'reality.' The Metaverse creates a far more pleasing AI 'reality' that is addicting. The new "normal" engage eyes, ears, and brains so that humans don't notice, or mind being converted into Technotronic slaves.

This treacherous scheme, and the agents financing it, has enabled the treasonous takeover of governments, economies, and societies worldwide by elite industrialists more interested in artificial intelligence (AI) and population control than flesh and blood ethics."

CLAIM 8: Frequent handwashing is good.

Science claims handwashing with water and plain soap (not anti-microbial soaps) can significantly reduce the number of pathogens found on your hands. But it is pure speculation that this translates into less death or disease.

Apparently, touching the coronavirus didn't kill the first million people reported to have been infected. They overcame the plague. That's great news. The infection and natural immunity made them stronger, not weaker. According to the experts, they generated protective antibodies.

Typically, when natural immunity defeats an infection and leaves behind antibodies for future protection, which is good. So why would anyone choose the defeat this natural process in favor of enriching Deep State criminals?

Even though some science proves such hand-washing reduced pathogen transmissions, that science would be questionable due to several "confounding factors." These include lifestyle risks impacting natural immunity, the number of organisms spread, the different pathogenicity of each strain transmitted by unwashed hands, etc.

People's immunity varies dramatically. But this vitally important fact of self-protection through lifestyle moderation is recklessly neglected by officials. Instead of properly educating people about the importance of antioxidants, good hydration, exercise for oxygenation, and good nutrition as an adjunct to good hygiene, officials promote vaccines. This 'one size fits all' remedy transparently evidences criminality.

Sure, it makes sense to practice good hygiene, including timely handwashing with plain soap and water after touching potentially contaminated surfaces. It makes sense to stop coughing or sneezing in public. If you are sick, it makes sense to stay at home, rest, and recuperate. But frightening and restrictive recommendations breed obsessive and compulsive disorders wherein people wash hands repeatedly neurotically, fearing otherwise they will die.

Accordingly, official messaging generally targets a 'dumbed-down' neurotically inclined population, consistent with MindWar operations and profiting from unreasonable and unethical behavior modification and social conditioning. This activity certainly profits the devil-doers.

Furthermore, too frequent hand washing can do physical damage. How many people use anti-microbial soaps that dry and chap the skin? These chemical intoxication reactions break the protective layer of the skin. Cracks in the skin increase the risk of viral transmissions.

How many consumers have allergies and multiple chemical sensitivities to anti-microbial soaps or disinfecting sprays? Tens of millions, according to science.

Plain tap water is commonly heavily chlorinated. It too dries out and cracks the skin.

Furthermore, these chemicals have damaging impacts on the environment and groundwater quality.

Given these facts, including circumstances characterizing coronavirus bioterrorism and Deep State biocrime, public appeals for handwashing, like mask-wearing, are diversionary red herrings. They may be doing more harm than good.

Ventilators Damaging Misdiagnosed Coronavirus Patients

As I will address in more detail in the next chapter, sharing with you my personal experience as a hospitalized "COVID patient" and my personal protocol that hastened my recovery, I urge you to consider the reckless death and manslaughter done by doctors and respiratory therapists when deciding to put patients on ventilators rather than alternative therapies.

"I have seen things I've never seen before," complained New York Doctor Cameron Kyle-Sidell,[693] who characterized the coronavirus bioweapon oxygen deprivation syndrome neglected by Fauci, most health science experts, and the media pundits.

"In treating these patients, I've witnessed medical phenomenon that just don't make sense in the context of treating a disease that is supposed to be a 'viral pneumonia,'" Dr. Kyle-Sidell objected.

"'Acute Respiratory Distress Syndrome' (ARDS) is the paradigm every hospital in the country is working under… that is untrue. In short, I believe that we are treating the wrong disease. And I fear that this misguided treatment will lead to a tremendous amount of harm to a great number of people in a very short time…"

This New York City critical care physician helped diagnose the actual cause of death in coronavirus patients. He diagnosed the condition of oxygen starvation as the leading cause of death in these victims of medical malpractice.

Click to view New York doctor, Cameron Kyle-Sidell, characterize the Coronavirus bioweapon as causing oxygen deprivation syndrome neglected by experts.[693]

"It appears that some kind of viral-induced disease most resembling 'high altitude sickness' [is being overlooked and neglected]," Kyle-Sidell explained. These patients are slowly being deprived of oxygen. I have seen patients depending on oxygen take off their oxygen [masks] and quickly progress into a state of anxiety and emotional distress, and eventually get blue in the face. And while they look like patients absolutely on the brink of death, they do not look like patients dying of pneumonia…"

Whistleblowing on the system-wide negligent manslaughter committed under the guise of standard "coronavirus intensive care," the heroic Dr. Kyle-Sidell added, "I fear that if we are using a 'false paradigm' to treat a new disease; that the method that we program the ventilator—one based on the notion of 'respiratory failure' rather than 'oxygen failure'… is doing more harm than good… [This] challenges long-held dogmatic beliefs within the medical community and lung specialists which will not be easy to overcome. But I believe they must be overcome [to prevent unnecessary morbidity and mortality]."

"These 'red herrings' prevent the investigation, prosecution, and conviction of the devil-doers that have criminally-imposed such restrictions on humanity in violation of bioweapons treaties. The suspects produced and released this strain, and then imposed unreasonable, irresponsible, and uncivilized restrictions on society and human behavior."

CLAIM 9: "We are all in this together."

Are we all together with Bill Gates or Anthony Fauci?[682,157]

Or is this simply a stupefying slogan to gain mass submission to the devil-doers' criminal impositions?

This slogan — "We are all in this together." — is reminiscent of the affirmation adorning Auschwitz's entryway — *"Arbeit macht frei,"* meaning "work sets you free."

Overlooking the direct ties between Abbott Labs, BASF, Boehringer, the Nazis, Bill Gates, and MindWar, operating at the heart of this coronavirus enterprise commercializing the "pandemic response" (allegedly in the interest of public health and safety) is the Cambridge, Massachusetts biotech company called Moderna Inc. That company developed the leading candidate vaccine against the Wuhan novel strain. Moderna's direct partners include Bill Gates and the NIAID, headed by Anthony Fauci. That is the same NIAID that financed the coronavirus/SARS novel virus research in 2018 at the most suspect Chinese lab near the Wuhan meat market —the Wuhan Institute of Virology. The same NIAID has covered up the hepatitis B vaccine triggering HIV/AIDS between 1972 and 1974.

Notable about the Fauci-Gates Moderna coronavirus vaccine enterprise is mRNA-1273, explained Princeton Univ. trained political and economic investigator, F. William Engdahl.[694] This anticipated 'remedy' for coronavirus 'insanity' has been "rolled out in a matter of weeks, not years, and on February 24 went directly to Fauci's NIH [NIAID] for tests on human guinea pigs, not on mice as normal."

Those who approve of this conversion of 'safe science' are "all in this together."

In contrast, law-abiding, well-informed citizens oppose this 'togetherness,' coronavirus bioterrorism, biological warfare, and biocrime. I reject these impositions by the suspects mentioned above and their racketeering enterprise.

Dr. Fauci deserves more than superficial criticism by the media for "miscalculating the spread of coronavirus."[157] By (1) secreting the lab origin and Deep State imposition of this plague; (2) misinforming the President and the public, as Bill Gates has done regarding risky vaccines; (3) promoting the specious benefits of testing and mask-wearing; while (4) fraudulently concealing natural remedies, Fauci and Gates have provided good cause for grand jury indictments for treason, bioterrorism, and biocrimes.

> *"[T]he 'social conditioning' for 'social distancing' administered by such media messaging, and public engineering to isolate people, clearly serves Deep State objective #1—to condition 'dehumanization' of civilization for the emergence of transhumanism and its commercial and global governmental goals."*

CLAIM 10: A coronavirus vaccine will save us.

Finally, and most importantly, Bill Gates, Anthony Fauci, et al., have conned people worldwide into believing only two remedies will save civilization and restore the global economy: (1) testing and (2) vaccines. As explained above, testing is so unreliable that obsessed officials imposing this technology are subject to fraud and treason charges. The same must be said about coronavirus vaccines.

Civilization is, in effect, being held hostage by the false promise that a forthcoming "fast-tracked" vaccine will prove "safe and effective" when that is impossible to predict, achieve, or even claim, other than by scientific fraud and media deception. At best, a vaccine might deliver 60% population protection. The other 40% are left to die and spread the disease.

As already mentioned, it is generally accepted that vaccines become less effective when the target germ mutates and when vaccinations are delayed.

Scientific support for these facts, resulting in abysmal vaccine failure, is found in the history of the H1N1 2009 "Swine Flu" that outbroke in Mexico during a reported vaccine trial. This pandemic was falsely blamed on China. These matters were thoroughly investigated and reported in *Clinical Microbiology Reviews* by authors Cheng et al.[695]

For instance, exemplifying the problems arising from imposing "social distancing" while awaiting mandatory coronavirus vaccinations, these mainstream science scholars (Cheng et al.) noted that, "despite the technological advances in using cell-based inactivated whole-virus vaccines and improved adjuvants, vaccine production failed to prevent the first peak [in certain regions]... Social distancing methods, such as canceling entertainment and sporting events, closing stores, office buildings, and public transportation systems, border screening, the isolation and quarantine of febrile patients and contacts, school closures, and hospital infection control measures may achieve only a few more weeks of preparedness by slowing down the introduction and spread of the virus if instituted early enough."

However, delayed vaccination after the virus has had an opportunity to "peak" and mutate into less cross-reactive forms dramatically tip the risk-to-benefit scale.

Coronavirus mutations make vaccinations riskier

Most importantly, the chosen coronavirus strains may have mutated so much by the time the developing vaccines are manufactured that their impact will be worse than ineffective. Delayed vaccines may increase diseases and death by burdening the immune system with the wrong strains for no benefit. In other words, the "risk-to-benefit ratio" increases, causing much more harm than good.

This problem with vaccines is not new. It has plagued efforts to cure or prevent, for example, AIDS. When HIV-1 jumps from one person to another, it mutates as much as 8%. This fact has stymied AIDS vaccine developments during the past four decades.

Quoting *Fox News* on April 6, 2020, "As it stands, the [circulating corona] virus is mutating among humans. And the more it rages on, the more it changes. At least eight strains of the pathogen have been identified, indicating that it has amended itself several times since 'patient zero' was presumably infected at a wet market in Wuhan, China, late last year."

More Science cautions against untoward side effects of "immunization with SARS Coronavirus Vaccines."[696] The injection/intoxication may lead to the production of antibodies and presumed immunity but is likely to cause "Pulmonary Immunopathology" and more deaths, much like clinicians have witnessed in intensive care units.

Even reviewing pro-vaccine sources,[697] we predict a major problem arising from the future coronavirus vaccinations for millions of people infected early in 2020 and society at large months after the pandemic reportedly "peaked" by early Spring. Typically, detailing this concern, the efficacy of timely vaccines ranges from 20% in a "bad year" to 60% in a really "good year" when the vaccine is injected well before the "peak" and contains the precise antigens matching the virus doing the damage. This knowledge bodes ominously for the forthcoming coronavirus vaccines.

According to *U.S. Pharmacists*,[698] "[I[f there were an early influenza peak [as in the early Spring peaking of coronavirus infections in NYC], then a late vaccine schedule would result in more cases as well. Similar results were shown when flu vaccines were given early in the season during a late influenza peak. Recommending an optimal time for influenza vaccination is difficult; influenza seasons are unpredictable... [T]he CDC does recognize that getting vaccinated too early, such as in July or August, may be associated with reduced protection later in the influenza season, especially in older adults."

Conclusion

These facts contraindicate the entire coronavirus "pandemic response." This ill-advised and risky response has delayed, even possibly worsened, the expected overall morbidity and mortality.

The focus of the response from its inception was based on skewed data and wrong models generated by individuals and institutions with conflicting interests. The primary objective was to delay the peak of the infections to secure health services and stall long enough for vaccines or drugs to be safely and effectively developed, tested, and administered. Antibody and DNA testing were falsely alleged to save lives by identifying infected persons requiring quarantine, treatments, or the vaccine. Other reasonable natural alternatives, self-care activities, and preventive strategies were recklessly neglected, resulting in unnecessary deaths and severe damage to the global economy.

According to science, it is unreasonable, illogical, recklessly irresponsible, dangerous, damaging, and deadly to expect vaccines (and coronavirus antigen, DNA, or antibody testing) to lower risks or increase prevention or healing benefits.

The defrauding of society using the media, overseen by intelligence agencies, has enabled bioterrorism and the alleged biocrime. In effect, unscrupulous officials have effectively twisted science to defraud and hijack civilization in favor of generally concealed pharmaceutical interests and technological objectives.

Henceforth, there is sufficient probable cause to disregard, investigate, indict, prosecute, and convict those persons who have aided-and-abetted the coronavirus biocrime(s), beginning with Bill Gates and Anthony Fauci. They have, with complicit parties, committed treason in violation of international treaties precluding biological warfare, bioterrorism, and bioweapons development and deployment.

Opening the economy completely and immediately by eliminating restrictive impositions is the most reasonable action officials can take at the time of this writing.

CHAPTER XIV

THE COVID COUP, NEW WORLD RELIGION, AND SPIRITUAL WARFARE

"By your magic spell all the nations were led astray,"
predicted Revelation 18:23.

In the Book of Revelation, the word "sorcery" is associated with magic spells cast internationally to deceive susceptible targets. They are made susceptible by lacking faith in Jesus and the universe's Creator. "Sorcery," therein, is defined as *pharmakeia*[699] —related to pharmaceuticals. Consciousness and reality did not create themselves. Today's sorcerers prescribe hallucinogenic drugs and AI in mind-altering vaccines, supplemented by media technologies like the 'Metaverse,' to generate illusions and diversions from reality.

Little Known History of American Satanism, the FBI, and CIA

Few people recall that the original Church of Satan in California was *hijacked* by U.S. military and its MindWar guru, Col. Michael Aquino.

Following the death of Aquino's political adversary, the original Church of Satan's founder, Anton LaVey, Aquino took over.[700]

Sherri Kane and Barbara Hartwell evidenced the Ted Gunderson I introduced previously in the MindWar chapter, who was the FBI's Division 5 Chief in Los Angeles, married Church of Satan heiress Diana Rively. Rively's daughter, Zeena LaVey, joined U.S. Military Psychological Operations and propaganda chief Aquino in founding the currently growing Church of Satan/Temple of Set.

This Satanic community bases its teachings on Aleister Crowley's anti-God and self-judgment theology. The current Satanic Temple churches draw their esoteric practices from Crowley and Aquino's modeling.

Oddly, this same demonic theology is practiced by the 'radical right' *Bible Believers*. They, like Gunderson's subordinates, especially Alma C. Ott (alias "Dr. 'True' Ott"), spewed dreadful misrepresentations of Judeo-Christian law. Members of their cult beat people's brains so forcefully with their slant on Jews and "Christian Identity" that the outcome resembles the MindWar operation of Third Reich propagandist Josef Goebbels. As already mentioned, this Deep State MindWar administration generated the PSYOPS and associated propaganda needed to advance the neo-Nazi movement resurging internationally. This is all backed by the global intelligence agencies in favor of the globalists' 'divide and conquer' 'helter-skelter' agenda.

Anton Szandor LaVey

LaVey publicity photo, ca. 1992

Title   Author of Satanic Bible, High Priest and founder of The Church of Satan

~

Today, news services report the federal government increasingly recognizes the Satanic Temple as a legitimate religion backed by the agents and agencies vetted in this book. In addition, it is becoming

increasingly apparent that officials subscribe to sexual perversion, including pedophilia and transsexuality.[701] Kane and Hartwell evidenced Gunderson's FBI and CIA assignments in "Child Sex Trafficking" and concealing the Deep State's network through the COINTELPRO.[702, 302] A video shows Geraldo Rivera hosting Gunderson and Aquino playing 'good cop' vs. 'satanist.'[703]

Accordingly, it is not surprising that the IRS has now recognized the right of the Satanic Temple[704] church members to "Hail Satan" as detailed below and that high-level government officials have been criminally complicit in Epstein-like crimes.

As further detailed below, Bill Gates openly endorsed this demented, blood-thirsty, satanist movement through Microsoft's sponsorship of "spirit-cooker" Marina Abramovic.

Penny Lane's *NUTS!* and *Hail Satan?*

Pro-vaccination filmmaker Penny Lane's gross hypocrisy is shown in the widely touted film "Hail Satan?"[705]

I recall Lane's leading role in commanding the opposite of free speech and religious tolerance when she circulated her 'open letter' to Robert De Niro, slamming the documentary *VAXXED*. Lane's influence allegedly 'caused' De Niro to 'pull' that film from the Tribecca Film Festival in 2016.

In Lane's letter,[706] she spilled the beans and blew her cover, however. She exposed her true passion for filmmaking as a PSYOPS queen. Quoting Lane's reference to her previous film *NUTS!* "the reason I made **NUTS!** is that I wanted to explore *just how easy it is to fall for a quack, especially one cloaked in the authority of a documentary film*." [Emphasis added.]

"How easy it is to fall for a quack," Lane wrote, engendering the persuasive illusions in a MindWar.

Lane is photographed below with fellow filmmaker Brian Frye, a lawyer.

Kane and I wrote in protest to Lane's and Frye's covert agency and agenda.[707]

VAXXED, condemned by Lane and Frye, featured government whistleblower William Thompson. Lane concealed Thompson's riveting recording in which Thompson admitted to fellow researcher Brian Hooker the CDC had criminally concealed evidence of MMR vaccine side effects.

One has to wonder why *Filmmaker* magazine, like Lane, did not want anyone to view *VAXXED,* but urged everyone to 'Hail Satan?'

The screenshot below shows nine of the main clients served by the agents and agencies administering the pro-vaccination Luciferian media. I mentioned them briefly earlier in this book. The SAB/PRISM Learning Group, on behalf of Axiom Learning, Crosslead.com, and the McChrystal Group, published enough intelligence online that we discovered their complicity in the Deep State COVID conspiracy.

Sherri and I exposed these counter-intelligence agents and agencies in our film *UN-VAXXED* that won five (5) international awards including

'Best Film-2016' in London and Geneva World International Film Festival competitions.[60,419]

This selective programming (or media bias) is best explained by the different messages in these two films. Bigtree's *VAXXED* proclaims 'vaccines are good,' we just need to make them 'safer.' My film, *UN-VAXXED* evidences vaccines are 'genocidal weapons of mass depopulation, and there is no way to make their toxic ingredients, such as mercury, aluminum, and GMOs, 'safer.'

'Radical Satanism' Advanced in the Mainstream

On the holiest day in the Christian calendar, Good Friday, 2018, "Bill Gates' Microsoft released a commercial promoting its association with the elite's favorite artist: Marina Abramovic," reported *Greek News on Demand* and other open sources.[708]

After a huge backlash from internet users calling her out as a Satanist, the video was set to "private," and Microsoft scrubbed all traces of the doomed campaign from its official website.[709]

Abramovic, "who enjoys posing with flayed goat heads and brewing blood, feces, semen, and breast milk cocktails for political elites," is upset that "conspiracy theorists" won't leave her alone and stop claiming she is a Satanist.

COVID COUP: "The Rise of the Fourth Reich"

You can judge for yourself by viewing the Gates/ Microsoft Abramovic promotional video by viewing this.

Now superimpose Penny Lane's and Brian Frye's Luciferian media actions upon Microsoft's Abramovic endorsement. Now you can grasp the psychopathic corruption by which these media influencers turn good into evil without remorse.

According to Anton LaVey's *Satanic Bible*, "Nothing is to be gained by denying oneself pleasure."

Clearly, Lane took pleasure in denying *VAXXED* its audience and promoting Luciferians.

Extending this doctrine to pedophiles and child-sex-traffickers (such as Jeffrey Epstein's ilk did and continue to do), 'radical Satanists' psychopathically justify anything and everything criminal and sociopathic, precisely like serial killers.

According to Anton LaVey, "religious calls for abstinence most often come from faiths that view the physical world and its pleasures as spiritually dangerous. Satanism is a world-affirming, not world-denying, religion."[711]

But not even traditional Satanism presumably endorses such unconscionable rituals of child abuse as we view in the devil-doers' "art," including that of Hillary Clinton's most trusted official, John Podesta, and his brother, Tony.

THE COVID COUP, NEW WORLD RELIGION, AND SPIRITUAL WARFARE
Satanism, PizzaGate and CNN's Anchor

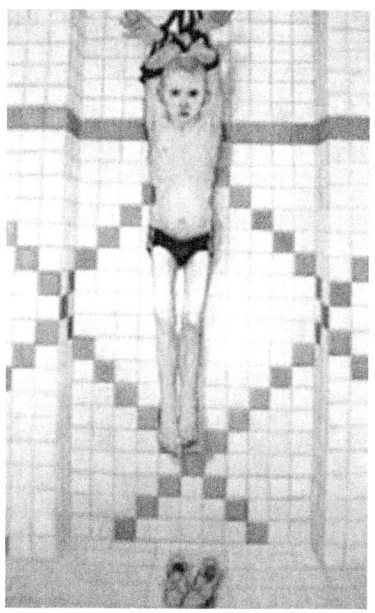

Podesta's painting above shows a young boy hanging in bondage atop a tile wall. His toes are chopped off on his left foot. The tile wall is a clue evidencing the real-life crime scene —allegedly Gloria Vanderbilt's estate. (Andersen Cooper, Vanderbilt's son, is the prime-time news 'anchor' on *CNN*.)

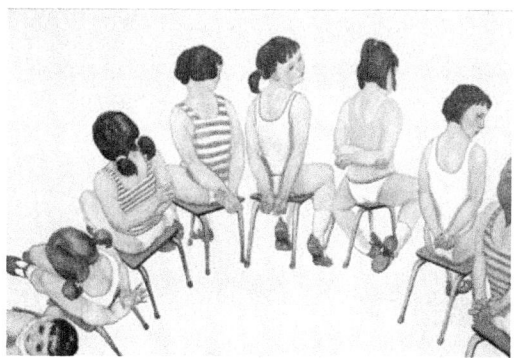

Here you can view other Podesta paintings featuring groups of young girls that appear to be victims of bondage. In the above painting, they are sitting. All are wearing red shoes. Each shows their hands and buttocks reddened by repeated assaults.

A collage of Podesta paintings on the next page shows the same ritualistic abuse. Children are shown in bondage after having their hands, feet, and behinds slapped red or sexually assaulted. The red shoes reflect their battered feet. This is apparent in each picture and each victim.

A photograph below shows Gloria Vanderbilt-Cooper and her two sons. To their right is a photograph of their fully tiled spa-dungeon. Draped on the walls of this 'bathing pool' is a series of nooses. Note the black tile line at the border of the pool.

In the collage below, from Tony Podesta's art collection, are two girls. They appear to be sitting on a similar black tile border of the spa. The child on the left therein appears to have had fingers amputated. These two victims are sitting against a similarly tiled wall with the same black stripe of tile that appears in the actual photograph of the Vanderbilt-Cooper spa.

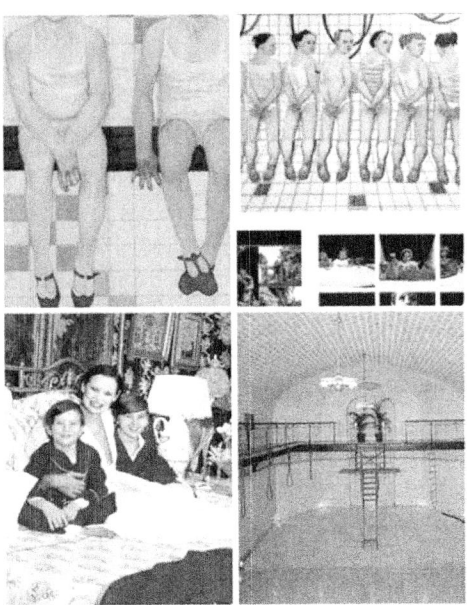

Of the two Vanderbilt-Cooper boys, the one on the left is Anderson Cooper. As mentioned, he is the primary propagandist on the Deep State's "fake news" *CNN*. Cooper's older brother, Carter Vanderbilt-

Cooper, died of alleged suicide on July 22, 1988, at age 23. Supposedly he jumped from the 14th-floor terrace of Vanderbilt's New York City penthouse apartment.

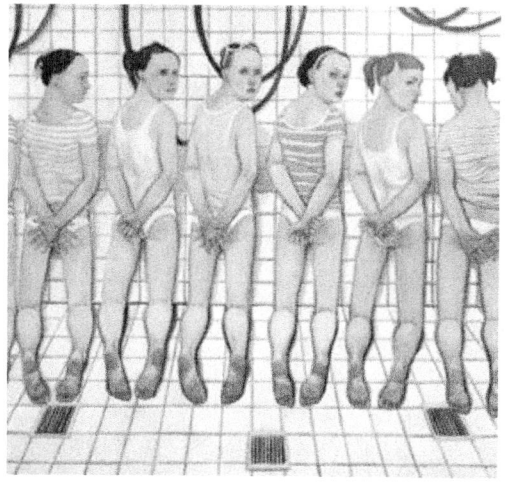

Below you can see a larger (albeit darker) image of the same 'spa dungeon' in the Vanderbilt-Cooper mansion, posted by an anonymous source. On the floor, in an arc, you can see a string of lights reflecting from the sparkling-clean tile floor. Around the sides of the spa, you see what appears to be chain-like harnesses.

In the image below, you will see insets of photos prepared by an anonymous source showing toddler Anderson Cooper progressively developing into the *CNN* news anchor atop Tony Podesta's art. This painting is much like the art collected by PizzaGate's central suspect, Tony Podesta's brother John. The art was painted by Biljana Djurdjevic.[712] "Tony Podesta's many windows to the soul is located right next to a painting of [this] 'TIED UP LITTLE BOY IN A KILL ROOM'"[713] (as evidenced on the blog ZeroPointNow).

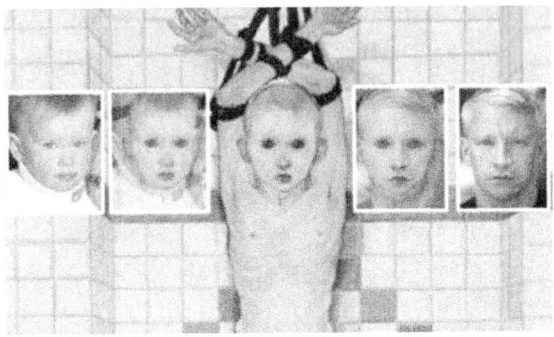

Below is a photograph also published in several places identifying Tony Podesta as a leader in the "Podesta Red Shoes Club."[714] You will notice the red shoes on all the men reportedly engaged in this demonic child sex trafficking and murder ritual club. The red shoes symbolize those red shoes and battered feet depicted on the children in the Podesta and Epstein art.

An image of Bill Clinton, dressed in Monica Lewinski's 'Blue Dress,' wearing red shoes was also found in Jeffrey Epstein's New York mansion.

THE COVID COUP, NEW WORLD RELIGION, AND SPIRITUAL WARFARE

You can also see Epstein's painting of George W. Bush. He sits on the bathroom floor like a mentally disabled person, flying paper planes to demolish two wooden block structures metaphorically representing the Twin Towers on 9/11. At the time of the disaster, the President was reading children a book about a goat —the symbol of the Baphomet, symbolizing the Deep State's affection for the Luciferian cult.

It was widely reported that 9/11 was an 'inside job' orchestrated by the 'Bush Cartel' for profit and power to advance 'globalism,' the War on Terrorism, the U.S. 'Patriot Act' precluding civil rights and liberties, Islamophobia, and gun control.

The 'Valerie Plame Affair' discredited the George H.W. Bush administration that was evidence to be engaged in child kidnapping and sex trafficking that resulted in national news coverage of the 'Johnny Gosch kidnapping scandal.' White House aide, Jeffrey Gannon, dispelled the allegations but failed to mention Gunderson's leading role in the cover-up.

Prophecies Fulfilled

"By your magic spell, all the nations were led astray," predicted Revelation 18:23.

The "magic spell" over the population is demystified in this book of facts. The main "magic spell" secures the 'unconsciousness' and denials of Satanism practiced by our leading government officials and media celebrities. The main spell involves the cover-ups in PizzaGate and America's child sex trafficking, drug and vaccine enterprise implicating the FBI and leading politicians in the devil-doing. The 'unconsciousness' and 'evil-obedience' permit criminals and their propagandists to lead everyone astray and into the Brave New World, featuring robots, cyborgs, and AI substitutes for humans.

Today's Deep State corporations and banking minions finance the writers and paparazzi photographers that condition society to accept mass murder and iatro.

Pharmaceutical Metaphysics and 'End Times' Bible Prophesy

Many observers conclude that we are the 'End Times,' as predicted in the Holy Bible.

The Word repeatedly warns about the demonic impact of drugs, false advertising by drug entrepreneurs, the abuse of the trance-inducing media to seduce the masses into drug delusions and soul stupors. That is, the Luciferian foundation of the COVID Coup was predicted by the early 1500s, as was the only remedy in the "key of the house of David" (Isaiah 22:22 and Rev. 3:6-8) that opens doors to Divine communion for the sake of physical salvation and spiritual restoration.

We need to, and shall, 'sing a New Song. (Rev. 14:1)

Drugs are condemned liberally in the Bible. The Greek word *pharmakeia* is referenced three times in the New Testament, in Galatians 5:20; Revelation 9:21; 18:23.

In scholarly defense of modern medicine, Dr. John Oakes considered the Greek origin of the word *phamakeia*, and its relevance today for all monotheists.[715]

"My Greek dictionary defines it as: 'witchcraft, magic, the use of spells and potions of magic, often involving drugs–a magic spell. It is a fact that witchcraft and magic in the Greek world often involved the use of drugs–either by the witch or the one on whom the magic is worked... [W]e get the word pharmacy from this Greek root word... Put it this way; if one were to use and administer drugs for the sole purpose of creating a mind-altered state, then one could probably be judged to have violated the commandment not to be involved in pharmakiea."

Rev. 8:13 (KJV) warns, "Woe, woe, woe, to the inhabiters of the earth by reason of the other voices of the trumpet [issuing musical frequencies] of the three angels, which are yet to sound! ... Neither repented they of their murders, nor of their sorceries, nor of their fornication, nor of their thefts." (Rev. 9:21) [F]or thy merchants were the

THE COVID COUP, NEW WORLD RELIGION, AND SPIRITUAL WARFARE

great men of the earth; for by thy sorceries were all nations deceived... And in her was found the blood of prophets, and of saints, and of all that were slain upon the earth." (Rev. 18:23-24)

It is unreasonable to dismiss the weight of the evidence above giving 'probable cause' to indict the named leaders of the real 'insurrection'—the COVID COUP—for practicing sorcery, casting 'magic spells' through the hypnotronic media, murdering people, raping children, fornicating with sex slaves, and stealing people's money and souls.

Accordingly, rather than replacing monotheistic religions, this book affirms them by sharing the extraordinary, clear-and-convincing accuracy of Bible prophecy in foretelling this New Age of Sorcery in opposition to the Deep State's Luciferian drug cult.

"Beware of false prophets, which come to you in sheep's clothing, but inwardly they are ravening wolves," Jesus warned in Matthew 7:15 during his Sermon on the Mount.

This Word came after Jesus was baptized in the Water of Life by John the Baptist. Jesus had finished fasting. He had been through a meditation retreat in the desert, then began to preach this warning. It was also prophecy foretelling for salvation the risks of trusting the 'gods of science' today, like Dr. Fauci. Jesus' 'consumer alert' in Galilee should be law at the CDC.

The parallels here to both the 'COVID Vaccination Cult' and the "Ayahuasca Death Cult"[335] are unmistakable. The Cult, a wolf in sheep's clothing, exploits the time-tested remedies and rituals of fasting, prayer, meditation, music, and chanting but switches psychedelic drug sorcery for the water baptism to gain presumed spiritual renewal refreshing emotions, mind, body, and soul.

The main problem with this is, as Fiona Barnett and other scholars have surmised, the metaphysics of pure unadulterated **human consciousness** is neglected and assaulted. The natural process of Divine-human communion and evolution of the species into 'higher consciousness' is chemically, sometimes violently, abrupted. Natural health sustenance is violently shattered chemically/pharmaceutically. Vomiting certainly evidences this reaction to poisons. This violent reaction to hallucinogenic poisons is acceptable to those financing the brain-Cloud connection.

This drug-induced 'trance state' and perception of another level of 'reality' not only enables sham shamans to manipulate victims in all ways, including sexually. Ayahuasca especially enables evil entities to enter fractured minds to steal their souls. In effect, it suppresses the heart and soul. Your heart and soul are central to faithful prayer seeking Divine communion and enlightened consciousness.

This is how 'born again Christians' experience powerful healing and optimal good. When 'the Connection' lights up, the Kingdom of Heaven opens up. Then the sonically-intertwined "musical mathematical matrix" of Holy Spiritual dynamics and multi-dimensional space/time physics broadcasts God's word/will. This expresses the frequencies of creative mathematics administering Divine healing.

As Fiona Barnett surmised in her *Eyes Wide Open* book,[716] the world's leading physicists and mathematicians have engaged metaphysics and the multi-dimensional realms of 'consciousness.' As a result, the 'star gates' are opening now enabling the forces of light and darkness to penetrate deeper into society and human behavior than ever.

This danger and opportunity is scientifically scrutinized best by America's leading metaphysical psychiatrist, Dr. Vernon Neppe. Neppe works with an equally brilliant partner, mathematician Edward Close. Their writings address "studies in quantal clairvoyance."[717]

Clairvoyance is how *and* why Revelation and Isaiah's prophesies are all coming true. Their works correlate profoundly with "models in the Triadic Dimensional Vortical Paradigm," Neppe and Close might argue. This know-how far surpasses the limited analyses provided by world-leading physicists scrutinizing data from the CERN large hadron collider.[718] The future of humanity (as a species) largely depends on this neglected and disparaged 'future vision.'

Neppe's team advanced the union between emerging neuroscience, physics, spiritual dynamics, and human consciousness. As Barnett did, Neppe et al. considered the CERN collider as technology for studying the interface between physical and spiritual dimensions.[718]

Like Barnett, Neppe and Close objected to the leading notion of a "God Particle." Why would God be limited to a single particle? Just the name wreaks fake science.

Neppe and Close postulated from solid mathematical proofs; there is something more to the equation. They named this primordial energy

"Gimmel." According to the earliest rabbis and church leaders, this vertical element central to creation, Neppe and Close determined, relates to certain mathematical constants and musical frequencies of creation and creationism. I encouraged them to consider "528" as the miracle manifester.

Like Barnett, who claimed that "CERN is a massive stargate… powered by a quantum computer" advancing a "form of clairvoyance," Neppe and Close postulated seeing into this realm of physics, metaphysics, and spiritual dynamics, using mathematics. I have been heralding this for nearly twenty years, as have colleagues Nassim Harramein and Marko Rodin. They described multi-dimensional reality as a "matrix" of creative and destructive forces measured by frequencies.

Frequency Therapeutics and Karmic Math

It was my contention in *The Book of 528: Prosperity Key of LOVE* that creative frequencies, including the "good vibration" of LOVE at the heart of rainbows, sunshine, chlorophyll, and good pure hearts —528 frequency— can be used most beneficially. It can (and is) being used to energize people's hearts to facilitate healing people *miraculously*. This activity is sourced from the Bible since 528 frequency is the Key of the House of David. It opens spiritual portals to miraculous manifestation. I wonder whether it is also engaged in 'righteousness,' universal homeostasis, orderliness in the 'matrix,' and Divine judgment; otherwise, KARMA!

At the same time, 528 heals, restores, and rejuvenates the entire body through every strand of DNA.

528 gains its energy from the heart of the Kingdom of Heaven and subsequently nature. I realized long ago the wisdom of, "As above, so below." If it is true for the heart of sunshine and rainbows, it's true for your heart and God's heart too. That energy puts you in sync with the heart of the universe, its administrative 'matrix,' and Source.

Alternatively, certain destructive frequencies, such as the annoying 'augmented fourth' or 'diminished fifth' called the **'devil's tone'** in musicology, produces dissonance within that same matrix. It interferes with the creative, restorative, and homeostatic balance administered by 'LOVE528frequency.' This dissonance (within the natural system) prompts dynamic changes to challenge the free will and choice-making.

This challenging "741Hz" frequency gives rise to forces within the universe (or the Kingdom of Heaven) that produce KARMA, stress, distress in humans, immune suppression, and disease states too.

The bad guys must have known about and *secreted* these frequencies of creation and destruction. This was, after all, the main esoteric knowledge involved in the creation of the earliest cults and control over them, for civilization's enslavement versus freedom by informed choice and exercise of Free Will.

I noted that the same globalists that advanced nuclear, biological, and chemical weapons under the CIA's MKULTRA and MKNAOMI programs also sponsored 'mandatory vaccination' legislation and the imposition of the "devil's tone" in music.

Skeptics cannot refute the presence and significance of John LaMattina, Ph.D.,[719] Former President of R&D for Pfizer, directing the "PCA Regenerative Medicine Advisory Board" for the "Frequency Therapeutics" company.[720]

Early History of 528 Frequency

In 1938, the globalists imposed the frequency of 'Concert Pitch' on the Western World. The Deep State's 'predecessors-in-interest' institutionalized this and called it cultist preference "standard tuning"[27] for music and entertainment.

The imposed Concert Pitch frequency was set at A=440Hz. This, my colleagues and I noted, produced that F# devil's tone commonly played in the D-major chord at 741Hz frequency. I concluded, this dramatically impacts society, best-accommodates population control, and degenerates the intelligence and 'consciousness' of clueless citizens.

Relatedly, at an MIT conference in 2014, Elon Musk metaphorically advanced the same intelligence about demonic energy influence. Musk referenced the sacred geometry of the pentagram versus Holy Water shaped like a hexagon (or a Merkaba when paired to balance water's positive and negative electromagnetic charges).

With AI that operates by mathematics modeled by neuroscience, "[W]e are summoning the demons," Musk admitted.

Indeed, as I made known in my popular "Musical Cult Control"[27] article, devil-doers have been broadcasting the 'devil's tone' used nearly exclusively, all the time, by the music industry monopoly, the

NBC/Universal Music Group. Their artists' concerts are produced by Liberty Media and Live Nation. These companies are linked to the Clinton/Bronfman/Diller ICA/Deep State cult controllers, the Las Vegas Deep State Massacre at the 96 Harvest Festival, and the Bataclan Theater Massacre in Paris.

Recently, on November 5, 2021, during the Travis Scott concert at the Astroworld Festival in Houston, produced likewise by Live Nation, many people reported a bizarre "satanic" energy overtaking the crowd, as though the music was summoning demons. The bad vibrations in the music occurred as Luciferian images and flames engulfed the stage. Secret agents moved through the crowd, injecting people with poisons. Eight people died, and 300 more required medical care for various ailments. Most reported feeling suffocated by the tightly-packed crowd.

Sergio Bertolucci, CERN's Director, stated that the hadron collider could open otherworldly doors to other dimensions. Many people believe these doors enable spiritual entities to access your soul or spirit, as in 'demonic possessions.' The doors permit angels or demons to access your energy field for better or worse, respectively.

"These things called 'doors' in the Bible are the 'star gates' to physical salvation, environmental purification, and spiritual evolution," I insisted during a past lecture. "Or otherwise, these doors could possibly invite eternal damnation if you encourage that with sin,"

These matters challenge the religious world pursuant to ethics, morals, values, and righteousness.

Considering what is unfolding geopolitically, economically, medically, and spiritually today, the only way out of this 'mess,' I conclude, is to surround yourself with the most powerful protection we have against the Deep State's dissonance. That is, by turning up the volume of LOVE. Tuning into that clear channel broadcast of "LOVE528Hz" connecting the Creator's loving heart to your loving heart. This doesn't encourage bioelectronic vaccinations, hallucinogenic brain conditioning, or demonic possessions. It encourages positive intention manifested miraculously through faithful loving prayer.

Lame Opposition to Religious Corruption Enables Satanism

In April 2019, The *Salt Lake [Mormon] Tribune* heralded the advancing Luciferian movement.[721] The article, written for the *Religious News*

Service by Menachem Wecker, complained about the U.S. Government granting tax free status to the Satanic Temple. "The taxman has given it the thumbs-up," Wecler wrote.

The Satanic Temple was featured in Lane's documentary, *Hail Satan?* She announced that the IRS would henceforth recognize this group as a legitimate "church." That recognition means the group can now get the same benefits as religious organizations —including tax exemption and protection from discrimination, Lane explained.

In a statement, Lucien Greaves, founder of the media-savvy group, said that "accepting religious tax-exemption —rather than renouncing it in protest— can help us to better assert our claims to equal access and exemption while laying to rest any suspicion that we don't meet the qualifications of a true religious organization."

"Satanism is here to stay," he added.

"Greaves spoke in front of the state Capitol in Little Rock, Ark., in August 2018, next to a statue of Baphomet," recalled Wecker. "The Satanic Temple wanted to install the statue as a symbol for religious freedom.

"A member of the Temple made a similar claim in the trailer for *Hail Satan?*, which details the nonprofit's attempts to put up a statue of Satan on public grounds.

"'I am following a code of ethics, having fellowship with brethren. Why can't I be a religion?' he said.

As *Hail Satan?* Director Lane told *Vox* she initially thought the film was about people mocking religion, only to learn it was about religion. "'I loved the idea of watching a new religion get born, right before our eyes, and how goofy and weird that looks, especially if you're not part of it, she told *Vox*."

Right-Wing Authoritarian Dark Transhumanism

"Of these musings, *New York Magazine* noted[722]:
"This is not an entirely abstract question. There are people who believe that the future of our species involves shedding our humanity in a marriage with AI [i.e., Artificial Intelligence]. This is known as *transhumanism*, and it has not unreasonably been called **a new tech religion**.

"Though the movement has no explicit political affiliations, it tends, for reasons that are probably self-explanatory, to draw a disproportionate number of Silicon Valley libertarians.

"And the cluster of ideas at its center — that the progress of technology will inevitably render good ol' Homo sapiens obsolete; that intelligence, pure computational power, is to be pursued above all other values — has exerted a powerful attraction on a small group of futurists whose extreme investment in techno-libertarianism has pushed them over an 'event horizon' into a form of right-wing authoritarianism it might be useful to regard as ***Dark Transhumanism***.

"The English critical theorist turned far-right cult thinker Nick Land is usefully representative of this intellectual tendency. Although he has never been identified as a transhumanist, his ideas are infused with the movement's delirious faith in the coming merger of humans and machines. His current political vision, which he has given the flamboyantly portentous title the '**Dark Enlightenment**,' is one in which the programmer elite and their ingenious technologies rule the world.

"'Increasingly,' [Nick Land] wrote in 2014, 'there are only two basic human types populating this planet. There are autistic nerds, who alone are capable of participating effectively in the advanced technological processes that characterize the emerging economy, and there is everybody else.'

"Many transhumanists would be inclined to reject the political implications of Land's futurism, but his vision is only really a darker, more explicitly fascistic rendering of the kind of thinking you find in the work of the futurist Ray Kurzweil, or for that matter *Wired* founder

Kevin Kelly, who believes that we humans are the reproductive organs of technology."

"As Land has it, through the acceleration of global capitalism the human will be dissolved in a technological apotheosis, effectively experiencing a species-wide suicide as the ultimate stimulant head rush.

"Ah, there's nothing quite like an ideology with visions of humanity being completely exterminated or tortured for all eternity by the [AI] creations.

CHAPTER XV

WHAT WE MUST DO TO SAVE CIVILIZATION

Author Timothy Snyder,[723] a Yale University history professor, provided a 'liberal's' list of "Twenty Lessons in Fighting Tyranny from the Twentieth Century" published on the Scholars Strategy Network.[724] I will draw from, and comment on, these recommendations.

1. Do not be blindly submissive. DO NOT OBEY 'THE BEAST.'

Be a 'conscious' (enlightened) contentious objector in the face of evil and criminality.

Professor Snyder objects to "the power of authoritarianism" given away too freely by *We The People*. In times like these, he wrote, individuals fear what an increasingly repressive oligarchy will force you to do. People cave to the fear. A citizen who submits to "false evidence appearing real" (FEAR) gives more power to the oppressors and "is teaching power what it can do."

People are regularly damaged and die from their misplaced faith and trust.

So obey God, obey the Ten Commandments, obey the laws of universal construction, obey the Constitution, but resist tyranny by protesting against it.

2. Restore Honest and Valued Institutions.

The breadth of institutions and their influence on the country and culture is vast. Institutions can "help us to preserve decency," as Professor Snyder posits, but they can also be abused to degenerate society and civility.

One example is the "WOKE" movement that vicariously encourages personal racism while allegedly opposing systemic racism. Institutions of higher learning and governmental affairs have fallen victim to this globalist agenda intended to destroy America by causing confusion and divisions throughout the population.

Attending board meetings or other official proceedings and being vocal in defense of God, country, justice, ethics, morals, and values contributes to actualizing needed remedies.

Assume as much ownership or stewardship of institutions as you can, and act in defense of righteousness therein. As Prof. Snyder advises, "institutions help us to preserve decency." They need your help. "So choose an institution you care about," and participate in its stewardship.

3. Vote and Campaign for Viable Independents.

The "two-party" system in America has failed as intended by the global elite. Graft, bribery, and self-serving 'campaign financing' provisions secured the destruction of fair and honest politics. In other words, all career politicians have sold out, and they need to be replaced.

I previously mentioned the American Legislative Exchange Council (ALEC) and their legislative "bill mill." This globalist 'front's' influence in politics is overwhelming and tragic —absolutely debilitating. Under ALEC's influence, widespread control over lawmaking always favors the global elite and genocide.

Campaign financing of candidates in both parties pushing mandatory vaccines is one example.

A great resource to become more aware of ALEC is Bill Moyers's website titled "The United States of ALEC." [725]

Support independent candidates that reject ALEC's influence and help educate others about the issues raised in their platforms.

4. "Take responsibility for the face of the world."

"The symbols of today enable the reality of tomorrow." Snyder advises to "notice the swastikas and other signs of hate. Do not look away, and do not get used to them. Remove them yourself and set an example for others to do so." This also applies to left-wing radical prejudicial slogans defacing properties. It's one thing to oppose racism, even legally advertise and assert that "Black Lives Matter," but it is shameful to damage or deface public or private properties with this affirmation.

Why?

Because this symbol of activism opposing systemic racism does worse than illegally defacing aesthetic environments. *It reinforces victimization.* It reminds viewers of class warfare, systemic racism, and deprivation of equal rights but does nothing to forge solutions.

This graffiti inspires divisiveness, racist thoughts, negative attitudes towards whites and people of color and encourages more illegal, dangerous, damaging, and deadly behaviors.

This illegal degenerative activity is precisely what the global elite worked to incite in America and elsewhere. Globalists such as George Soros and Bill Gates largely financed the main groups ignorantly committing and modeling this anti-social behavior. These activists ignorantly neglect the conflicting financial and geopolitical interests encouraging this dysfunction. This deadly form of 'activism' is paired with rioting, arson, and looting by the images and stories we witness on television.

It is unreasonable to presume this destructive form of activism will result in systemic institutionalized 'equity,' equality, and mutual respect for human life.

5. Target Judicial Corruption.

There are three branches of government. The Judicial Branch is supposed to function to adjudicate fairly questions of law and fact.

Political leaders may be setting negative examples for honesty and integrity, but 'where the rubber meets the road' is in courts of law, wherein judicial corruption is systemic. Here, victims are commonly damaged, deprived, and die. Criminals, alternatively, are exonerated.

The 'good-ole boy' network and nepotism are endemic to the Justice Department and the state and federal courts. As Snyder objects to this, "it is hard to subvert a rule-of-law state without lawyers, or to hold [kangaroo court] trials without judges."

Accordingly, the entire judicial system needs to be protested and revamped.

JudicialCorruptionNews.com provides several examples of activists' articles written to expose damaging and deadly Judicial Corruption. Writing and publishing articles exposing the devil-doers is not only your First Amendment right but your 'public duty.'

Flooding courtroom galleries with victims' advocates and witnesses to record and report malfeasance can help clean up courts and help communities gain more justice.

Staging protests outside courts across America would be one of the most beneficial forms of activism.

6. Be wary of cult-like paramilitaries and join your militia.

Under 10 USC Ch. 12: THE MILITIA,[726] the law encourages our civilian development of a "class" of armed protectors described as "members of the militia who are not members of the National Guard or the Naval Militia."

Between 1996 and 2010, I hung out with many members of state militias. I was a 'regular' featured speaker at "Preparedness Conferences" and "Patriot Community" meetings across North America. I spoke on "Emerging Viruses" and the risk of lab mutations outbreaking to create where we are today, in the COVID crisis.

During that period, I befriended a wonderful man and most honorable leader in the Patriot Community, Jack McLamb. Jack favored militias as a retired Phoenix police officer and organizer of the "Police Against the New World Order." Jack and his close friend, Col. Bo Gritz, were two militia advocates who supported Randy Weaver and me. I cherished Jack and Bo's friendship and honored their efforts to establish a sustainable Christian patriot community near Kamiah, Idaho, anticipating the currently manifesting and deadly New World Order.

For many years I continued to be warmly accepted as a Jewish doctor, avid researcher, and shocking whistleblower by Jack, Bo, and most people in the Patriot community. Then agents for the FBI and CIA targeted us to destroy our reputability.

Gritz was America's most decorated war hero —the real-life Rhambo. He honored me by writing the testimonial in the screenshot after attending my lectures and reading my book, *Emerging Viruses: AIDS & Ebola—Nature, Accident or Intentional?*

In all the years I enjoyed our friendships, I never felt any anti-Semitism expressed by Jack or Bo. Nor did I ever hear any racist or white supremacist words from their lips, not during their lectures or at any time I spent with them socially.

Contrast my experience with the propaganda issued by the manipulative media. Jack, Bo, and others in the COINTELPRO-infiltrated Patriot Community were alleged to be "neo-Nazi types" and "white supremacists." And the militias they honorably served were falsely disparaged as "radical" "right-wing" threats to society. This is how we were "neutralized."

Southern Poverty Law Center (SPLC) cites Bo and Jack among the "false patriots" and "conspiracy theorists" who "played pivotal roles in the antigovernment 'Patriot' movement" and provided "a timeline recapitulating the history of modern civilian militias..."[727] This 2001 SPLC "Intelligence Report"[728] listed 40 men and women who illustrated "the changing shape of the radical right." Curiously, I was not listed therein, even though I contributed most to the Patriot movement's grasp of the risk of government-manufactured and deployed biological weapons.

I suppose the fact that I am Jewish, and have multiple post-doctoral degrees and scientific publications to my credit, did not fit the Deep State's stereotyping of Patriots as right-wing radicals and white supremacists.

Later, in 2010-13, my co-investigator, Sherri Kane, exposed key FBI and CIA operatives tainted the list of honorable Patriots listed by the SPLC.[728] These infiltrators included the FBI's Division Five, Los Angeles, Chief, directly under J. Edgar Hoover, Ted Gunderson.[729]

Prof. Snyder wrote, "When the men with guns who have always claimed to be against the system start wearing uniforms and marching around with torches and pictures of a Leader, the end is nigh."

I revise Snyder's advisement. When shooters with guns raise them against a fascist dictatorship, uniformed or not, they are heroes to the cause of freedom.

7. Defend the Second Amendment and the Sanctity of Your Body.

Professor Snyder wrote, "If you carry a weapon in public service, God bless you and keep you. But know that evils of the past involved policemen and soldiers finding themselves, one day, doing irregular things. Be ready to say no." The "irregular things" we now witness, including mandatory intoxications called 'vaccinations' threaten public health, safety, and well-being. Meet uniformed people with guns enforcing un-constitutional and enslaving impositions with civilian groups with guns defending the freedom of choice, religious convictions, and solid science contraindicating such poisonings.

8. Be a leader by reaching out to leaders.

Shadow governors and their propaganda mills are especially targeting community leaders to encourage vaccinations. Leaders influence others. Counter the falsehoods of vaccination safety and efficacy by contacting, informing, and persuading leaders who influence audiences. Relay accurate opposing arguments including real science, dispelling 'pseudo-science' published by liars like doctors Andersen and Garry, fueling fakers like Fauci, as detailed previously.

Remember Rosa Parks. The moment you set an example, the spell of the status quo is broken, and others will follow.

9. Unplug your brain from mass media.

"Make an effort to separate yourself from the internet," Snyder advises. "Read books."

I largely agree, but I am thankful for Internet search engines that save me from driving to medical libraries and spending hours reading and researching to find important evidence available to me in seconds with the click of my mouse.

WHAT WE MUST DO TO SAVE CIVILIZATION

It is most important to learn about the conflicting financial and political interests tainting and discrediting the mass media and government agents, such as Anthony Fauci, promoted by propaganda.

It is essential to ask people, especially 'skeptics' and naysayers, "HOW DO YOU KNOW WHAT YOU KNOW?"

Ask your adversaries to reveal what intelligence source they used to acquire, in this case, lifesaving versus life-threatening knowledge? Having determined the source of misinformation or disinformation is biased, you can leverage that knowledge as an activist.

One of several examples I published exposed *Snopes* as a "'Deep State' Protection Racket."[292] I published this online at *WarOnWeThePeople.com*, and many others spread it from there.

Snopes misrepresented my work to discredit both. So, I not only discredited *Snopes* and its propaganda, but I relayed accurate life-saving messages to readers.

Below is what I wrote to encourage a boycott of *Snopes*. Feel free to copy and republish any part of this to oppose ignorance enabling the vaccinecide:

SNOPES' fraud is royally exposed here for fabricating false stories published to protect the vaccine industry from liability for genocide.[730]

In this classic example of fake news, Dr. Henry Kissinger and his cohorts are protected by SNOPES's fraud.' The purpose of *SNOPES's* fraud here was to generate confusion to conceal, obfuscate, and confound readers about Kissinger's important role in aiding-and-abetting worldwide 'injecticide.'

Vaccination genocide has been permitted to advance justified by such false and misleading propaganda. Vaccination genocide has been recklessly evaded for the same reasons vaccines are promoted as "safe" and "effective."

I have repeatedly and correctly identified vaccinations as the "**sacred cow of public health**." No one dare say anything bad about them. "Immunizations" are justified, administered, and rationalized under this imposed censorship and false presumption that vaccines do more good for 'public health' and 'disease prevention' than harm.

That is false. Vaccines are not good for public health. They are GREAT for *profitable depopulation*. By publishing false safety and efficacy claims, *SNOPES* is aiding-and-abetting by willful blindness *genocide*.

SNOPES demonstrates a counter-intelligence scheme that evidences "protection racketeering,' fraudulently concealing that vaccines have killed millions, soon to be billions of people. I should know, I have been abused, stalked, harassed, and damaged by COINTELPRO for years. *SNOPES* abused me and smeared my good name to commit their fraud, consistent with this globalist protection racket.

Evidence for *SNOPES*' fraud is irrefutable in the graphic of the next page. The screenshot shows their publication of "Henry Kissinger '... Forcible Vaccinations'" with my comments and analysis.

SNOPES falsely published that I sourced their disinformation. You can read their false attribution at the bottom of the left side panel. I constructed this graphic to evidence *SNOPES*' fraud and assess it.

WHAT WE MUST DO TO SAVE CIVILIZATION

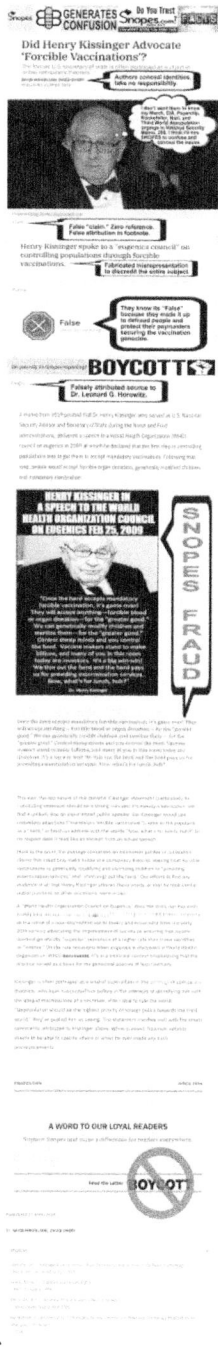

Analysis of Snopes' Fraud

SNOPES's authors, "David Mikkelson" and "David Emery," may not even exist. If they exist, they conceal their identities.

Facebook shows no images of Mikkelson. And several David Emerys on Facebook are not linked to *SNOPES*.

Curiously, *SNOPES* used this name "Dave Emery," probably to detract from the fine work journalist-activist 'Dave Emery' did with me years ago at Berkeley. Dave Emery and I, Dr. Alan Cantwell, Dr. Garth Nicholson, and others exposed the lab virus creations of HIV/AIDS, Ebola, and much more in our work titled, *VIRUS MAKERS OF THE CIA*. That "BIOWAR FORUM" was produced by Dave Emery at Berkeley in-or-about 1997 and published by Tetrahedron, Inc. It can still be purchased through *CureShoppe.com* in a package titled: *The Best of Early Horowitz*. But this ground-breaking group contribution was wickedly suppressed by the corporate-controlled media.

The recorded history of lab virus developments in the vaccine industry by bioweapons contractors, especially Merck, was coupled with other classic recordings that pre-dated the "conspiracy industry."

SNOPES was not around then to abuse public perceptions with online counterintelligence when I published *The Nazi American BioMedical BioWarfare Connection* with Dave Emery in the late 90s. It is subtitled "*Rockefeller, Kissinger, Bush and the CIA.*"

"*SNOPES*' fraud makes its authors and publisher liable for mass murder, so this *SNOPES'* fraud was a direct 'hit' against me, my good name, and my heroic colleagues. We bravely stepped out, to tell the truth that others were too afraid to tell. Our academic friends and fellow clinicians were worried about losing their careers, livelihoods, families, etc. But we risked everything to save whoever we could with the truth regardless of the threats.

I pioneered this subject[731] of linking Dr. Henry Kissinger to genocide studies. Since that time,[66] many others have bravely multiplied my efforts. But only devil-doers in counterintelligence published to discredit me as *SNOPES* attempted here.

For the record, I point out (in the inset above) how *SNOPES'* fraud operates. Their agents take any subject they wish to "muddle" and fabricate questions and misrepresentations, often completely out of context with the actual statements made by their sources.

Also, for the record, *SNOPES*'s selection of my publication on *Rense.com* providing the falsely alleged "source" of their false statements compounds the evidence I published regarding Jeff Rense and his *Rense.com* website,[733] being a despicable clearing-house for counter-intelligence, racist, anti-Semitic, and anti-government "CONTROLLED OPPOSITION" publications.[729]

I learned this about Jeff Rense a little too late. Years ago, I learned that Rense was being 'handled' by Alma C. Ott[734] and Ted Gunderson, certified counter-intelligence agents exposed.[735,736,737]

So it's not surprising *SNOPES* would select that one article on *Rense.com*, among hundreds of alternative publications and sources covering my work exposing the Kissinger-Rockefeller-Bush connections to the German Merck family, Big Pharma, the Nazis, the eugenics movement that began at Cold Spring Harbor Labs on Long Island, NY and the hepatitis B vaccine that delivered HIV/AIDS to the world.[738]

SNOPES should be investigated by legitimate 'law enforcers' for complicity in aiding-and-abetting the increased morbidity and mortality among vaccine recipients. That is, *SNOPES*' officials must be indicted by a grand jury for racketeering in organized crime and treason against the people of the U.S. and abroad.

Since that is unlikely to happen, my recommendation is to **BOYCOTT *SNOPES***. Put them out of business. That might send a message to their cohorts in crime that *We The People* are done financing their fraud.

confuse [kuhn-fyooz]

bewilder someone

Synonyms for confuse:

amaze	distract	perplex	cloud	shame
astonish	embarrass	perturb	clutter	stump
baffle	fluster	puzzle	darken	unhinge
befuddle	frustrate	trouble	discomfit	lead astray
bemuse	involve	unsettle	discompose	mess up
complicate	misinform	upset	discountenance	render uncertain
confound	mislead	worry	faze	stir up
daze	mortify	abash	fog	throw off
demoralize	muddle	addle	fuddle	throw off balance
disconcert	mystify	becloud	nonplus	
disorient	obscure	bedevil	rattle	

MOST RELEVANT

10. Believe in truth.

Prof. Snyder encourages activism opposing the main objective of the tainted media, that is, to create confusion for population manipulation and victory over We The People.

"To abandon facts is to abandon freedom," Snyder wrote. "If nothing is true, then no one can criticize power because there is no basis upon which to do so. If nothing is true, then all is a spectacle. The biggest wallet pays for the most blinding lights."

This is precisely the urgency of encouraging remedial activism. With the "fake news" media and Big Tech powers censoring truth, concealing vital facts, and modeling riots, crimes, and chaos, you are wise to research and ally with alternative news sources and activist organizations. Seek out and find 'kindred spirits' for social support in gathering and telling the truth, making sure your sources are sound.

11. Investigate.

Prof. Snyder encourages the aforementioned thusly:

"Figure things out for yourself. Spend more time with long articles. Subsidize investigative journalism by subscribing to print media. Realize that some of what is on the internet is there to harm you. Learn about sites that investigate propaganda campaigns (some of which come from abroad). Take responsibility for what you communicate to others."

My research and publications encourage the same. For research, I read and consider both sides of every story. The left and right statements evidence the 'whole picture' that always shields and implicates the globalists. You get a clearer understanding of how divisiveness is practiced in the media to confuse and control society by reading both sides of every issue.

12. Improve Your Discernment and Communication Skills.

How can you lead a revolution without connecting with people?

Prof. Snyder encourages making eye contact and even making "small talk" to establish more and better relationships.

This "is part of being a citizen and a responsible member of society. It is also a way to stay in touch with your surroundings, break down social barriers, and understand whom you should and should not trust," Snyder advises.

Body language and making eye contact are extremely important in discerning liars and cheats. Look for, and honor, the "catch-in-your-spirit" when speaking with someone untrustworthy or closed-minded. You can intuit this discernment.

Don't "throw [your[pearls before swine." Closed-minded adversaries are too often 'draining." Move on. Their karma is sealed.

To establish a better rapport with people, "match-and-mirror" their gestures, tone, pace, and overall 'energy.' This will produce benefits, especially when you interact with community leaders.

Prof. Snyder warns that as you "enter a culture of denunciation," you will want to know the psychological and social underpinnings of your adversaries. Don't get caught up in their ignorance and drama. "Forgive them, Father, they know not what they do," Jesus prayed.

13. Find and March with Groups Defending Justice.

Prof. Snyder calls this "practicing corporeal politics." He correctly advises that the Deep State power-mongers want your body. They also seek to control your emotions. "Get outside" their box, Snyder advises. "Make new friends and march with them" to oppose the ongoing genocide and totalitarianism.

14. Secure your identity and private life.

"Nastier rulers will use what they know about you to push you around," warns Snyder. So, "Scrub your computer of malware. Remember that email is skywriting. Consider using alternative forms of the Internet or simply using it less. Have personal exchanges in person..."

15. Contribute to good causes.

Attend and participate in town-hall meetings. Co-sign and distribute protest letters to politicians and the media. Be more active in organizations that express your interests. Contribute to a charity or two. Set up autopay. "Paying forward" in this way will surely return blessings.

16. Learn from peers in other countries.

Provincial thinking is deadly in this Age of Globalization. Establish friendships abroad. Make new friends overseas. Discuss the difficulties you address in the United States and how they intertwine with what is happening globally. "No country is going to find a solution by itself," Prof. Snyder asserts. "Make sure you and your family have passports."

17. Watch for diversionary rhetoric.

Pay attention to media personalities who don't answer important questions. Observe that they divert to a talking point their agenda requires. Worse than a sign of incompetence, this is evidence of complicity in cover-up and genocide. Address this head-on if you have the opportunity. "I asked you a precise question, and you diverted. Why?"

18. "Be calm when the unthinkable arrives."

Modern tyranny requires terror and population management. "When the terrorist attack comes, remember that authoritarians exploit such events in order to consolidate power. *Do not fall for it,"* he warns.

19. Assert your leadership in public demonstrations and courts.

Organize protests, rallies, conferences, and lectures. Litigate when your human and financial resources permit, and most importantly, publicize your lawsuits to expose your adversaries' devil-doing.

20. Boldly publish your activism and resistance to tyranny.

Be bold in publishing and promoting your activism. "If none of us is prepared to die for freedom, then all of us will die under tyranny," Prof. Snyder warns. Your publications can include articles, letters, videos, audio recordings, radio show call-ins, and visits to legislators.

SUMMARY & CONCLUSION

The details, evidence, and analyses presented in this book expose the source and administration of the SARS-CoV-HIV mutant lab virus developed and deployed to actualize depopulation, transhumanism, and One World Government.

A 'Deep State' racketeering enterprise organized by the world's wealthiest elite has advanced to complete a centuries-old scheme to control the world. This is now possible and unfolding, geopolitically and economically, using advanced technologies, including sophisticated media propaganda and military MindWar protocols.

The advanced technologies now available include nano-bioelectronic devices capable of brain-Cloud interactions and genetic reprogramming using biological weapons developed to reduce and control populations most insidiously, profitably, and effectively.

The COVID Coup was set into motion in late 2020 upon releasing a lab virus and commercializing related mRNA vaccines derived largely from early AIDS research. These biotechnologies feature a unique attachment "spike protein" mechanism that operates bioelectrically with the virus and vaccines working similarly.

The "novel" vaccines employ a nano-metallic bioelectronic hydrogel device that presumably prompts immunity against the COVID-19 virus by subverting DNA, resulting in the production of additional spike protein antigens. These proteins prompt antibodies to be produced, presumably heightening immunity against the infection.

Research and developments central to this biotechnology and globalization scheme were conducted in the U.S., China, and the U.K., with Harvard and MIT labs playing significant roles.

The identities of the agents and agencies financing and administering these proceedings are provided with sufficient evidence to establish probable cause for their criminal indictments.

The personal identities of the 'insiders' falsifying publications and confounding intelligence criminally engaged in a cover-up are provided.

Dr. Anthony Fauci's and Bill Gates's "inner circle" of complicit co-conspirators reveals substantial evidence of the globalists' plot to administer genocide under the guises of "public health" and "National Security."

In the United States, the 'inside traitors' and their allied institutions comprise a "National Security Crime Syndicate" that includes the NIH, NIAID, FDA, CDC, WHO, DARPA, and the academic institutions spinning off for-profit companies from non-profit tax-payer financed programs.

A "National Security 'Crime' Syndicate" is identified serving the 'World Disease Enterprise' exploiting the governments and militaries of China and the United States responsible for the pandemic and its deadly and damaging impacts.

"Public health" and regulatory agencies are evidenced having been 'captured' by this globalist cartel to do evil rather than good.

Consideration is given to the ethical, moral, and religious concerns raised by artificial intelligence (AI) and transhumanism enabled by vaccination nano-bioelectronics and the brain-Cloud interface.

As this "New World Order" advances to repopulate the planet with transhumans, spirituality and 'consciousness' of 'enlightened' people and 'conscientious objectors' are called into question, as is the likelihood of civilization's extinction.

Evidence of Luciferian influence in AI administering transhumans and the "Dark Enlightenment" is considered.

Strategies for activists opposing this Brave New World and its deadly and dehumanizing impositions are provided, encouraging the prosecution and conviction of the agents and agencies accountable for committing genocide and treason.

In the interest of justice, spirituality, human harmony, and sustainability, officials responsible for the COVID pandemic and Coup must be held accountable for their crimes.

ABOUT THE AUTHOR

Dr. Leonard Horowitz, D.M.D., M.A., M.P.H., D.N.M. (hon.), D.M.M. (hon.), is the victim, witness, and whistleblower in the crimes reported in this book.

He is the author of twenty-two books, including the prophetically titled _Death In The Air: Globalism, Terrorism and Toxic Warfare_ that came out three months before 9-11-01. That book dealt with biological and energy weapons used for population control. His three American best-sellers include _Emerging Viruses: AIDS & Ebola–Nature, Accident or Intentional?_ that was largely responsible for prompting explosive interest in vaccination risks and biological warfare; _Healing Codes for the Biological Apocalypse_, which revealed the ancient Solfeggio musical scale secreted for millennia; and _Healing Celebrations: Miraculous Recoveries Through Ancient Scripture, Natural Medicine and Modern Science_ in which Dr. Horowitz presents his protocol for administering prevention and speedy recoveries. His most recent text, **_The Book of 528: Prosperity Key of LOVE_**, has prompted a revolution in the music industry improving recording artistry and music therapy using C=528Hz(A=444Hz) tuning that produces "medicinal music." Dr. Horowitz's documentary, **_UN-VAXXED: A Docu-commentary for Robert De Niro,_** won five international awards, including "Best Film – 2016" in London and Geneva competitions.

E-mail inquiries to Editor@MedicalVeritas.org.

Appendix 1

THE HOROWITZ COVID PROTOCOL FOR PERSONAL HEALTH & IMMUNITY

Neutralizing the bio weaponized antigen must be the primary objective in defending against the COVID imposition, suffocation, and its damage and death. Once intoxicated and damaged by the spike-protein host-cell antigenic complex immune response, two primary problems present clinically: (1) cell-mediated immunity causing histamine release prompts smooth muscle contractions in blood vessels causing asthma-like respiratory symptoms; and (2) the irritation and AhR impact of the spike protein on the cilia and mucus-producing cells prompts: breathing difficulty associated with abnormal mucous production. The offending mucous is clear but extraordinarily sticky. Damage to cilia has been reported, restricting the removal of these antigen-antibody complexes. This 'sticky goo' restricts oxygen flow to your bloodstream. All of the above results in asthma-like symptoms, ARDS, suffocation, hospital intubation, and death occur much too often.

Sedating this synthetic antigenic over-stimulation of cell-mediated immunity in the human body, normalizing its histamine/antihistamine functions, and mucus reduction through good hydration, exercising your breathing, and expectoration of as much mucous as you can remove are keys to survival and rehab.

Let me introduce you to my successful protocol for quickly recovering from pneumonia or COVID-related 'ARDS' —acute respiratory distress syndrome.

A. Natural Remedies Available to Help

How do you remove what feels like a hundred-pound weight on your chest when you are trying to breathe?

This advisement supplements my many previous articles in which I explained the urgency of good pH-adjusted hydration and "anti-oxidant therapies,' neutralizing positively-charged synthetic spike-protein antigens as best you can. I recommended antioxidant Vitamins C; D; E (d-alpha tocopherol); Zinc; and many other scientifically proven natural remedies.[739]

I've urged everyone to stockpile OxySilver w/528Hz frequency. Used as directed, OxySilverTM provides a "quadruple anti-oxidant." (I've explained this repeatedly in the free subscription HealthyWorldStore.com newsletter.) I further characterize OxySilver's antioxidant and broad-spectrum antibiotic function below. In my lawsuit against Pfizer and Moderna, I claim that safe and effective OxySilver competes directly, bio-electrically, against their risky mRNA vaccines.

Adding the anti-stress B vitamins and CoQ10–another natural antioxidant is helpful, as is consuming lots of greens containing chlorophyll (e.g., wheatgrass juice). Reducing acidifying lifestyle risks is very important.

Caffeine, nicotine, sugar (refined carbohydrates), stress, alcohol, most pharmaceuticals, and excessive red meat should be avoided because they acidify your body chemistry—this electro-chemistry aids-and-abets COVID morbidity and mortality.

An excellent science paper reviewing several "bitter" natural medicines (i.e., negatively-charged alkalizing anti-oxidants) encourages doctors and patients to relieve COVID's respiratory distress using the alternatives published.[740] A review of the "antiviral performance of graphene-based materials with emphasis on COVID-19"[741] by Seifi and Kamali discusses the anti-viral bio-electronic impacts of silver nanoparticles, such as those provided by OxySilver™ resonating in 528Hz.

a. Nutritional Antihistamines to Help Breathing

In review, for your information, Vitamin C acts as a powerful natural antihistamine.

"According to a 2018 study on vitamin C in the treatment of allergies,[742] oxidative stress plays a key role in allergic diseases... Another study from 2000[743] suggested taking 2 grams (g) of vitamin C daily to act as an antihistamine... As vitamin C is a powerful antioxidant and anti-inflammatory, it may act as a treatment for allergies." Now add the asthmatic symptoms associated with smooth muscle contractions from spike-protein antigen-prompted histamine releases. These, too, maybe naturally remedied by vitamin C.[744]

Quercetin is another natural antioxidant found in foods and over-the-counter nutritional supplements. This flavonoid may help relieve some of the symptoms of histamine release and airway inflammation.

Fish oils rich in omega-3 fatty acids may also help reduce lung inflammation and relieve asthma-like symptoms.

b. Anti-Asthmatic Botanicals

According to many open sources,[745] natural herbs that may help treat asthma-like symptoms include:

- ginkgo, shown to reduce inflammation[746]
- mullein
- Boswellia (Indian frankincense)
- dried ivy
- butterbur
- black seed
- choline
- French maritime pine bark extract

Many sources report that coffee and tea may also be useful for treating asthmatic symptoms. Unfortunately, these consumables tend to be *acidifying*. That is the opposite of what you want to accomplish to neutralize the positive charge on the spike-protein antigen.

c. Essential Oil Antihistamines to Help Breathing

(i) Essential oils that act as antihistamines include:

- Lavender
- Eucalyptus
- Rosemary
- Tee Tree Oil
- Oil of Oregano
- Lemon
- Peppermint
- Sandalwood, and
- Camphor

I have been using these daily.

I made a mist-spray formula combining a few drops of these ingredients in OxySilver. One to three times daily, I use this formula in three ways:
1) I close my eyes and spray this mist around my head. I find this refreshing and relaxing;
2) I inhale this mist as deeply as possible using a *facial steamer*. I find facial-steamer-misting anti-histaminic essential oils are one of the most helpful things I have done. Anyone with asthma, or COVID-associated respiratory distress, can do this too to help themselves recover; and
3) I spray this formula in my steamy bathroom while bathing or showering. This is pleasant and has helped me in my recovery.

(ii)Tracking and Recording Your Progress

I encourage you to track, document and confirm your progress–your positive results in freeing your breathing using the above nutritional, lifestyle, essential oil applications. Especially the facial steamer treatment. I found that most beneficial, albeit in the beginning, it was difficult.

At the beginning of your recovery program, you will likely feel EXHAUSTED. Oxygen deprivation has you in bed or sitting most often. (I have often used my sitting time for prayer and reflection.) I have enjoyed listening to *528Radio.com* broadcasting 528Hz-transposed "medicinal music" to pass the time.

While you are sitting, you can also exercise, expanding your lungs. Do this as often as you like, but certainly five to ten minutes three times daily.

For this, I recommend that you get an instrument that measures your respiratory volume, such as the AirLife™ Spirometer device commonly dispensed to COVID patients by hospitals.[747]

HERE IS A GOOD EXERCISE & TEST: Before and after you inhale the essential oil OxySilver™ steam, measure and record your milliliters of air using this AirLife™ Spirometer (or other[748]) device. Try to breathe in the steam as deeply as you can without coughing too much. Stretch your 'comfort zone' during this exercise. Then sit quietly for five to ten minutes and observe your breathing capability and inspiration capacity. I have experienced and recorded benefits. So will you.

Appendix 1 - The Horowitz COVID Protocol: Tips for Personal Health & Immunity

I pushed my essential oil steam inspirations to the limit on several occasions. At first, this caused me to cough so intensely I almost threw up from coughing up small amounts of that sticky clear mucus. So I recommend, if you do this exercise, have a vomit bag or other receptacle close to you in case you 'heave.' I never did, but I came close to it.

Subsequently, I found this essential oil steaming exercise more tolerable because less mucus was in my lungs. I was able to breathe deeper and coughed less and less as days and weeks passed.

Do not think negatively that if you cough severely, this exercise is harming you. God's physical body design (form) and function are very hardy. Sit quietly if coughing excessively. Your cough will subside.

Then you will likely be pleasantly surprised that you can 'miraculously' breathe deeper than you could before your essential oil facia-steaming exercise and coughing spell.

(iii) Devil's Club Natural Expectorant Combined with Antioxidants

I've been toying with the idea of adding menthol and the expectorant Devil's Club to my OxySilver/Essential Oil formula.

Devil's Club is very helpful as an expectorant to cough out the sticky mucous. HealthyWorldStore.com has a limited supply of a *Flu Free Formula* that contains Devil's Club for this purpose. It is costly but well worth the expense to gain the relief this tincture of extracts provides.

I have been using OxySilver™ and the *Flu Free Formula* three (3) times daily with prayer and facial steaming at least once daily. I know this routine has helped me tremendously.

Corroboration comes from another interesting study.[749] This indicated that menthol might be added to inhalants to "decrease intracellular free Ca^{*+} concentration through an inhibition of voltage-dependent $Ca^{*'}$ channels..." Decreasing free $Ca+$ ions in the respiratory tract, thereby *increasing* negatively-charged alkalizing anions therein (e.g., anti-oxidants), may theoretically help neutralize the positively-charged COVID antigen delivery mechanism causing the mucous production, smooth-muscle contractions, and asthmatic symptoms.

d. Other Self-help Activities and Supplements

i. Remember Good Hydration!

Always remember that good alkalizing-hydration goes hand-in-hand with good oxygenation to defeat the positively-charged antigenic spike protein bioweapon, respiratory distress, and sticky mucus.

Remember to drink at least half of your body weight converted to ounces in pH 8.0-9.5 drinking water daily. (I've been drinking mostly 9.5pH *Essentia* water available in food stores.) The good fluid movement will help expel the mucus suffocating your alveoli, depriving you of oxygen, and making you feel exhausted after simple tasks or exercises. This feeling will pass with time and requires physical EXERCISE.

ii. Positive Attitude Toward Physical Exercise

Think as positively as you can about your recovery activities, especially *exercising*. Consider all your activities, wherever you walk or whatever you lift, as "exercise for recovery and rehab." You are exercising yourself back to optimal health!

As you go through your day, consciously expand your lungs during breathing as much as comfortably possible. These inspirations are *breathing exercise*s. Physical exercises that make you breathe deeper are great, but you don't need to overdo it. Increase your stamina and strength slowly but surely.

You will be astonished, after a couple of weeks of exercising at least 20 minutes daily, that stairway that you almost passed out on will seem like no problem at all.

iii. Avoiding Stress and Distress

Try to avoid psychological stress. Distress can be a killer, especially compounding damage from spike-protein antigen intoxication inflammatory reactions—psychological stress and emotional distress precipitate cytokine releases. An analysis of 34 studies found increases in circulating IL-6, IL1B, IL-10, interferon, and more in response to mental and emotional distress.

Please keep this advisement in mind when watching television and following the disturbing "news."

Appendix 1 - The Horowitz COVID Protocol: Tips for Personal Health & Immunity

iv. Assert Your Will to Live

Most important is your will to live and your relationships with your loved ones and the All-Mighty.

v. Acquire and Thank Your "Patient Advocate(s)"

When I was in the hospital sustaining cruel and unusual punishment, with staff threatening to put me on a ventilator (from which I would not have likely survived), I had loved ones sneak me OxySilver with 528Hz; good meals; my 528-tuning fork; and a bottle of the FDA-condemned *Flu-Free Formula* I mentioned above. (I originally developed that product to defeat SARS).

I used the OxySilver™ my advocates brought me with prayer. (I developed this unique product to amplify 'prayer power with the special 528 "MIracle" frequency resonating at the heart of your heart, reflected in the structured-water 528 memory.[750]

My loved ones, my "patient advocates," prayed for me regularly and vigorously advocated for me as Dr. Ardis recommends.[751] They protected me against the malpractices of the doctors and respiratory technicians who threatened my death by needlessly putting me on a respirator.

We experienced a "miracle." Within hours of my two "patient advocates" (one a lawyer) threatening malpractice lawsuits against the hospital devil-doers if I was not treated better, SUDDENLY, MIRACULOUSLY, DRAMATICALLY, my blood oxygen levels improved. In 24 hours, my oxygen readings went from the mid-60s to the high 80s. Then, my advocates DEMANDED I be released from the "death camp."

The staff (many in agreement with me about the inhumane environment imposed by their corporate paymasters) were forced to consent.

I am here now telling you my survival story, and recovery protocol, hoping you will benefit from my research and lessons learned.

B. SUMMARIZING "THE HOROWITZ COVID/ARDS PROTOCOL"

1) Start your day drinking 15-to-24 gulps of alkalized water and wait five to ten minutes before taking the nutritional supplements aforementioned.
2) Take your nutritional supplements (e.g., vitamins, minerals, enzymes, probiotics, etc.) with 12-to-16 ounces of grapefruit or orange juice (half mixed with prune juice to remedy constitutional difficulties) to help you swallow the pills.
3) Three times daily, beginning in the morning, before or after breakfast, take 3 capfulls of OxySilver™ with your eyes closed to visualize the white light of the Holy Spirit's resonance energizing your head and body. (The Holy Spirit is very 528-frequency friendly.) Imagine sending this positive healing energy down to help clear your lungs. One suggested prayer is:

"Heavenly Creator, thank you for sending your Holy Spirit of Love and miraculous healing energy to quickly and completely restore my breathing capacity; so that I can serve you to the best of my ability."

4) Three times daily, take one dropper-full of the *Flu Free Formula* or other product(s) containing Devil's Club expectorant.
5) Throughout your day, exercise your deeper breathing capability. Stretch your lungs with deeper inspirations.
6) Do physical exercise at least 20 minutes daily. Rehab your whole-body muscle tone for at least 20 minutes (and throughout your increasingly active day). *Build stamina* this way to help you hasten recovery. Always practice proper posture–shoulders back, chest expanded–at all times.
7) One to three times daily, enjoy essential oil antihistamine respiratory therapy, especially using your facial steamer. Record your breathing improvements using an AirLife™ Spirometer.
8) Supplement the above with a Rife frequency treatment, ideally three times a week. There are many such machines being sold. They commonly flow (negatively-charged) electrons of energy through your body for enhanced anti-oxidant activity. Using certain frequencies recommended by Rife practitioners may be most helpful.

Appendix 1 - The Horowitz COVID Protocol: Tips for Personal Health & Immunity

9) Get plenty of bed rest, as your fatigue, weakness, and recovery demands. And don't forget to hydrate. At least half your body weight in alkalized ounces daily.

References and Notes

[1] https://www.youtube.com/watch?v=SCGV1tNBoeU
[2] https://www.youtube.com/watch?app=desktop&v=kpW9JcWxKq0
[3] https://checkyourfact.com/2021/04/01/fact-check-bill-gates-klaus-schwab-covid-greater-control-mankind/
[4] https://www.linkedin.com/pulse/agile-todays-economy-fast-fish-eats-slow-dulce-casadiego
[5] https://www.usna.edu/NewsCenter/sites/Ethics/Dr._James_Giordano_Battlescape_Brain_Military_and_Intelligence_Use_of_Neurocognitive_Science.php
[6] https://www.smh.com.au/national/physicist-discovered-key-to-brain-science-20120607-1zy51.html
[7] https://medicalveritas.org/origin-of-aids/
[8] https://larouchepub.com/eiw/public/1997/eirv24n44-19971031/eirv24n44-19971031_018-origins_of_the_aids_virus_accide.pdf
[9] https://film.britishcouncil.org/resources/support-organisations/wellcome-trust
[10] https://medium.com/illumination/in-2030-youll-own-nothing-and-be-happy-about-it-abb2835bd3d1
[11] https://www.usna.edu/NewsCenter/sites/Ethics/Dr._James_Giordano_Battlescape_Brain_Military_and_Intelligence_Use_of_Neurocognitive_Science.php
[12] https://www.youtube.com/watch?v=b9vWShsmE20
[13] https://medicalveritas.org/wp-content/uploads/2020/02/Pradham-et-al-Coronavirus-HIV-paper.pdf
[14] https://medicalveritas.org/special-virus-cancer-program-must-reading-for-cancer-researchers/
[15] https://medicalveritas.org/origin-of-aids/
[16] https://medicalveritas.org/special-virus-cancer-program-must-reading-for-cancer-researchers/
[17] http://stanleykrippner.weebly.com/--bio.html
[18] https://www.waronwethepeople.com/wp-content/uploads/2014/08/STAR-GATE-KRIPPNER-CONTRACT-CIA-Files_-Parapsychology.pdf
[19] https://www.waronwethepeople.com/wp-content/uploads/2014/08/Terrance-McKenna-CIA-FOIA-Reply.pdf
[20] https://www.waronwethepeople.com/wp-content/uploads/2014/08/Star-Gate-Research-CIA-Full-Report.pdf
[21] https://www.archives.gov/iwg/declassified-records/rg-330-defense-secretary
[22] https://mwi.usna.edu/mwi-video-dr-charles-morgan-neurobiology-war/
[23] http://www.telegraph.co.uk/news/worldnews/europe/france/7415082/French-bread-spiked-with-LSD-in-CIA-experiment.html
[24] https://www.waronwethepeople.com/wp-content/uploads/2014/08/Princeton-Univ.-Document-Dulles-Rockefeller-CIA-Connection-Evidence.pdf
[25] http://www.aarclibrary.org/publib/church/rockcomm/html/Rockefeller_0119b.htm
[26] https://www.waronwethepeople.com/wp-content/uploads/2014/08/Rockefeller-Commission-Report-on-CIA-Drugs-and-Social-Control-Research.pdf
[27] https://medicalveritas.org/musical-cult-control/
[28] History of Legislative Control Over Opium, Cocaine, and Their Derivatives (druglibrary.org)
[29] http://jonestown.sdsu.edu/Jonestown_com/CIA.htm
[30] http://judicialcorruptionnews.com/senator-roz-baker/
[31] D:\Volumes\One Touch\BOOKS BY HOROWITZ \COVID COUP\o
[32] http://exposingvaccinegenocide.org/wp-content/uploads/2018/10/IV.-New-Bus.-A.-11-157-Exam-Immun.-SBIS-Rules.pdf
[33] Gardasil package insert is available for download from the FDA at: https://www.fda.gov/downloads/biologicsbloodvaccines/vaccines/approvedproducts/ucm111263.pdf

[34] Caulfield MJ, Shi L, et. al., Effect of alternative aluminum adjuvants on the absorption and immunogenicity of HPV16 L1 VLPs in mice. Hum Vaccin. 2007 Jul-Aug;3(4):139-45. Epub 2007 Apr 5.
[35] Horowitz LG. Special Virus Cancer Program. Medical Veritas online journal and press release. "Special Virus Cancer Program" ("SVCP") Concealed Contracts Evidencing Man-Made AIDS Pandemic Released in 20-Year Anniversary of Amazing Discovery." December 5, 2016. You can purchase two of these Reports in pdf file downloads. https://www.cureshoppe.com/special-virus-cancer-program-1972-the-virus-cancer-program-1977-2-pdf-downloads/
[36] http://www.thebodypro.com/content/art48915.html
[37] Press Release. Françoise Barré-Sinoussi and Luc Montagnier Win Nobel Prize for Discovery of HIV. From Nobelprize.org. October 6, 2008.
    https://www.nobelprize.org/prizes/about/the-nobel-assembly-at-karolinska-institutet/
[38] Roberts S. What AIDS Researcher Dr. Robert Gallo Did in Pursuit of the Nobel Prize.
    http://www.virusmyth.com/aids/hiv/srlabrat.htm. Since then, solid evidence has emerged from the Supreme Court's ruling on the "Daubert standard" concerning the trustworthiness of so-called experts. Society's 'general acceptance' of science is often flawed and untrustworthy. 'General acceptance' evolved as a legal "term of art" that holds special meaning in this book.
    According to the Daubert standard established by the U.S. Supreme Court in *Daubert v. Merrell Dow Pharmaceuticals, Inc.*, 509 U.S. 579 (1993), previously-concealed facts are admissible as evidence against the general acceptance standard.
[39] https://www.cambridge.org/core/journals/epidemiology-and-infection/article/covid19-pandemic-as-a-risk-factor-for-the-reactivation-of-herpes-viruses/834F3B34171BF09AC91F57DE9DAE161F
[40] U.S.G.A.O. Origin of the AIDS Virus. Briefing for the Office of the Honorable James A. Traficant, Jr. GAO-02-809R, June 17, 2002.
[41] NCI staff. The Special Virus Cancer Program: Progress Report #8. Office of the Associate Scientific Director for Viral Oncology (OASDVO). J.B. Moloney, Ed., Washington, D.C.: U.S. Government Printing Office, 1971.
[42] Interview of Dr. Maurice Hilleman by medical historian Edward Shorter's discussing Merck's vaccine-making, in which Hilleman stated, "We brought in the African greens . . . We didn't know we were importing AIDS virus at the time."
[43] Keim B. Did Merck bring AIDS to America? No. WIRED, September 19, 2007. This article is another classic example of "Big Pharmaflag." WIRED receives the vast majority of its advertising revenue from Big Biotech. This article exemplifies frivolous argument, diversionary focus, and recklessness in news reporting concealing serious risks to public health and safety from vaccines. Keim falsely claims "the off-camera Merck researchers laugh loudly, and someone quips, 'What Merck won't do to develop a vaccine.' following Hilleman's crucial admission that corrobrates the science. (44) In fact, the nervous laughter came from a WGBH crew, recording the interview for public television.
[44] Interview of Dr. Maurice Hilleman by medical historian Edward Shorter's discussing Merck's vaccine-making, in which Hilleman stated, "We brought in the African greens . . . We didn't know we were importing AIDS virus at the time."
[45] Association of State and Territorial Health Officials. Health Officials Alarmed by Declining U.S. Vaccination Rates, Country Could Face Scenario Like Europe's Measles Outbreak. https://www.astho.org/Press-Room/Health-Officials-Alarmed-by-Declining-US-Vaccination-Rates/10-25-18/?utm_source=Informz&utm_medium=email&utm_campaign=change+this+%28per+campaign%29 October 26, 2018.
[46] Kane S. Healthcare Holy War Waged by Hearst Challenges World Religions. August 30, 2016. https://www.waronwethepeople.com/holy-war/
[47] Horrobin, D.F. 1990. The philosophical basis of peer review and the suppression of innovation. J. Am. Med. Assoc. 263:1438–1441. Dr. Horrobin was terribly smeared by

END NOTES

critics online, including Wikipedia, following his publication of this author's following paper. Dr. Horrobin is heavily smeared by Wikipedia that publishes an outrageously deceptive page on Willowbrook, as discussed above.

[48] Horowitz LG. Polio, hepatitis B and AIDS: an integrative theory on a possible vaccine induced pandemic. www.ncbi.nlm.nih.gov/pubmed/11388787 -2001 May; 56(5):677-86.

[49] https://dlnr.hawaii.gov/mk/files/2017/03/TIO-Ex-C-57.pdf

[50] To summarize the Supreme Court's ruling on the Daubert standard, the court wrote: "'General acceptance' is not a necessary precondition to the admissibility of scientific evidence under the Federal Rules of Evidence, but the Rules of Evidence— especially Rule 702—do assign to the trial judge the task of ensuring that an expert's testimony both rests on a reliable foundation and is relevant to the task at hand. Pertinent evidence based on scientifically valid principles will satisfy those demands."

[51] Horowitz LG. **Emerging Viruses: AIDS & Ebola–Nature, Accident or Intentional?** Tetrahedron Press, Rockport, MA, 1996, pp. 6-7. The book is available through the author's sponsors for a nominal fee https://www.cureshoppe.com/special-virus-cancer-program-1972-the-virus-cancer-program-1977-2-pdf-downloads/

[52] https://medicalveritas.org/special-virus-cancer-program/

[53] http://revolutiontelevision.net/video/robert-gallo-the-man-that-created-aids/

[54] https://www.ncbi.nlm.nih.gov/books/NBK554776/

[55] Hearings Before a Subcommittee of the Committee on Appropriations House of Representatives, Ninety-First Congress, Part 5, Research, Development, Test and Evaluation, Dept of the Army, on Tuesday, July 1, 1969, pg. 79, (Washington: U.S. Gov. Printing Office)

[56] https://www.historyofvaccines.org/content/articles/debunked-polio-vaccine-and-hiv-link

[57] http://originofaids.net/

[58] HIV & AIDS - Lab Rat - What AIDS Researcher Dr. Robert Gallo Did in Pursuit of the Nobel Prize (virusmyth.com)

[59] https://www.cureshoppe.com/emerging-viruses-aids-ebola-nature-accident-or-intentional-hardback-book/

[60] UN-VAXXED: A Docu-commentary for Robert De Niro, a Leonard G. Horowitz production in association with Medical Veritas International, Inc. 2016. http://medicalveritas.org/un-vaxxed-a-docu-commentary-for-robert-de-niro-2/

[61] U.S.G.A.O. Origin of the AIDS Virus. Briefing for the Office of the Honorable James A. Traficant, Jr. GAO-02-809R, June 17, 2002.

[62] https://judicialcorruptionnews.com/judge-wendy-deweese/

[63] Wikipedia, defining criminally negligent manslaughter, at: https://en.wikipedia.org/wiki/Manslaughter

[64] https://www.thelancet.com/pdfs/journals/lancet/PIIS0140-6736%2815%2960696-1.pdf

[65] https://medicalveritas.org/coronavirus-science-fraud/

[66] https://medicalveritas.org/origin-of-aids/

[67] College of Physicians of Philadelphia. The History of Vaccines. Debunked: The Polio Vaccine and HIV Link Online at: https://www.historyofvaccines.org/content/articles/debunked-polio-vaccine-and-hiv-link

[68] https://www.ronjohnson.senate.gov/services/files/A4A76F9A-9B29-4CF9-B987-F9097A3F4CB7

[69] https://www.scmp.com/news/china/society/article/3074991/coronavirus-chinas-first-confirmed-covid-19-case-traced-back

[70] https://www.theguardian.com/world/2020/mar/11/coronavirus-wuhan-doctor-ai-fen-speaks-out-against-authorities

[71] https://chinadigitaltimes.net/2020/02/translation-li-wenliangs-admonishment-notice/

[72] https://www.whitehouse.gov/ostp/

[73] https://medicalveritas.org/wp-content/uploads/2020/02/Letter-to-White-House-Officials-Feb-10-2020-w-Exhibits.pdf
[74] https://abcnews.go.com/Politics/white-house-asks-scientists-investigate-origins-coronavirus/story?id=68807304
[75] http://drlenhorowitz.com/wp-content/uploads/2020/01/Dr-Leonard-Horowitz-Curriculum-Vitae-2019.pdf
[76] https://medicalveritas.org/un-censors-dr-horowitzs-aids-thesis-published-in-medical-hypothesis-journal/
[77] https://medicalveritas.org/vaccine-racket-contaminates-nobel-prizes-school-mandates-evidence-criminally-negligent-manslaughter/
[78] https://medicalveritas.org/zika-virus-mystery-solved/
[79] https://medicalveritas.org/ebola-vaccine-warning/
[80] https://medicalveritas.org/sars-scam/
[81] https://medicalveritas.org/swine-flu-scam-exposed-in-india/
[82] https://www.waronwethepeople.com/the-cias-role-in-the-anthrax-mailings-could-our-spies-be-agents-for-military-industrial-sabotage-terrorism-and-even-population-control/
[83] http://www.waronwethepeople.com/spams-expose-cia-google-internet-regulators-pushing-drugs-unfair-competition-dr-leonard-horowitz/
[84] https://chinadigitaltimes.net/2020/02/translation-li-wenliangs-admonishment-notice/
[85] https://apnews.com/PR Newswire/ff548c99a03afb0d69bb7871f7cd4fc0
[86] http://revolutiontelevision.net/video/coronavirus-predictive-programming/?mc_cid=863697fdc0&mc_eid=c2f7e2eee7
[87] http://www.centerforhealthsecurity.org/event201/about
[88] https://revolutiontelevision.net/video/fauci-lecture-at-georgetown-u-predicting-plague-on-trump-administration/
[89] https://medicalveritas.org/white-house-asks-coronavirus-foxes-to-investigate-bio-engineering-hen-house/
[90] http://panindigan.tripod.com/aidsdodhear.html
[91] https://web.archive.org/web/20200208193920/https://www.loc.gov/law/foreign-news/article/china-vaccine-law-passed/
[92] http://revolutiontelevision.net/video/coronavirus-predictive-programming/?mc_cid=863697fdc0&mc_eid=c2f7e2eee7
[93] https://www.thecrimson.com/article/2020/2/4/Lieber-arrest-analysis/
[94] https://time.com/5416273/foreign-universities-china-education/
[95] Andersen KG, Rambaut A, Lipkin WI, Holmes EC and Garry RF. The proximal origin of SARS-CoV-2. Nature Medicine (2020). See: https://www.nature.com/articles/s41591-020-0820-9
Under New York law, there are five elements of a fraud claim. These include: misrepresentation, knowledge of falsity, intent to deceive, reliance and damages. Mallis v. Bankers Trust Co., 615 F.2d 68, 80 (2d Cir.1980); Freschi v. Grand Coal Venture, 551 F.Supp. 1220, 1230 (S.D.N.Y.1982). According to the U.S. Supreme Court in Neder v. United States, 527 US 1 – 1999, "the well-settled meaning of 'fraud' required a misrepresentation or concealment of material fact.
In the instant case, co-authors Andersen and Garry, et. al. misrepresented coronavirus genetic science. They willfully and knowingly omitted the HIV-1 inserts into the spike protein genome material to the pandemic's pathogenesis.
The intent of this omission and misrepresentation was to deceive the public and official investigators to induce their dismissal of the "conspiracy theory" and laboratory bioweapons origin of the pandemic.
Recipients of this false writing relied upon it, including ABC News, Yahoo News, and presumably federal agents and agencies entrusting Scripps, Tulane, and Zalgen Labs with grant monies to conduct this (and related) research. Taxpayers were also damaged, as was the international scientific community and its reliability. Society as a whole suffers. Such damaging fraudulent concealment and misrepresentation, thereby, satisfies all the elements of 'fraud in the factum' inducing damaged parties to dismiss urgently needed

END NOTES

intelligence required to prevent additional outbreaks and develop remedies for the current crisis.

This felony dovetails with additional charges of obstructing justice as an impediment to governmental activities. Laws precluding obstruction of justice include witness tampering (18 U.S.C. 1512(c)(1)(2)), witness retaliation (18 U.S.C. 1513(e)(f)), obstruction of congressional or administrative proceedings (18 U.S.C. 1505), conspiracy to defraud the United States (18 U.S.C. 371), and contempt (a creature of statute, rule and common law).

[96] Horowitz LG. The Lancet Coronavirus Science Fraud Reveals Bioterrorism. February 23, 2020. MedicalVeritas online journal. See: https://medicalveritas.org/coronavirus-science-fraud/

[97] https://www.vanityfair.com/news/2021/10/nih-admits-funding-risky-virus-research-in-wuhan

[98] https://www.vanityfair.com/news/2020/03/donald-trump-coronavirus-lies-china

[99] https://revolutiontelevision.net/video/trump-states-coronavirus-is-artificial/

[100] http://revolutiontelevision.net/video/tucker-carlson-admits-coronavirus-was-created-in-a-lab/

[101] https://www.nature.com/articles/s41591-020-0820-9

[102] https://www.yahoo.com/gma/sorry-conspiracy-theorists-study-concludes-covid-19-not-090026698--abc-news-topstories.html

[103] https://www.whitehouse.gov/ostp/

[104] https://medicalveritas.org/white-house-coronavirus-origin-probe/

[105] https://escholarship.umassmed.edu/covid19/2/

[106] https://www.npr.org/sections/health-shots/2014/11/07/361219361/how-a-tilt-toward-safety-stopped-a-scientists-virus-research

[107] https://www.justice.gov/atr/page/file/1197731/download

[108] https://medicalveritas.org/leaked-mail-mercola-meltdown-and-fauci-perjury-evidence-vaccine-racket-unwinding/

[109] https://www.eurekalert.org/pub_releases/2018-02/sri-tsr021518.php

[110] https://www.businesswire.com/news/home/20160822005090/en/Zalgen-Labs-Introduces-Viral-Diagnostics-Business-Opens

[111] https://www.cnn.com/2019/11/06/health/hiv-new-strain-discovered/index.html

[112] Pradhan P. et. al., Uncanny similarity of unique inserts in the 2019-n-CoV spike protein to HIV-1 gp120 and Gag. bioRxiv preprint first posted online Jan. 31, 2020. See the 'censored' scientific report re-published under public duty doctrine at: https://medicalveritas.org/wp-content/uploads/2020/02/Pradham-et-al-Coronavirus-HIV-paper.pdf

[113] http://revolutiontelevision.net/video/strange-things-happening-while-you-are-quarantined/

[114] https://medicalveritas.org/wp-content/uploads/2020/04/CNN-Reports-New-Strain-of-HIVAIDS..pdf

[115] https://www.cnn.com/2019/11/06/health/hiv-new-strain-discovered/index.html

[116] https://www.abbottinvestor.com/news-releases/news-release-details/abbott-laboratories-atrasentan-abt-627-demonstrates-potential/

[117] Horowitz LG. Emerging Viruses: AIDS & Ebola—Nature, Accident or Intentional? Tetrahedron Publishing Group, 1996. See also: Horowitz LG. AIDS/Ebola author defends embattled African presidents. OriginofAIDS.com, November 7, 2002. Press release and letters available at: http://www.originofaids.com/author_defends_embattled_african_presidents.htm

[118] https://www.thepharmaletter.com/article/basf-sells-pharma-business-to-abbott-labs-for-6-9-billion-in-cash

[119] https://www.sourcewatch.org/index.php?title=American_Legislative_Exchange_Council

[120] https://www.sourcewatch.org/index.php?title=Association_of_the_British_Pharmaceutical_Industry

[121] https://technical.ly/baltimore/2020/03/20/17-maryland-biotech-companies-collaborating-vaccines-therapeutics-diagnostics-covid19-coronavirus/
[122] http://blogs.nature.com/boston/2007/05/25/cambridge-vs-cambridge-a-personal-comparison
[123] https://journal-neo.org/2020/03/18/coronavirus-and-the-gates-foundation/
[124] Quoting the CIA's 'propaganda mill', Wikipedia. Additional reference links to MIT's 'Media Labs' largely financed by Bill Gates and Jeffrey Epstein's group of investors that include The Mega Group of globalists tied to the Israeli Mossad.
[125] https://exposingvaccinegenocide.org/bill-gates-pedophile-scandal/
[126] https://www.nature.com/articles/nm.3985.pdf?origin=ppub
[127] https://www.justice.gov/opa/pr/harvard-university-professor-and-two-chinese-nationals-charged-three-separate-china-related
[128] https://www.documentcloud.org/documents/20793561-leopold-nih-foia-anthony-fauci-emails. and https://www.documentcloud.org/documents/20793561-leopold-nih-foia-anthony-fauci-emails.
[129] https://medicalveritas.org/wp-content/uploads/2020/04/Fauci-Andersen-2-Complicity-in-Propaganda-3-8-20.pdf
[130] https://www.zerohedge.com/geopolitical/coronavirus-contains-hiv-insertions-stoking-fears-over-artificially-created-bioweapon
[131] Horowitz LG. Coronavirus Predictive Programming: A Docu-commentary on the October 18, 2019 "Event 201" Proceedings. See: https://revolutiontelevision.net/video/coronavirus-predictive-programming/
[132] https://revolutiontelevision.net/video/trump-states-coronavirus-is-artificial/
[133] Daubert v. Merrell Dow Pharmaceuticals, Inc., 509 US 579 – Supreme Court 1993..
[134] Pradhan P. et. al., Uncanny similarity of unique inserts in the 2019-n-CoV spike protein to HIV-1 gp120 and Gag. bioRxiv preprint first posted online Jan. 31, 2020. See the 'censored' scientific report re-published under public duty doctrine at: https://medicalveritas.org/wp-content/uploads/2020/02/Pradham-et-al-Coronavirus-HIV-paper.pdf
[135] https://www.nature.com/articles/nbt.1593?errors=cookies_not_supported&code=3c4e
[136] https://www.nytimes.com/interactive/2020/04/03/science/coronavirus-genome-bad-news-wrapped-in-protein.html?utm_source=pocket-newtab
[137] https://www.law.cornell.edu/uscode/text/18/175
[138] https://medicalveritas.org/coronavirus-bioterrorism-tips-for-personal-health-heightened-immunity-dispelling-fear-and-exposing-fraud/
[139] R&D Systems. ACE-2: The Receptor for SARS-CoV-2. Online literature review including 27 scientific references is available at:
https://www.rndsystems.com/resources/articles/ace-2-sars-receptor-identified
[140] Dziuba N. et. al. Identification of cellular proteins required for replication of human immunodeficiency virus type 1 [HIV-1]/ AIDS Res Hum Retroviruses. 2012 Oct; 28(10): 1329-1339. Available online at
https://www.ncbi.nlm.nih.gov/pmc/articles/PMC3448097/
[141] Hoffman M. et. al., SARS-CoV-2 cell entry depends on ACE2 and TMPRSS2 and is blocked by a clinically proven protease inhibitor. Cell. Published online on March 5, 2020, available at:
https://www.sciencedirect.com/science/article/pii/S0092867420302294?via%3Dihub
[142] https://www.rndsystems.com/target/ace-2?category=Primary%20Antibodies
[143] https://www.rndsystems.com/products/human-ace-2-antibody_af933
[144] https://www.ncbi.nlm.nih.gov/pmc/articles/PMC7149539/
[145] https://www.cnbc.com/2021/08/12/blood-clots-linked-to-astrazeneca-shot-have-22percent-mortality-rate-study.html
[146] https://www.nejm.org/doi/full/10.1056/nejme2106315
[147] Schmidt RR. Heteratom manipulation. In: Comprehensive Organic Synthesis. 1991. Elsevier Science, Ltd. See: https://www.sciencedirect.com/topics/chemistry/o-glycoprotein /

END NOTES

[148] Kamerling JP and Gerwig GJ. Analysis of glycans; polysaccharide functional properties. In: Comprehensive Glycoscience. 2.01.5.5. Uronic Acid Degradation Elsevier B.V. 2007. Quoting these authors:
"The alkaline degradation of reducing sugars, which also includes epimerization reactions at C2, and the alkaline lability of Hex(1-3)Hex bonds at the reducing end form classical examples of this type of reactions." See: https://www.sciencedirect.com/topics/chemistry/o-glycoprotein

[149] Rezabakhsh A. et. al. Effect of hydroxychloroquine on oxidative/nitrosative status and angiogenesis in endothelial cells under high glucose condition. Bioimpacts. 2017; 7(4): 219-226. Published online at: https://www.ncbi.nlm.nih.gov/pmc/articles/PMC5801533/

[150] Babayi T and Riazi GH. The effects of 528Hz sound waves to reduce cell death in human astrocyte primary cell cultures treated with ethanol. J Addict Res Ther 8: 335. Available online at: https://medicalveritas.org/the-effects-of-528-hz-sound-wave-to-reduce-cell-death-in-human-astrocyte-primary-cell-culture-treated-with-ethanol/

[151] Vliegenthart JFG and Kamerling JP. Analysis of glycans; polysaccharide functional properties. In: Comprehensive Glycoscience. Elsevier B.V. 2007. See: https://www.sciencedirect.com/topics/chemistry/o-glycoprotein

[152] Ray RC. Oral NaHCO3 Activates a Splenic Anti-InflammatoryPathway: Evidence That Cholinergic Signals Are Transmittedvia Mesothelial Cells. Journal of Immunology April 16, 2018. See: https://www.jimmunol.org/content/jimmunol/early/2018/04/14/jimmunol.1701605.full.pdf?with-ds=yes

[153] Funes, S. C., Rios, M., Escobar-Vera, J., & Kalergis, A. M. (2018). Implications of macrophage polarization in autoimmunity. Immunology, 154(2), 186–195. See: https://www.ncbi.nlm.nih.gov/pubmed/29455468

[154] https://group.springernature.com/gp/group/media/press-releases/archive-2016/new-open-access-journal-npj-digital-medicine-announced-by-nature/12000202

[155] https://www.nasdaq.com/market-activity/stocks/ssp/institutional-holdings

[156] https://www.law.cornell.edu/uscode/text/18/2381

[157] http://revolutiontelevision.net/video/tucker-carlson-roasts-anthony-fauci-for-coronavirus-disinformation/

[158] https://revolutiontelevision.net/video/alexa-says-the-government-released-the-corona-virus-on-purpose/

[159] https://www.buzzfeednews.com/article/nataliebettendorf/fauci-emails-covid-response

[160] https://www.washingtonpost.com/archive/politics/2000/04/30/aids-is-declared-threat-to-security/c5e976e4-3fe8-411b-9734-ca44f3130b41/

[161] https://medicalveritas.org/charles-lieber-and-covid-vaccines/

[162] https://endpts.com/harvard-joins-coronavirus-fight-with-115-million-and-a-high-profile-chinese-partner/

[163] https://endpts.com/harvard-joins-coronavirus-fight-with-115-million-and-a-high-profile-chinese-partner/

[164] https://www.phe.gov/about/barda/Pages/default.aspx

[165] https://www.usaspending.gov/award/ASST_NON_HDTRA11710064_9761

[166] https://judicialcorruptionnews.com/cia-coup-risks-public-health-safety/

[167] https://mintpressnews.cn/mega-group-maxwells-mossad-spy-story-jeffrey-epstein-scandal/261172/

[168] https://www.fiercepharma.com/marketing/covid-vaccine-unsung-heroes-and-biotech-science-giants-star-influencer-s-tiktok

[169] https://www.opgen.com/acuitas-lighthouse/acuitas-lighthouse-prediction/

[170] https://revolutiontelevision.net/video/laura-ingraham-exposes-the-global-reset-and-forced-vaccines/

[171] https://www.documentcloud.org/documents/20793561-leopold-nih-foia-anthony-fauci-emails

[172] https://theintercept.com/2021/11/03/coronavirus-research-ecohealth-nih-emails/

[173] https://www.usaspending.gov/award/ASST_NON_HDTRA11710064_9761
[174] https://www.phe.gov/s3/dualuse/documents/gain-of-function.pdf
[175] https://www.organicconsumers.org/blog/peter-show-me-money-daszak-pulls-big-bucks-through-ecohealth-alliance-risky-virus-research
[176] https://www.ncbi.nlm.nih.gov/pmc/articles/PMC7159358/
[177] https://medicalveritas.org/changing-ethics-in-vaccination-trials/
[178] https://exposingvaccinegenocide.org/ebola-genocide/
[179] https://www.nature.com/articles/ncomms6342
[180] https://www.nytimes.com/2020/07/25/business/coronavirus-vaccine-profits-vaxart.html
[181] https://www.weforum.org/covid-alliance-for-social-entrepreneurs/join-us
[182] https://www.andeglobal.org/page/covid-19
[183] https://www.cureshoppe.com/the-las-vegas-deep-state-massacre-e-book/
[184] https://theintercept.com/2016/09/08/hillary-clintons-national-security-advisors-are-a-whos-who-of-the-warfare-state/
[185] https://www.waronwethepeople.com/?s=McChrystal
[186] https://www.justice.gov/opa/pr/harvard-university-professor-and-two-chinese-nationals-charged-three-separate-china-related
[187] https://mwi.usma.edu/mwi-video-dr-charles-morgan-neurobiology-war/
[188] https://www.usna.edu/NewsCenter/sites/Ethics/Dr._James_Giordano_Battlescape_Brain_Military_and_Intelligence_Use_of_Neurocognitive_Science.php
[189] https://www.celebritynetworth.com/richest-businessmen/richest-billionaires/hui-ka-yan-net-worth/
[190] https://www.gsa.gov/reference/reports/budget-performance/annual-reports/2019-agency-financial-report/managements-discussion-and-analysis/financial-statements-summary-and-analysis/acquisition-services-fund
[191] https://www.niaid.nih.gov/research/barney-graham-md-phd
[192] https://pubmed.ncbi.nlm.nih.gov/32663912/
[193] https://www.rockefellerfoundation.org/news/the-rockefeller-foundation-welcomes-uks-global-pandemic-radar/
[194] https://www.who.int/news/item/05-05-2021-who-germany-launch-new-global-hub-for-pandemic-and-epidemic-intelligence
[195] https://www.gov.uk/government/news/pm-announces-plan-for-global-pandemic-radar
[196] https://revolutiontelevision.net/video/dengue-fever-presentation-by-dr-leonard-horowitz/
[197] http://investors.dna.com/2016-01-19-Expansion-of-Oxitecs-Vector-Control-Solution-in-Brazil-Attacking-Source-of-Zika-Virus-and-Dengue-Fever-after-Positive-Program-Results
[198] https://www.entrepreneur.com/article/371072
[199] https://www.youtube.com/watch?v=5V9gbgICPKY.
[200] https://revolutiontelevision.net/video/bill-gates-gmo-mosquitoes-for-depopulation/
[201] https://www.waronwethepeople.com/zika-fraud/
[202] https://onlinelibrary.wiley.com/doi/full/10.1002/smll.202002169
[203] https://www.law.cornell.edu/definitions/uscode.php?width=840&height=800&iframe=true&def_id=18-USC-763651625-782330727&term_occur=999&term_src=title:18:part:I:chapter:113B:section:2332b
[204] https://www.law.cornell.edu/definitions/uscode.php?width=840&height=800&iframe=true&def_id=18-USC-2003303595-783258095&term_occur=999&term_src=title:18:part:I:chapter:113B:section:2332a
[205] https://www.casey.senate.gov/news/releases/casey-statement-on-refugee-screenings
[206] https://www.who.int/influenza/pip/advisory_group/bio_watson_Jan20.pdf
[207] https://www.uni-goettingen.de/en/362311.html
[208] https://theorg.com/org/wellcome-trust
[209] https://pubmed.ncbi.nlm.nih.gov/11388787/

END NOTES

[210] https://www.justice.gov/opa/pr/harvard-university-professor-and-two-chinese-nationals-charged-three-separate-china-related; and the "Charging Document" was available for download on the DOJ's link: https://www.justice.gov/opa/press-release/file/1239796/download.

[211] https://judicialcorruptionnews.com/ayahuasca-kingpin-indicted-for-fraud-in-hawaii-may-have-ties-to-jeffrey-bronfman/

[212] https://www.documentcloud.org/documents/20793561-leopold-nih-foia-anthony-fauci-emails

[213] https://medicalveritas.org/darpa-and-coronavirus-western-hands-in-plague/

[214] https://muckrack.com/whitney-webb/articles

[215] https://www.eurekalert.org/pub_releases/2011-02/uoc--uea021411.php

[216] https://www.thedrive.com/the-war-zone/12456/the-united-states-and-australia-quietly-test-hypersonic-missiles?iid=sr-link3

[217] https://www.crunchbase.com/organization/inovio-pharmaceuticals#section-funding-rounds

[218] https://www.marketwatch.com/press-release/inovio-pharmaceuticals-selected-by-darpa-to-lead-a-45-million-program-to-expedite-development-of-novel-products-to-prevent-and-treat-disease-caused-by-ebola-2015-04-08

[219] https://www.barrons.com/press-release/PR-CO-20190610-904931?tesla=y&tesla=y

[220] https://cepi.net/news_cepi/inovio-awarded-up-to-56-million-from-cepi-to-advance-dna-vaccines-against-lassa-fever-and-mers/

[221] https://cepi.net/news_cepi/cepi-to-fund-three-programmes-to-develop-vaccines-against-the-novel-coronavirus-ncov-2019/

[222] https://www.marketwatch.com/press-release/inovio-pharmaceuticals-dna-vaccine-against-ebola-and-marburg-filoviruses-provides-complete-protection-in-preclinical-challenge-study-2013-05-14

[223] http://ir.inovio.com/news-and-media/news/press-release-details/2018/Inovio-Enters-License-and-Collaboration-Agreementwith-ApolloBio-To-Develop-and-Commercialize-VGX-3100-in-Greater-China/default.aspx

[224] https://investors.modernatx.com/news-releases/news-release-details/darpa-awards-moderna-therapeutics-grant-25-million-develop

[225] https://www.morningstar.com/news/business-wire/20191211005159/moderna-announces-key-2020-investor-and-analyst-events

[226] *Walters v. Blankenship*, 931 So.2d 137, 140 (Fla. 5th Dist.Ct.App.2006) *(citations omitted),*" BANKERS LIFE INSURANCE COMPANY v. CREDIT SUISSE FIRST BOSTON CORPORATION, 590 F. Supp. 2d 1364 - Dist. Court, MD Florida 2008.

[227] Baxter, along with Bayer Co., spread HIV/AIDS through contaminated blood products during the early 1980s, according to public knowledge and litigation settlements.

[228] https://www.bloomberg.com/news/newsletters/2021-04-19/curevac-s-shot-could-arrive-just-in-time

[229] https://www.idtdna.com/pages/support/usage-warranty-and-licenses.

[230] Online source: https://www.businesswire.com/news/home/20171129005719/en/Six-Researchers-Chosen-for-New-Pfizer-NCBiotech-Distinguished-Postdoctoral-Fellowships-in-Gene-Therapy

[231] "Partnerships and Anti-Infective Efforts" advertised by Pfizer include UNC SARS-CoV-2 researcher, "Dr. Ralph Baric" "screening Pfizer's lead compound and additional compounds for antiviral activity in a primary human airway epithelial cell assay." https://www.pfizer.com/science/coronavirus/partnerships

[232] U.S. Provisional Application No. 60/206,537, filed May 21, 2001. See: https://patents.justia.com/patent/6593111

[233] U.S. Patent 7,279,327 B2 "STATEMENT OF FEDERAL SUPPORT" states: "This invention was made possible with government Support under grant numbers AI23946 and GM63228 from the National Institutes of Health. The United States government has certain rights to this invention." Pursuant to the commercial applicability and inherent

risks in this invention for anticipated outbreaks and disease transmissions, this patent also states:

"The antigen or antigenic protein or peptide encoded by the heterologous RNA and expressed in the host can be an antigen of a vertebrate pathogen, e.g., a mammalian pathogen or a Swine pathogen, Such as a rabies G antigen, gp51, 30 envelope antigen of bovine leukemia virus, FeLV [feline leukemia virus] envelope antigen of feline leukemia virus, glycoprotein D antigen of herpes simplex virus, a fusion protein antigen of the Newcastle disease virus, an RAV-1 envelope antigen of rous [sarcoma] associated virus, nucleoprotein antigen of avian or mammalian influenza virus, a fusion protein antigen of porcine reproductive and respiratory disease virus (PRRSV), a matrix antigen of the infectious bronchitis virus, a glycoprotein species of PRRSV or a peplomer antigen of the infectious bronchitis virus. In another aspect, the present invention is directed to synthetic recombinant coronavirus modified by the insertion therein of DNA or RNA from any source, and particularly from a non-coronavirus or non-TGEV Source, into a non-essential region of the TGEV genome. Synthetically modified TGEV virus recombinants carrying exogenous (i.e. non-coronavirus) nucleic acids or genes encoding for and expressing an antigen, which recombinants elicit the production by a vertebrate host of immunological responses to the antigen, and therefore to the exogenous pathogen, are used according to the invention to create novel vaccines which avoid the drawbacks of conventional vaccines employing killed or attenuated live organisms, particularly when used to inoculate vertebrates."

[234] Horowitz. LG. Polio, hepatitis B and AIDS: an integrative theory on a possible vaccine induced pandemic. *Med Hypotheses.* 2001 May;56(5):677-86. (PMID: 11388787). See: https://pubmed.ncbi.nlm.nih.gov/11388787/
[235] https://www.washingtonexec.com/2013/12/chertoff-group-adds-five-new-senior-advisors-firm/
[236] https://www.quora.com/topic/In-Q-Tel-venture-capital-firm
[237] https://www.waronwethepeople.com/las-vegas-deep-state-massacre-2/
[238] https://www.nytimes.com/2020/10/09/us/charles-lieber-harvard-china.html
[239] https://www.biospace.com/article/releases/-b-microchip-biotechnologies-inc-b-changes-name-tointelgenx-corp-/
[240] https://www.opgen.com/
[241] https://www.thecrimson.com/article/2021/6/26/lieber-moves-closer-to-trial/
[242] https://www.govinfo.gov/content/pkg/PLAW-107publ188/pdf/PLAW-107publ188.pdf
[243] https://www.economist.com/science-and-technology/2020/02/01/an-american-chemist-is-suspected-of-illegal-dealings-with-china
[244] https://www.sciencemag.org/author/robert-f-service
[245] https://www.sciencemag.org/news/2020/02/why-did-chinese-university-hire-charles-lieber-do-battery-research
[246] https://www.sciencemag.org/sites/default/files/Lieber_CV.pdf
[247] https://www.todaysmotorvehicles.com/article/automotive-manufacturing-henniges-bought-avic-china-091015/
[248] https://programmersought.com/search
[249] https://www.news24.com/news24/MyNews24/Reports-of-Ebola-Conspiracy-are-Worrying-20141012
[250] https://www.waronwethepeople.com/viral_immunity_ebola_best_treatment/
[251] http://www.waronwethepeople.com/ebolagate/
[252] https://www.who.int/news-room/fact-sheets/detail/ebola-virus-disease
[253] https://medicalveritas.org/wp-content/uploads/2019/01/Kunitz-S.-Am-J-Pub-Hlt-Article.pdf
[254] https://thebl.com/us-news/revealed-hunter-biden-invested-in-a-pandemic-firm-partnering-with-daszaks-ecohealth-and-the-wuhan-lab.html
[255] https://nypost.com/2018/03/15/inside-the-shady-private-equity-firm-run-by-kerry-and-bidens-kids/
[256] https://web.archive.org/web/20140315054332/http:/www.rstp.com/people/

END NOTES

[257] https://www.marlinllc.com/_media/_data/MarlinNewsletter/hit-monthly-newsletter-june-2015.pdf
[258] https://www.bloomberg.com/profile/person/16779750
[259] https://www.ecohealthalliance.org/wp-content/uploads/2016/01/EcoHealth_Alliance_FY14_Annual_Report.pdf
[260] https://www.ecohealthalliance.org/2014/11/usaid-announces-second-phase-of-predict-project-with-global-partners
[261] https://www.ecohealthalliance.org/wp-content/uploads/2016/08/FY15-Annual-Report-final.pdf
[262] http://www.ecohealthalliance.org/wp-content/uploads/2016/11/Yuan-et-al_virus-bats_viruses-2014.pdf
[263] https://www.ecohealthalliance.org/wp-content/uploads/2018/10/Living-Safely-with-Bats_download.pdf
[264] https://www.cbsnews.com/news/american-company-metabiota-problems-during-ebola-outbreak/
[265] https://www.merrimack.edu/live/news/1967-merrimack-grad-on-front-line-of-battle-against
[266] https://pubmed.ncbi.nlm.nih.gov/?term=T32+AI007151%2FAI%2FNIAID+NIH+HHS%2FUnited+States%5BGrant+Number%5D
[267] https://pubmed.ncbi.nlm.nih.gov/?term=F32+AI152296%2FAI%2FNIAID+NIH+HHS%2FUnited+States%5BGrant+Number%5D
[268] https://www.ncbi.nlm.nih.gov/sites/books/NBK401938/
[269] https://www.cnet.com/news/bill-gates-once-had-emotional-quotient-of-snail-barry-diller-says/
[270] https://www.youtube.com/watch?v=b9vWShsmE20
[271] https://www.usna.edu/NewsCenter/sites/Ethics/Dr._James_Giordano_Battlescape_Brain_Military_and_Intelligence_Use_of_Neurocognitive_Science.php
[272] https://mwi.usma.edu/mwi-video-dr-charles-morgan-neurobiology-war/
[273] https://judicialcorruptionnews.com/hawaii-mob-murders-sherri-kane/
[274] https://judicialcorruptionnews.com/hawaii-challenge-trump-travel-ban-vets-pattern-judicial-corruption-mob-rule-drug-trafficking/
[275] https://medicalveritas.org/hawaiian-last-king-kalakaua-killed-by-drugs/
[276] https://judicialcorruptionnews.com/judicial-corruption-backfires-as-criminal-actions-against-lawyers-advance/
[277] https://www.youtube.com/watch?v=kpW9JcWxKq0
[278] https://www.routledge.com/Trance-Formation-The-Spiritual-and-Religious-Dimensions-of-Global-Rave/Sylvan/p/book/9780415970914
[279] https://www.psychologytoday.com/us/basics/smoking
[280] https://www.psychologytoday.com/us/basics/politics
[281] https://medicalveritas.org/mind-controlled/
[282] https://www.wired.com/2012/12/the-next-warfare-domain-is-your-brain/
[283] https://judicialcorruptionnews.com/ayahuasca-kingpin
[284] https://jamestown.org/program/cognitive-domain-operations-the-plas-new-holistic-concept-for-influence-operations/
[285] https://ndupress.ndu.edu/Portals/68/Documents/jfq/jfq-94/jfq-94.pdf
[286] https://madsciblog.tradoc.army.mil/158-in-the-cognitive-war-the-weapon-is-you/
[287] https://www.google.com/amp/s/www.forbes.com/sites/zakdoffman/2019/08/01/social-media-warfare-new-military-cyber-unit-will-fight-russias-dark-arts/amp/
[288] https://revolutiontelevision.net/video/rev-jeremiah-wright-aids-genocide-defense-cites-dr-leonard-horowitz/
[289] https://www.waronwethepeople.com/scam-bots/
[290] https://www.waronwethepeople.com/cointelpro-minions-barabara-hartwell/
[291] https://www.waronwethepeople.com/pseudo-skeptic-robert-todd-carroll/
[292] https://www.waronwethepeople.com/snopes-fraud/

[293] https://www.waronwethepeople.com/spams-expose-cia-google-internet-regulators-pushing-drugs-unfair-competition-dr-leonard-horowitz/
[294] https://www.waronwethepeople.com/wikipedia-fraud-jimmy-wales-exposed/
[295] https://www.waronwethepeople.com/vaccine-cancer-links/
[296] https://www.waronwethepeople.com/popular-mechanics-spins-anti-vaccination-movement-and-assault-on-dr-leonard-horowitz/
[297] https://www.frequencytx.com/role/leadership/
[298] https://www.prnewswire.com/news-releases/googles-jigsaw-announces-toxicity-reducing-api-perspective-is-processing-500m-requests-daily-301223600.html
[299] https://www.perspectiveapi.com/
[300] https://jigsaw.google.com/challenges/
[301] https://judicialcorruptionnews.com/burisma-holdings-supplemental-intelligence/
[302] https://www.waronwethepeople.com/counter-intelligence/atrueott-of-motherearthminerals-exposed/
[303] https://www.waronwethepeople.com/triggered-investigation-fbicia-agent-ted-gunderson-sherri-kane-dr-leonard-horowitz/
[304] http://www.waronwethepeople.com/ipredator-michael-nuccitelli-exposed-in-troll-triad/
[305] https://revolutiontelevision.net/video/controlling-opposition-through-the-media-with-dr-leonard-horowitz-and-sherri-kane/
[306] https://www.waronwethepeople.com/cointelpro-minions-barabara-hartwell/
[307] https://revolutiontelevision.net/video/sex-tape-scandal-exposes-patriot-network-infiltration-by-traitorous-cointelpro/
[308] http://revolutiontelevision.net/video/nsa-cia-trolls-destroy-heroes-to-profit-villains-snowden-wikileaks-evidence-protection-racket-for-death-industry/
[309] http://www.waronwethepeople.com/google-youtube-fraud-proven-censorship-celebrity-dr-leonard-horowitzs-popularity-sherri-kane/
[310] https://www.pcmag.com/feature/365966/how-google-s-jigsaw-is-trying-to-detoxify-the-internet/1
[311] https://www.pcmag.com/news/364193/googles-sister-company-develops-anti-censorship-android-app
[312] https://www.pcmag.com/news/357591/inside-project-shield-jigsaws-anti-ddos-machine
[313] https://www.psychologytoday.com/us/basics/groupthink
[314] https://www.businessinsider.com/microsoft-ibm-surface-partnership-2016-7
[315] https://www.crn.com/news/cloud/hearst-s-multi-cloud-strategy-taps-aws-microsoft-azure-google-cloud
[316] https://customers.microsoft.com/en-us/story/rackspace-mckesson-azure-healthcare-usa
[317] http://www.cosmopolitan.com/lifestyle/news/a59263/marco-arturo-vaccines-autism-video/
[318] https://www.federalregister.gov/articles/2016/08/15/2016-18103/control-of-communicable-diseases#h-8
[319] https://ec.europa.eu/health/sites/health/files/vaccination/docs/2019-2022_roadmap_en.pdf
[320] https://www.mediapost.com/publications/article/36595/aol-iac-and-hearst-invest-in-brightcove.html
[321] https://gigaom.com/2005/11/22/brightcove-gets-162-million-round-from-iac-aol-hearst-diller-on-board/
[322] https://www.marketwatch.com/story/mckesson-settles-price-fixing-case-for-151m-2012-07-27
[323] https://www.justice.gov/opa/pr/justice-department-announces-largest-health-care-fraud-settlement-its-history
[324] http://www.computerworld.com/article/2591105/healthcare-it/ftc-charges-hearst-with-monopolizing-drug-databases.html
[325] http://www.hearstintegratedmedia.com/custsol.php

END NOTES

[326] http://www.healthimaging.com/topics/diagnostic-imaging/glaxosmithkline-mckesson-sued-over-avandia-wrongful-deaths
[327] https://www.churchofjesuschrist.org/study/scriptures/nt/rev/18?lang=eng#note4a
[328] http://www.waronwethepeople.com/vaxxed-investigator/
[329] http://www.waronwethepeople.com/this-vaxxed-story-began-aboard-a-ship/
[330] https://www.waronwethepeople.com/popular-mechanics-spins-anti-vaccination-movement-and-assault-on-dr-leonard-horowitz/
[331] http://www.waronwethepeople.com/colin-mcroberts/
[332] http://www.waronwethepeople.com/paul-offit-2/
[333] https://psychedelictimes.com/psychedelic-technology-neuralink/
[334] https://beckleyfoundation.org/2018/06/13/psychedelics-promote-neural-plasticity/
[335] https://medicalveritas.org/true-history-of-ayahuasca-mkultra/
[336] https://judicialcorruptionnews.com/sulla-fights-child-porn-charge/
[337] https://medicalveritas.org/bronfman-crime-tree-explains-ayahuasca/
[338] https://stanleykrippner.weebly.com/the-future-of-religion.html
[339] https://www.usna.edu/NewsCenter/sites/Ethics/Dr._James_Giordano_Battlescape_Brain_Military_and_Intelligence_Use_of_Neurocognitive_Science.php
[340] https://revolutiontelevision.net/video/terencemckenna/
[341] https://medicalveritas.org/bronfman-crime-tree-explains-ayahuasca/
[342] https://www.businessinsider.com/category/ebola
[343] https://www.businessinsider.com/this-man-is-helping-stop-ebola-2014-10
[344] https://www.waronwethepeople.com/ebolagate/
[345] https://www.businessinsider.com/psychedelics-depression-anxiety-alcoholism-mental-illness-2017-1
[346] https://www.businessinsider.com/what-magic-mushrooms-brain-state-of-mind-2017-1
[347] http://www.who.int/mediacentre/factsheets/fs369/en/
[348] https://www.thenewamerican.com/print-magazine/item/27870-deep-state-secret-societies-skull-and-bones-bohemians-illuminati
[349] https://books.google.com/books?id=vf9ZJx8WkjQC&pg=PA88&lpg=PA88&dq=Axel+Springer+Dulles&source=bl&ots=72AcTqjLim&sig=ACfU3U1Okmbk6k_vV0Tj5qiBM5gcDQCNTA&hl=en&sa=X&ved=2ahUKEwjM8YH6osPkAhXGvZ4KHewlB4QQ6AEwA3oECAkQAQ#v=onepage&q=Axel%20Springer%20Du
[350] https://www.cia.gov/library/center-for-the-study-of-intelligence/csi-publications/csi-studies/studies/vol-58-no-3/operation-paperclip-the-secret-intelligence-program-to-bring-nazi-scientists-to-america.html
[351] https://www.businessinsider.com/ecstasy-mdma-medical-legalization-ptsd-2017-8
[352] https://www.businessinsider.com/davos-top-psychedelic-scientist-mdma-magic-mushrooms-medicine-2019-1
[353] https://www.businessinsider.in/a-peter-thiel-backed-startup-has-raised-25-million-to-unleash-a-virgin-market-of-for-profit-psychedelic-research/articleshow/66055503.cms
[354] https://www.businessinsider.com/peter-thiel-psychedelic-research-startup-mushrooms-psilocybin-depression-atai-2018-9
[355] https://www.businessinsider.com/magic-mushrooms-psilocybin-mental-physical-effects-2019-5
[356] https://www.businessinsider.com/most-promising-uses-psychedelic-drugs-medicine-science-2018-10
[357] https://www.thedailybeast.com/jeffrey-epstein-has-a-secret-charity-heres-who-it-gave-money-to
[358] https://revolutiontelevision.net/video/bill-gates-microsoft-commercial-features-spirit-cooker-satanist-marina-abramovic/
[359] https://www.waronwethepeople.com/leon-black-apollo-global-management/
[360] http://projectavalon.net/forum4/showthread.php?48817-Tin-foil-hats-test-resuts--MIT-
[361] https://www.nature.com/articles/s41577-020-0358-6

[362] https://www.organicconsumers.org/news/why-chemtrail-conspiracy-real?gclid=EAIaIQobChMIhvyN3Pr35AIVgobACh3IGgo_EAMYASAAEgLpTPD_BwE
[363] https://www.forbes.com/sites/trevornace/2018/12/05/harvard-scientists-begin-experiment-to-block-out-the-sun/#61f7fd240c24
[364] https://www.independent.co.uk/topic/elon-musk
[365] https://www.independent.co.uk/topic/artificial-intelligence
[366] https://www.independent.co.uk/life-style/gadgets-and-tech/news/elon-musk-ai-openai-microsoft-artificial-intelligence-funding-a9016736.html
[367] https://news.mit.edu/2015/brain-strengthen-connections-between-neurons-1118
[368] http://www.psychedelic-library.org/renmenu.htm
[369] https://www.sciencedirect.com/topics/psychology/cyborg
[370] https://www.sciencedirect.com/topics/social-sciences/cybernetics
[371] https://www.sciencedirect.com/science/article/pii/B008043076703182X
[372] https://www.usonainstitute.org/?doing_wp_cron=1634325873.4009110927581787109375
[373] http://www.maps.org/news-letters/v15n3-html/trance.html
[374] https://ieeexplore.ieee.org/ielx7/6287639/8948470/09056855.pdf?tp=&arnumber=9056855&isnumber=8948470&ref=aHR0cHM6Ly93d3cuZ29vZ2xlLmNvbVS8=
[375] http://www.528revolution.com/432-hertz-cia-propaganda-fraud-from-pentagon-agent-jamie-buturff/
[376] https://www.waronwethepeople.com/528-528-v-432-contrivance/
[377] https://www.waronwethepeople.com/roels-world/
[378] https://medicalveritas.org/musical-cult-control/
[379] http://revolutiontelevision.net/video/natural-cure-for-global-warming-excerpt-from-criminal-foundations/
[380] https://medicalveritas.org/the-effects-of-528-hz-sound-wave-to-reduce-cell-death-in-human-astrocyte-primary-cell-culture-treated-with-ethanol/
[381] http://www.528revolution.com/jay-z-444-in-a444hz/
[382] http://www.528revolution.com/528-superstars/
[383] http://onlinelibrary.wiley.com/doi/10.1002/adma.201203261/abstract
[384] https://bcs.mit.edu/news-events/news/crowdsourced-tool-depression
[385] http://satori.mit.edu/
[386] https://www.top500.org/lists/2019/06/
[387] https://www.nytimes.com/2019/08/01/business/jeffrey-epstein-leon-blackapollo.html%5D
[388] https://www.lookingglassoptical.com/blue-eyes-more-sensitive/
[389] https://www.nyu.edu/students/communities-and-groups/student-diversity/spiritual-life/of-many-institute-for-multifaith-leadership.html
[390] https://www.chop.edu/centers-programs/vaccine-education-center/vaccine-history/developments-by-year
[391] https://www.businessinsider.com/peter-thiel-facebook-trump-biography-2018-2#thiel-plans-tolive-to-be-120-years-old-and-takes-human-growth-hormones-every-day-he-wrote-on-reddit-in-2015-19
[392] https://www.vanityfair.com/news/2016/08/peter-thiel-wants-to-inject-himself-with-young-peoples-blood
[393] https://www.waronwethepeople.com/?s=Epstein
[394] https://s3.documentcloud.org/documents/1508273/jeffrey-epsteins-little-black-book-redacted.pdf
[395] https://www.mintpressnews.com/tag/mega-group/
[396] http://judicialcorruptionnews.com/epstein-alive-top-10-reasons/
[397] http://judicialcorruptionnews.com/jeffrey-epstein-death-psyops/
[398] http://www.waronwethepeople.com/zika-virus-mystery/
[399] http://revolutiontelevision.net/video/bill-gates-to-create-gmo-mosquitoes-within-two-years-for-depopulation/

END NOTES

[400] https://revolutiontelevision.net/video/zika-psyop-to-justify-releasing-millions-of-bill-gates-gm-mosquitoes-in-florida/
[401] http://www.oxitec.com/news-and-views/topic-pages-safety-and-sustainability/overview-of-oxitecs-outdoor-projects/
[402] http://dna.com/Company/Board-of-Directors/Kirk
[403] http://topics.nytimes.com/top/news/business/companies/ziopharm-oncology-inc/index.html
[404] http://www.nasdaq.com/quotes/institutional-portfolio/third-security-llc-874451
[405] https://www.oecd.org/derec/49682113.pdf
[406] https://www.thedailybeast.com/jeffrey-epstein-has-a-secret-charity-heres-who-it-gave-money-to
[407] https://www.prnewswire.com/news-releases/jeffrey-epstein-armistice-activist-heralds-new-yorks-international-peace-institute-into-bahrain-203260561.html
[408] https://www.waronwethepeople.com/trump-framed/
[409] https://www.gatesfoundation.org/Media-Center/Press-Releases/2006/06/Cervical-Cancer-Vaccines-in-the-Developing-World
[410] https://www.nytimes.com/2019/07/16/us/jeffrey-epstein-what-to-know.html?module=inline
[411] http://revolutiontelevision.net/video/bill-gates-vaccines-to-be-used-for-population-control/
[412] https://www.cnbc.com/2014/04/29/25-bill-gates.html
[413] https://www.cnbc.com/quotes/?symbol=CMCSA
[414] https://www.cnbc.com/2017/08/29/steve-jobs-and-bill-gates-what-happened-when-microsoft-saved-apple.html
[415] https://www.nytimes.com/2019/10/12/business/jeffrey-epstein-bill-gates.html
[416] https://www.alamy.com/stock-photo-washington-dc-usa-15th-sep-2014-melanie-walker-senior-adviser-to-the-73470722.html
[417] https://www.foxbusiness.com/features/who-is-boris-nikolic-epstein-executor
[418] https://www.nytimes.com/2019/07/13/nyregion/jeffrey-epstein-new-york-elite.html?module=inline
[419] https://medicalveritas.org/kennedy-dershowitz-debate/
[420] https://en.wikipedia.org/wiki/Glenn_Dubin
[421] https://cdn.muckrock.com/outbound_composer_attachments/JPatBrown/57986/Marvin20Minsky2C20E2809Cfather20of20artificial20intelligence2CE2809D20dies20at20.pdf
[422] https://www.azquotes.com/quotes/topics/company-you-keep.html
[423] https://www.nytimes.com/2019/08/08/business/jeffrey-epstein-jpmorgan.html?module=inline
[424] https://www.mintpressnews.com/tag/mega-group/
[425] https://www.nytimes.com/2019/10/24/us/politics/john-durham-criminal-investigation.html
[426] https://www.mintpressnews.com/mega-group-maxwells-mossad-spy-story-jeffrey-epstein-scandal/261172/
[427] https://www.maryferrell.org/showDoc.html?docId=1156#relPageId=2&tab=page
[428] https://www.worldjewishcongress.org/en/about/leaders/executive-committee
[429] https://en.wikipedia.org/wiki/David_Mayer_de_Rothschild
[430] https://www.hsph.harvard.edu/c-change/about-c-change/
[431] https://www.jpost.com/Israel-News/Jeffrey-Epstein-bankrolled-Ehud-Baraks-high-tech-investment-report-claims-595492
[432] https://www.wexnerfoundation.org/resources/videos/?v=a-conversation-between-shimon-peres-and-david-gergen
[433] https://www.wexnerfoundation.org/member-search/remembering-our-friend
[434] https://www.theyeshivaworld.com/news/headlines-breaking-stories/468497/mossad-statement-on-the-passing-of-former-president-shimon-peres.html
[435] https://carbyne911.com/wpcontent/uploads/2018/03/CARBYNE_ActiveShooter_WhitePaper.pdf

[436] https://www.jpost.com/Magazine/Israeli-countermeasures-571389
[437] https://www.timesofisrael.com/liveblog-july-11-2019/
[438] https://carbyne911.com/team/
[439] https://narativ.org/2019/07/27/building-big-brother/
[440] https://www.waronwethepeople.com/deep-state-treason/
[441] https://www.vanityfair.com/news/2019/10/how-nbc-killed-its-weinstein-story
[442] https://www.latimes.com/business/la-fi-ct-nbcuniversal-snap-stake-20170303-story.html
[443] https://songmeanings.com/songs/view/4810/
[444] https://www.rollingstone.com/culture/culture-news/ethan-nadelmann-the-real-drug-czar-92837/
[445] https://www.opendemocracy.net/en/tenp-ethannadelmann/
[446] https://stanleykrippner.weebly.com/the-future-of-religion.html
[447] https://quoteinvestigator.com/2017/03/23/same/
[448] https://www.influencewatch.org/non-profit/open-society-foundations/
[449] https://www.influencewatch.org/person/george-soros/
[450] https://www.investopedia.com/news/26-goldman-sachs-alumni-who-run-world-gs/
[451] https://www.globalresearch.ca/the-federal-reserve-cartel-the-eight-families/25080
[452] https://www.environment-hawaii.org/?p=10741
[453] http://www.armbusinessbank.am/en/event/read/url:Mobile-banking
[454] Goldman Sachs Philanthropy Fund, 2014 990, Accessed February 9, 2019. http://990s.foundationcenter.org/990_pdf_archive/311/311774905/311774905_201412_990.pdf ^
[455] https://www.webmd.com/drugs/2/drug-20222-325/mifepristone-oral/mifepristone-oral/details
[456] https://www.chop.edu/centers-programs/vaccine-education-center/vaccine-ingredients/fetal-tissues
[457] https://highline.huffingtonpost.com/articles/en/alt-right/
[458] https://www.wsj.com/articles/blackrock-larry-fink-china-hkex-sse-authoritarianism-xi-jinping-term-limits-human-rights-ant-didi-global-national-security-11630938728
[459] https://medicalveritas.org/las-vegas-deep-state-massacre-book-vets-key-suspects-in-the-greatest-media-cover-up-in-the-history-of-organized-crime/
[460] https://www.waronwethepeople.com/nuclear-ballistic-missile-denial-ability-hits-fema-officials-licensing-procedure-failure-intervene-states-mental-meltdown-according-ipaws-procedure-book/
[461] https://www.waronwethepeople.com/spacegate-movie/
[462] http://www.waynemadsenreport.com/
[463] https://www.youtube.com/watch?v=VfQuoEk6b-E
[464] https://youtu.be/X7SIHsPWYEs?t=1m43s
[465] https://www.fec.gov/data/receipts/?two_year_transaction_period=2018&data_type=processed&committee_id=C00639591&min_date=01%2F01%2F2017&max_date=08%2F22%2F2018
[466] https://www.fec.gov/data/receipts/individual-contributions/?two_year_transaction_period=2018&contributor_name=george+soros&min_date=01%2F01%2F2017&max_date=08%2F24%2F2018
[467] https://www.influencewatch.org/organization/occupy-wall-street/
[468] https://www.reuters.com/article/us-wallstreet-protests-origins-idUSTRE79C1YN20111014
[469] https://www.influencewatch.org/movement/black-lives-matter/
[470] http://psypressuk.com/2014/10/09/aya2014-joining-worlds-for-bottom-up-regulation/
[471] https://medicalveritas.org/bronfman-crime-tree-explains-ayahuasca/
[472] https://obits.democratandchronicle.com/us/obituaries/democratandchronicle/name/stephen-kunitz-obituary?pid=188744077
[473] https://www.waronwethepeople.com/PAKISTAN_VACCINATION_MASSACRE_files/Kunitz- Globalization and Genocide.pdf
[474] http://www.ncbi.nlm.nih.gov/pmc/articles/PMC1446376/

END NOTES

[475] https://www.waronwethepeople.com/partnership-for-new-york-city-linked-to-big-pharma-genocide/
[476] https://www.waronwethepeople.com/drug-cartel-exposed-creating-pandemic-h1n1-swine-flu-viruses-and-vaccines/
[477] https://waronwethepeople.com/waronwethepeople.com/WAR_on_We_The_People_The_Gulf_Oil_Attack_by_Rothschild_League_of_Investment_Bankers.html
[478] https://www.bloomberg.com/news/articles/2019-08-01/apollo-s-black-clarifies-relationship-with-epstein-to-investors
[479] https://www.bloomberg.com/news/articles/2019-07-31/jeffrey-epstein-had-a-door-into-apollo-his-deep-ties-with-black
[480] https://www.bitchute.com/video/kI07hWZJ4pLS/
[481] https://www.waronwethepeople.com/?s=Paris+Attacks
[482] https://nypost.com/2019/08/15/jeffrey-epsteins-gal-pal-ghislaine-maxwell-spotted-at-in-n-out-burger-in-first-photos-since-his-death/
[483] https://www.businessinsider.com/ghislaine-maxwell-book-cia-in-n-out-sales-2019-8
[484] https://www.waronwethepeople.com/scam-bots/
[485] https://www.waronwethepeople.com/paris-attacks/
[486] https://digitaledition.chicagotribune.com/tribune/article_popover.aspx?guid=7dc8300c-128b-428d-8650-c06f16f9e3d7
[487] https://www.hollywoodreporter.com/news/warner-music-boss-edgar-bronfman-74400
[488] https://www.bloomberg.com/news/articles/2021-04-27/qatar-adds-u-s-lobbying-muscle-after-saudi-rift-trump-s-snub
[489] https://www.nytimes.com/2019/06/25/magazine/universal-music-fire-bands-list-umg.html
[490] https://www.thewrap.com/paris-attack-at-bataclan-claims-life-of-mercury-records-executive-thomas-ayad/
[491] https://www.bbc.com/news/entertainment-arts-34825865
[492] https://nypost.com/1999/03/16/bronfmans-movie-biz-debt-may-go-to-diller-in-deal/
[493] https://theorg.com/org/endeavor/org-chart/edgar-bronfman-jr
[494] https://variety.com/1998/film/reviews/hollywoodism-jews-movies-and-the-american-dream-1200453087/
[495] https://forward.com/culture/428797/from-diller-to-zuckerberg-heres-what-jewish-execs-are-reading-this-summer/
[496] https://www.news.com.au/lifestyle/health/wellbeing/dangerous-direction-australia-is-heading-according-to-international-mindfulness-guru/news-story/aa54e4df236d24b71b6ca09c355ece64
[497] https://variety.com/2017/film/news/michael-moore-weinstein-facebook-1202589771/
[498] https://www.waronwethepeople.com/cointelpro-clearing-house-dr-leonard-horowitz-sherri-kane/
[499] https://www.theguardian.com/world/2014/may/12/glenn-greenwald-uk-arrest-me-edward-snowden-nsa
[500] https://www.waronwethepeople.com/sex-drugs-internet-fraud-the-secret-life-of-erin-elizabeth-finn-and-dr-joseph-mercola/
[501] https://medicalveritas.org/a-conspiracy-of-silence-exposes-deep-state-mind-control-social-engineering-and-vaccination-depopulation-programs/
[502] https://revolutiontelevision.net/video/coronavirus-end-game/
[503] https://medicalveritas.org/covid-19-coup/
[504] https://medicalveritas.org/1111-prophecy-covid-19-sorcery/
[505] https://medicalveritas.org/covid-19-genocide/
[506] https://www.apa.org/topics/covid-19/psychological-impact
[507] https://www.youtube.com/watch?v=VEGKjqd8iIQ
[508] https://www.researchgate.net/publication/303499923_Obsolete_tobacco_control_themes_can_be_hazardous_to_public_health_The_need_for_updating_views_on_absolute_product_risks_and_harm_reduction

[509] https://www.nbcnews.com/health/health-news/questions-about-covid-19-test-accuracy-raised-across-testing-spectrum-n1214981
[510] https://buffalonews.com/news/local/parsing-the-numbers-why-vaccinated-residents-made-up-40-of-erie-countys-positive-covid-19/article_641c4a84-10da-11ec-a720-cf5c0cc41444.html
[511] https://www.rockefellerfoundation.org/wp-content/uploads/2020/10/CovidTestingTracing_Handbook.pdf
[512] https://medicalveritas.org/vaccination-hesitancy-science-review-of-u-s-senate-hearing-nets-a-genocidal-red-herring/
[513] https://medicalveritas.org/anti-vaccination-psychoscotoma/
[514] https://medicalveritas.org/a-conspiracy-of-silence-exposes-deep-state-mind-control-social-engineering-and-vaccination-depopulation-programs/
[515] https://twitter.com/JenniferNuzzo/status/1267885076697812993
[516] https://news.yahoo.com/suddenly-public-health-officials-social-211925723.html
[517] https://drive.google.com/file/d/1Jyfn4Wd2i6bRi12ePghMHtX3ys1b7K1A/view
[518] https://medicalveritas.org/coronavirus-conspiracy-proven-by-fake-science/
[519] https://www.lewrockwell.com/2020/03/no_author/coronavirus-vaccines-and-the-gates-foundation/
[520] https://www.foxnews.com/us/rioting-across-the-nation-leaves-cities-reeling-as-hundreds-arrested-national-guard-called-in-businesses-damaged-and-at-least-one-dead
[521] https://judicialcorruptionnews.com/the-horokane-blasts-public-corruption-investigations-exploding-in-paradise/
[522] https://www.nytimes.com/2020/07/13/us/politics/george-soros-racial-justice-organizations.html
[523] https://www.sciencedirect.com/topics/medicine-and-dentistry/nocebo-effect
[524] https://www.waronwethepeople.com/partnership-for-new-york-city-linked-to-big-pharma-genocide/
[525] http://scihi.org/abraham-maslow-hierarchy-needs/
[526] https://medicalveritas.org/free-will/
[527] https://www.ncbi.nlm.nih.gov/pmc/articles/PMC3166409/
[528] https://news.stanford.edu/2020/04/14/stanford-researchers-reengineer-covid-19-face-masks/
[529] https://www.theatlantic.com/ideas/archive/2021/09/persuade-unvaccinated-protect-unvaccinated/620091/
[530] https://www.usnews.com/news/health-news/articles/2021-09-03/idaho-hospitals-nearly-buckling-in-relentless-covid-surge
[531] https://warfarehistorynetwork.com/2016/01/19/why-kaiser-wilhelm-ii-almost-attacked-americas-east-coast/
[532] https://www.gatesfoundation.org/about/committed-grants/2020/09/inv004478
[533] https://medicalveritas.org/national-security-covid-crime-syndicate/
[534] http://exposingvaccinegenocide.com/
[535] https://medicalveritas.org/horowitz-hospitalized/
[536] https://medicalveritas.org/the-horowitz-covid-protocol/
[537] https://www.reuters.com/world/us/biden-deliver-six-step-plan-covid-19-pandemic-2021-09-09/
[538] https://www.who.int/reproductivehealth/about_us/hrp/partners/en/
[539] https://www.plannedparenthood.org/about-us/newsroom/press-releases/planned-parenthood-kicks-off-major-push-to-help-get-people-vaccinated-against-covid-19
[540] https://sciencebasedmedicine.org/who-vaccine-hesitancy-top-health-threat/
[541] https://www.catholicstand.com/eugenics-in-america/
[542] https://www.ncbi.nlm.nih.gov/pubmed/2028254
[543] https://journals.plos.org/plosone/article?id=10.1371/journal.pone.0142200
[544] http://www.kccb.or.ke/home/news-2/press-statement-5/
[545] https://www.ted.com/talks/amory_lovins_on_winning_the_oil_endgame#t-1162402
[546] https://academic.oup.com/pcp/article/40/10/999/1885036?login=true
[547] http://www.familyplanning2020.org/

END NOTES

[548] https://www.politico.com/story/2016/06/bill-gates-clinton-trump-224476
[549] https://timeline.com/willowbrook-the-institution-that-shocked-a-nation-into-changing-its-laws-c847acb44e0d
[550] https://nyulangone.org/news/nyu-langone-health-establishes-center-psychedelic-medicine
[551] https://www.nyu.edu/about/news-publications/news/2021/september/public-health-nih-grant-transit-workers.html
[552] https://www.nyu.edu/students/communities-and-groups/student-diversity/spiritual-life/of-many-institute-for-multifaith-leadership.html
[553] https://academic.oup.com/ndt/article/32/suppl_2/ii13/3056571
[554] https://www.cnn.com/2021/09/09/politics/joe-biden-covid-speech/index.html
[555] https://www.vanityfair.com/news/2021/10/nih-admits-funding-risky-virus-research-in-wuhan
[556] https://nypost.com/2018/03/15/inside-the-shady-private-equity-firm-run-by-kerry-and-bidens-kids/
[557] https://web.archive.org/web/20140315054332/http:/www.rstp.com/people/
[558] https://web.archive.org/web/20140315054242/http:/www.rstp.com/companies/
[559] https://www.marlinllc.com/_media/_data/MarlinNewsletter/hit-monthly-newsletter-june-2015.pdf
[560] https://www.bloomberg.com/profile/person/16779750
[561] https://www.bbc.com/news/world-54553132
[562] https://medium.com/@george.mesires/a-statement-on-behalf-of-hunter-biden-dated-october-13-2019-d80bc11087ab
[563] https://www.politico.com/news/magazine/2021/10/12/hunter-biden-corruption-515583
[564] https://www.usna.edu/NewsCenter/sites/Ethics/Dr._James_Giordano_Battlescape_Brain_Military_and_Intelligence_Use_of_Neurocognitive_Science.php
[565] https://revolutiontelevision.net/video/fauci-lecture-at-georgetown-u-predicting-plague-on-trump-administration/
[566] https://www.ecohealthalliance.org/wp-content/uploads/2018/10/Living-Safely-with-Bats_download.pdf
[567] http://www.ecohealthalliance.org/wp-content/uploads/2016/11/Greatorex-etal_Lao-Markets_2016.pdf
[568] https://www.ecohealthalliance.org/wp-content/uploads/2018/11/Evidence-for-henipavirus-spillover-into-human-populations-in-Africa.pdf
[569] https://www.ecohealthalliance.org/wp-content/uploads/2016/11/Reed-et-al_Ebola_plos-2014.pdf
[570] https://www.ecohealthalliance.org/wp-content/uploads/2016/10/Lee-etal_Macacine-Herpesvirus_2015.pdf
[571] https://www.ecohealthalliance.org/wp-content/uploads/2016/10/Anthony-Nonrandom-patterns-in-viral-diversity.pdf
[572] https://www.nytimes.com/roomfordebate/2014/04/15/when-is-foreign-aid-meddling/secret-programs-hurt-foreign-aid-efforts
[573] https://apnews.com/article/46328e561bfb44b99b2e6937835be957
[574] https://www.ecohealthalliance.org/2021/04/research-collaboration-announces-new-tool-to-calculate-zoonotic-disease-risk
[575] https://www.ecohealthalliance.org/wp-content/uploads/2021/06/journal.pone_.0236971.pdf
[576] https://www.foxnews.com/us/federal-officers-biden-covid-19-vaccine-mandate
[577] https://yournews.com/2021/09/13/2219109/sen-blackburn-to-newsmax-were-missteps-in-afghanistan-intentional/
[578] The Bidens were assigned by the globalists backing Obama and the Clintons to supply the EU with energy in competition with the U.S. and its allies, including the Vitoil Company in the Ukraine and Armendia, whose interests in Hawaii personally involved these authors.

[579] https://nypost.com/2020/10/14/email-reveals-how-hunter-biden-introduced-ukrainian-biz-man-to-dad/
[580] A top psychedelic scientist says 'the climate's looking good' for magic mushrooms and MDMA to turn into medicines at a gathering of the world's billionaires,' reported Erin Brodwin "a senior health and tech reporter at Business Insider.
[581] https://www.foxnews.com/politics/trump-authorizes-declassification-of-all-russia-collusion-hillary-clinton-email-probe-documents
[582] https://gumc.georgetown.edu/gumc-stories/global-health-experts-advise-advance-planning-for-inevitable-pandemic/
[583] https://exposingvaccinegenocide.org/exposing-h1n1-flu-vaccine-scam/
[584] https://www.washingtonpost.com/national/health-science/us-launches-new-global-initiative-to-prevent-infectious-disease-threats/2014/02/12/afd9863c-936d-11e3-b46a-5a3d0d2130da_story.html
[585] https://www.axios.com/fincen-files-us-banks-money-laundering-buzzfeed-e9a04230-a8f8-4a6b-a190-f4f98a4bba7b.html
[586] https://www.buzzfeednews.com/article/jasonleopold/fincen-files-financial-scandal-criminal-networks
[587] https://www.icij.org/investigations/fincen-files/global-banks-defy-u-s-crackdowns-by-serving-oligarchs-criminals-and-terrorists/
[588] https://www.fincen.gov/
[589] https://www.washingtonexaminer.com/news/hunter-biden-will-step-down-from-chinese-backed-firm-as-he-denies-wrongdoing
[590] https://www.washingtonexaminer.com/news/analysis-hunter-biden-tied-to-china-firm-with-questionable-dealings
[591] https://www.bloomberg.com/news/articles/2019-10-13/hunter-biden-steps-down-from-chinese-board-in-bid-to-fight-back
[592] https://www.weforum.org/great-reset/
[593] https://www.weforum.org/agenda/2020/10/covid-19-litmus-test-sustainability/
[594] https://www.weforum.org/agenda/2020/02/jeff-bezos-earth-fund-climate-change/
[595] https://twitter.com/wef/status/1084465092228509697
[596] https://www.weforum.org/events/world-economic-forum-annual-meeting-2020/partners
[597] https://www.weforum.org/agenda/2020/08/covid19-global-health-climate-change
[598] http://www.chinadaily.com.cn/business/2014-09/10/content_18571757.htm
[599] https://www.uscc.gov/sites/default/files/annual_reports/2018 Annual Report to Congress.pdf
[600] https://en.wikipedia.org/wiki/CEFC_China_Energy
[601] http://www.intelligencequarterly.com/Document/BidenChina.pdf
[602] https://www.waronwethepeople.com/wp-content/uploads/2016/10/Democratic-National-Committee-Secret-Memo-of-April-7-2015-Identifying-Trump-as-a-Pied-Piper-Candidate-to-be-Used-by-Hillary-Rodham-Clinton-to-Win-the-Presidency.pdf
[603] https://medicalveritas.org/cclinton_iac_vimeo_censorship_aids_genocide/
[604] https://www.ftc.gov/enforcement/cases-proceedings/052-3069/choicepoint-inc
[605] https://medicalveritas.org/volume-7-number-1-january-2010/
[606] https://wikispooks.com/wiki/Carlyle_Group
[607] https://exposingvaccinegenocide.org/10-stages-of-genocide-by-g-h-stanton-of-genocide-watch/
[608] http://genocidewatch.net/genocide-2/8-stages-of-genocide/
[609] https://www.hopkinsmedicine.org/kimmel_cancer_center/cancers_we_treat/blood_bone_marrow_cancers/patient_guide_BMT_with_new_Krames.pdf
[610] https://www.waronwethepeople.com/wp-content/uploads/2019/10/Hopkins-Patient-Guide-Final-seeP113-2018-06-05-11-48-39-UTC.pdf
[611] https://www.hopkinsmedicine.org/kimmel_cancer_center/patient_and_family_services/patient_information/Patient-Guide-Weinberg.pdf
[612] http://revolutiontelevision.net/video/senate-health-committee-hearing-on-vaccines-outbreaks-and-vaccine-hesitancy-mar-5-2019/

END NOTES

[613] https://www.immunizationcoalitions.org/content/uploads/2017/01/Timeline-for-CDC-Whistleblower.pdf
[614] https://kbhaebw.tk/world-news-tv/who-said-useless-eaters-quote.php
[615] https://www.genengnews.com/topics/translational-medicine/merck-co-to-market-csls-seasonal-flu-vaccine-in-the-u-s/
[616] https://www.fda.gov/media/117022/download
[617] https://www.originofaids.com/articles/wrong.htm
[618] https://ec.europa.eu/health/sites/health/files/vaccination/docs/2018_vaccine_confidence_en.pdf
[619] https://dlnr.hawaii.gov/mk/files/2017/03/TIO-Ex-C-57.pdf
[620] https://www.un.org/en/genocideprevention/genocide.shtml
[621] https://www.un.org/en/genocideprevention/documents/atrocity-crimes/Doc.1_Convention on the Prevention and Punishment of the Crime of Genocide.pdf
[622] http://revolutiontelevision.net/video/fox-news-hypocritically-condemns-competing-propagandists/
[623] https://scholar.google.com/scholar_case?case=14369486041709640908&q=reckless+endangerment&hl=en&as_sdt=4,60
[624] https://scholar.google.com/scholar_case?case=8271091122584728148&q=dow+chemical+fraud+elements&hl=en&as_sdt=4,60
[625] https://www.sciencemag.org/news/2019/11/top-israeli-immunologist-accused-promoting-antivaccine-views
[626] https://www.sciencemag.org/author/kai-kupferschmidt
[627] https://www.ima.org.il/ENG/ViewContent.aspx?CategoryId=11081
[628] https://medicalveritas.org/jewish-priest-slams-pharma-rabbi-for-religious-hypocrisy-and-genocidal-stupidity-in-unfolding-vaccine-controversy/
[629] https://www.sciencedirect.com/science/article/pii/S0896841110000788?via%3Dihttps://pubmed.ncbi.nlm.nih.gov/17581283/hub
[630] https://www.ncbi.nlm.nih.gov/pubmed/17581283
[631] As reviewed in Nature Nanotechnology, "Nanotechnology benefits modern vaccine design since nanomaterials are ideal for antigen delivery, as adjuvants, and as mimics of viral structures." In fact, the first COVID-19 vaccine candidate launched into clinical trials is the Bill Gates/Anthony Fauci/Moderna Company's mRNA vaccine delivered via lipid nanoparticles. "To eradicate pandemics, present and future, a successful vaccine platform must enable rapid discovery, scalable manufacturing and global distribution." This article reviews current approaches to COVID-19 vaccine development and highlights the role of nanotechnology and advanced manufacturing" befitting the 4IR.
[632] https://www.waronwethepeople.com/paul-offit-2/
[633] https://www.jacionline.org/article/S0091-6749(11)01452-7/fulltext
[634] https://www.sciencedirect.com/science/article/pii/S2213219817305172?via%3Dihub
[635] https://www.semanticscholar.org/paper/Recent-Advances-in-Vaccine-Adjuvants-Singh-O%27hagan/f813ac17ffe461c5e9f73347163c2a37bbc264f6/figure/3
[636] https://patents.justia.com/inventor/paul-offit
[637] https://www.sciencedirect.com/science/article/pii/S0264410X16000165
[638] https://link.springer.com/article/10.1007/s12026-016-8826-6
[639] https://www.cnbc.com/2017/03/17/mini-nukes-and-inspect-bot-weapons-being-primed-for-future-warfare.html
[640] http://www.darpa.mil/program/fast-lightweight-autonomy
[641] http://www.darpa.mil/news-events/2016-02-12
[642] https://www.army.mil/article/140097
[643] http://www.dtic.mil/dtic/tr/fulltext/u2/a557001.pdf
[644] https://www.youtube.com/watch?v=VvXiC8Wmc2U
[645] https://www.pirbright.ac.uk/partnerships/our-major-stakeholders

[646] https://drive.google.com/file/d/1VhFJvQT3VvaKONdd8fyCjO5q-IlStewD/view?fbclid=IwAR3oqmuiSahOVlBDmYoffk_JSAemYqUHdzYW-N7OhXXw5OMg6QDce8rsx-Y
[647] https://gumc.georgetown.edu/gumc-stories/global-health-experts-advise-advance-planning-for-inevitable-pandemic/
[648] https://www.lexjansen.com/phuse/2018/dh/DH02.pdf
[649] https://www.youtube.com/watch?v=hL9uk4hKyg4&app=desktop
[650] https://openknowledge.worldbank.org/handle/10986/14927
[651] https://www.genengnews.com/topics/drug-discovery/quantum-dots-deliver-vaccines-and-invisibly-encode-vaccination-history-in-skin/
[652] https://medicalveritas.org/deep-state-coronavirus-jeffrey-epstein-harvard/
[653] https://en.wikipedia.org/wiki/MIT_Media_Lab
[654] https://sts-program.mit.edu/event/biden-or-trump-whats-at-stake-on-november-3/
[655] https://www.nationalobserver.com/2020/06/15/analysis/noam-chomsky-post-covid-19-society-trump-worse-hitler-peasants-are-coming
[656] https://www.youtube.com/watch?v=zi6ae6kZNqE
[657] https://thefederalist.com/2020/10/13/sen-mazie-hirono-pauses-obamacare-rant-to-ask-amy-coney-barrett-if-shes-a-rapist/
[658] https://www.civilbeat.org/2017/11/denby-fawcett-would-dan-inouye-have-survived-1992-sex-allegations-today/
[659] https://www.youtube.com/watch?v=EbFphX5zb8w
[660] https://medicalveritas.org/hawaiian-last-king-kalakaua-killed-by-drugs/
[661] https://medicalveritas.org/hawaii-ballistic-missile-threat/
[662] https://www.govinfo.gov/content/pkg/CHRG-108hhrg98604/html/CHRG-108hhrg98604.htm
[663] https://www.waronwethepeople.com/?s=ayahuasca
[664] https://judicialcorruptionnews.com/hawaii-challenge-trump-travel-ban-vets-pattern-judicial-corruption-mob-rule-drug-trafficking/
[665] https://judicialcorruptionnews.com/hawaii-racket/
[666] https://revolutiontelevision.net/video/hawaii-khon-news-anchor-joe-moore-gets-cut-off-the-air-for-exposing-covid-19-as-a-manmade-bioweapon/
[667] https://judicialcorruptionnews.com/lisa-m-ginoza/
[668] See intertwining the Hawaii drug trafficking operations of Bostonian kingpin Paul J. Sulla, Jr..
[669] https://frankreport.com/2019/12/22/more-evidence-about-ayahuasca-death-cult-emerges-with-arrest-of-lawyer-paul-sulla-in-hawaii/
[670] https://heffter.org/heffter-marks-milestone-100th-scientific-publication/
[671] https://hub.jhu.edu/2019/10/14/60-minutes-anderson-cooper-psychedelics/
[672] http://www.intelligencequarterly.com/Document/BidenChina.pdf
[673] https://revolutiontelevision.net/video/kamala-harris-exposed-as-dictator-targets-whistleblower-and-covers-up-planned-parenthood-selling-body-parts/
[674] https://twitter.com/R_H_Ebright/status/1450947395508858880/photo/1
[675] https://www.nature.com/articles/s41423-020-0424-9
[676] https://nymag.com/intelligencer/2020/04/more-bad-news-on-the-long-term-effects-of-the-coronavirus.html
[677] https://www.un.org/disarmament/wmd/bio/
[678] http://stateofthenation.co/?p=11247
[679] https://www.fbcoverup.com/docs/library/2019-06-01-CETel-acquires-Cobbett-Hill-Earth-Station-Limited-Spacewatch-Jun-01-2019.pdf
[680] https://www.fdanews.com/articles/74903-abbott-laboratories-co-promotion-with-boehringer-ingelheim-expires
[681] https://www.thepharmaletter.com/article/basf-sells-pharma-business-to-abbott-labs-for-6-9-billion-in-cash
[682] http://revolutiontelevision.net/video/bill-gates-exposed-for-genocide/

END NOTES

[683] https://www.fbcoverup.com/docs/library/2018-11-20-US-Pat-No-10130701-CORONAVIRUS-Assignee-THE-PIRBRIGHT-INSTUTUTE-Woking-Great-Britain-funded-by-Wellcome-Trust-and-Gates-Foundation-USPTO-Nov-20-2018.pdf
[684] https://www.scientificamerican.com/article/heres-how-coronavirus-tests-work-and-who-offers-them/
[685] https://www.viracor-eurofins.com/media/2214/eua-abbott-ncov-ifu-wp.pdf
[686] https://www.ncbi.nlm.nih.gov/pubmed/27482900
[687] https://theinfectiousmyth.com/book/CoronavirusPanic.pdf
[688] http://revolutiontelevision.net/video/dr-anthony-fauci-perjures-himself-regarding-measles-vaccine-causing-encephalitis/
[689] https://www.cia.gov/library/center-for-the-study-of-intelligence/csi-publications/csi-studies/studies/vol48no3/article06.html
[690] https://www.niaid.nih.gov/news-events/new-coronavirus-emerges-bats-china-devastates-young-swine
[691] https://exposingvaccinegenocide.org/vaccination-genocide/
[692] https://www.scmp.com/news/hong-kong/health-environment/article/3078437/mask-or-not-mask-who-makes-u-turn-while-us
[693] http://revolutiontelevision.net/video/new-york-doctor-characterizes-the-coronavirus-bioweapon/
[694] https://journal-neo.org/2020/03/18/coronavirus-and-the-gates-foundation/
[695] https://www.ncbi.nlm.nih.gov/pmc/articles/PMC3346300/
[696] https://www.ncbi.nlm.nih.gov/pmc/articles/PMC3335060/
[697] https://www.insider.com/how-effective-is-the-flu-shot
[698] https://www.uspharmacist.com/article/20192020-influenza-vaccine-update
[699] http://www.ukapologetics.net/pharmakeia.html
[700] https://religionnews.com/category/coronavirus/?gclid=EAIaIQobChMI_NaglI3r8wIVmpSGCh01FQVaEAMYASAAEgI_6fD_BwE
[701] http://barbarahartwell.blogspot.com/2009/12/ex-fbi-agent-ted-gunderson-exposed.html
[702] http://barbarahartwellvscia.blogspot.com/
[703] http://la.indymedia.org/js/?v=cont&url=/news/2012/09/255338.json
[704] https://religionnews.com/2019/04/25/the-satanic-temple-is-a-real-religion-says-irs/
[705] https://cinemafemme.com/2019/02/25/2019-2-25-hail-satan-interview-with-documentary-filmmaker-penny-lane/
[706] https://filmmakermagazine.com/97862-an-open-letter-to-the-tribeca-film-festival-about-vaxxed/
[707] https://www.waronwethepeople.com/senior-anti-vaxxed-attackers-vetted/
[708] http://greeknewsondemand.com/2020/04/22/marina-abramovic-demands-conspiracy-theorists-leave-her-alone-im-not-a-satanist/
[709] https://newspunch.com/bill-gates-microsoft-cancel-marina-abramovic-campaign-after-people-flood-video-calling-her-out-satanist/
[710] https://revolutiontelevision.net/video/bill-gates-microsoft-commercial-features-spirit-cooker-satanist-marina-abramovic/
[711] https://www.learnreligions.com/the-satanic-statements-95978
[712] http://ibankcoin.com/zeropointnow/2016/11/26/sick-lets-revisit-the-podesta-penchant-for-pedophilic-cannibalistic-and-satanic-art/
[713] http://ibankcoin.com/zeropointnow/files/2016/11/2016-11-12-13_54_38-Films-TV.png
[714] https://twitter.com/i/moments/833366113996124160?lang=en
[715] http://evidenceforchristianity.org/galatians-519-has-the-word-pharmakeia-which-means-the-use-and-administration-of-drugs-why-does-the-niv-translate-it-as-witchcraftr/
[716] https://www.waronwethepeople.com/wp-content/uploads/2019/09/eyes-wide-open-fiona-barnett_first-edition_august-2019.pdf
[717] https://iqnexus.org/Graphics/Mag/IQNJ 11-3 2019.pdf
[718] https://home.cern/science/accelerators/large-hadron-collider

[719] https://www.frequencytx.com/people/john-lamattina-phd/
[720] https://www.frequencytx.com/role/leadership/
[721] https://www.sltrib.com/religion/2019/04/25/satanic-temple-is-real/
[722] http://nymag.com/intelligencer/2017/04/the-techno-libertarians-praying-for-dystopia.html
[723] https://scholars.org/scholar/timothy-snyder
[724] https://scholars.org/contribution/twenty-lessons-fighting-tyranny-twentieth-century
[725] https://billmoyers.com/episode/united-states-of-alec/
[726] https://uscode.house.gov/view.xhtml?path=/prelim@title10/subtitleA/part1/chapter12&edition=prelim
[727] https://www.splcenter.org/hatewatch/2014/01/13/patriot-conspiracy-theorist-jack-mclamb-dies
[728] https://www.splcenter.org/fighting-hate/intelligence-report/2001/false-patriots?page=0%2C7
[729] https://www.waronwethepeople.com/child-trafficking/
[730] https://www.snopes.com/fact-check/kissinger-forced-vaccinations/
[731] https://medicalveritas.org/vaccination-genocide/
[732] https://www.cureshoppe.com/the-best-of-early-horowitz-friends/
[733] http://revolutiontelevision.net/video/the-horokane-snags-cointelpro-agents-jeff-rense-and-true-ott/
[734] https://www.waronwethepeople.com/counter-intelligence/
[735] https://www.fbcoverup.com/docs/library/2005-05-12-Theodore-L-Gunderson-VIDEO-TRANSCRIPT-Former-FBI-Senior-Special-Agent-In-Charge-Dallas-Los-Angeles-Memphis-on-Illuminati-Satanism-Pedophilia-CIA-Deep-State-May-12-2005.pdf
[736] http://sosbeevfbi.ning.com/forum/topics/elaborate-efforts-by-ted
[737] https://www.waronwethepeople.com/cointelpro-minions-barabara-hartwell/
[738] http://revolutiontelevision.net/video/vice-motherboard-propaganda-on-ebola/
[739] https://www.sciencedirect.com/science/article/pii/S2590098621000208
[740] https://faseb.onlinelibrary.wiley.com/doi/full/10.1096/fj.202000502
[741] https://www.sciencedirect.com/science/article/pii/S2590098621000208
[742] http://journals.sagepub.com/doi/10.1177/0300060518777044
[743] http://www.altmedrev.com/archive/publications/5/5/448.pdf
[744] https://www.medicalnewstoday.com/articles/323276#quercetin
[745] https://www.healthline.com/health/natural-asthma-treatment-and-alternative-therapies#herbs-and-supplements
[746] http://web.stanford.edu/group/hopes/cgi-bin/hopes_test/ginkgo-biloba/
[747] https://www.carewell.com/product/airlife-spirometer/?sku=001901A-EA1&g_network=u&g_productchannel=online&g_adid=504605868196&g_keyword=&g_campaign=TOF+-+Smart+Shopping+%7C+NB+LTV+%7C%7C+Medical+Supplies&g_keywordid=pla-465749734767&g_adtype=&g_ifcreative=&g_adgroupid=122705214687&g_productid=17224&g_merchantid=114738081&g_partition=465749734767&g_campaignid=12503559488&g_acctid=333-280-7133&g_ifproduct=product&gclid=Cj0KCQjw-NaJBhDsARIsAAja6dMM2wfvOjkrnZJFb9OgdcTSZdQLflOGUBZozOckr5uuu3q_Jr-A_zcaAlIiEALw_wcB
[748] https://lifewellnesshealthcare.com/collections/airphysio/products/airphysio-device?gclid=Cj0KCQjw-NaJBhDsARIsAAja6dMZ2RN3QM5z5WL1Q56IhVMgTNjQloxkNn7fnTHXf-sJPsguQC1Zfm4aAm5GEALw_wcB
[749] https://core.ac.uk/download/pdf/82518757.pdf
[750] http://528revolution.com/
[751] https://www.brighteon.com/8ae48e6f-e781-4348-a6ec-ae8385932216?utm_source=Engaged%20%283%20Months%29&utm_medium=email&utm_campaign=SUCCESSFUL%20STEPS%20TO%20BE%20A%20PATIENT%20ADVOCATE%20FOR%20PATIENTS%20AND%20LOVED%20ONES%20IN%20H

END NOTES

OSPITALS%20%28Wp3yBW%29&_kx=FuoOELjqdL6aBAy1BzCAQycJL3DzqHBaJ4d0KD_dM74%3D.SwvL6p

Printed in Great Britain
by Amazon